Sensors Update

Volume 11

Sensor Technology –
Applications – Markets

Related Wiley-VCH titles:

Sensors Vol. 1–9
edited by W. Göpel, J. Hesse, J.N. Zemel

ISBN 3-527-26538-4

Sensors Applications
edited by J. Hesse, J.W. Gardner, W. Göpel

Vol. 1

Sensors in Manufacturing
edited by H.K. Tönshoff, I. Inasaki

ISBN 3-527-29558-5

Vol. 2

Sensors in Intelligent Buildings
edited by O. Gassmann, H. Meixner

ISBN 3-527-29557-7

Upcoming Volumes:

Sensors in Automotive Technology
Sensors in Household Appliances
Sensors in Medicine and Health Care
Sensors in Aerospace Technology
Sensors in Environmental Technology

Sensors Update

Edited by
H. Baltes, G.K. Fedder, J.G. Korvink

Volume 11

Series Editors

Prof. Dr. H. Baltes
Physical Electronics Laboratory
ETH Hoenggerberg, HPT-H6
8093 Zürich
Switzerland

Prof. Gary K. Fedder
ECE Department &
The Robotics Institute
Carnegie Mellon University
Pittsburgh, PA 15213-3890
USA

Prof. Dr. Jan G. Korvink
Institute for Microsystem
Technology
University of Freiburg
Georges-Köhler-Allee 103
79110 Freiburg
Germany

The series Sensors Update was founded by
Prof. Dr. H. Baltes, Prof. Dr. W. Göpel and Prof. Dr. J. Hesse.

Library of Congress Card No.: applied for

British Library Cataloguing-in-Publication Data:
A catalogue record for this book is available from the British Library

Bibliographic information published by Die Deutsche Bibliothek
Die Deutsche Bibliothek lists this publication in the Deutsche Nationalbibliografie; detailed bibliographic data is available in the Internet at <http://dnb.ddb.de>.

ISBN 3-527-30601-3

Printed on acid-free paper.

Composition: K+V Fotosatz GmbH, Beerfelden
Printing: betz-druck GmbH, Darmstadt
Bookbinding: J. Schäffer GmbH & Co. KG, Grünstadt
Printed in the Federal Republic of Germany

Preface

Improvements in sensor technology are empowering new embedded systems to not just process information, but also to interact with people and the environment. To this end, systems developers are ramping up demand for miniature sensors with low-power operation and computer-compatible outputs. It is widely predicted that within the next few years such sensor-augmented systems will be wireless, interconnected, distributed and networked. The importance of sensors is reflected in a myriad of seedling business activities, even as the overall global economy is still struggling. Embedded sensors are poised to make a large impact as enabling components, and will no doubt play a key role in the next economic upswing. In light of these trends, it remains critical to stay abreast of the latest developments in sensor technologies and applications. *Sensors Update* plays a unique role by providing timely comprehensive reviews in the sensors area.

Sensors Update presents review articles that place new developments within a historical perspective while including the critical technical depth necessary for independent evaluation. Generally, each volume in the series is partitioned into three major sections. 'Sensor Technology' reviews key areas in applied and basic research, 'Sensor Applications' provides overviews of new or improved applications, and 'Sensor Market' covers emerging trends, reviews of suppliers, or patent status for a particular market sector.

This 11th volume of the series, following the 9-volume series *Sensors*, includes a number of complementary articles related to biochemical and chemical sensing. Research continues to expand in these areas, and has garnered even more heightened interest within the past year. Articles in other key areas are on microsystem compact modeling, microelectronic bonding processes and earthquake sensors.

Timely reviews that are a pleasure to read require the effort of many people. The series editors extend a heartfelt thanks to the many colleagues who have committed their time and effort in writing, reviewing or editing articles. The editors are grateful to and thank the publishers, Wiley-VCH, and their staff Dr. Jörn Ritterbusch and Hans-Jochen Schmitt for their support. Special thanks are due to the series publishing editor, Dr. Martin Ottmar, who replaced Dr. Claudia Barzen on the *Sensors Update* team this past year.

H. Baltes, Zürich G. K. Fedder, Pittsburgh J. G. Korvink, Freiburg

Editorial Advisory Board

Contents

List of Contributors

H. Furukawa
Yamatake Corporation
Kawana, Fujisawa-shi
Kanagawa
Japan

C. Hagleitner
Physical Electronics Laboratory
ETH Hoenggerberg
Zurich
Switzerland

S. Ichida
Yamatake Corporation
Kawana, Fujisawa-shi
Kanagawa
Japan

K. Koganemaru
Tokyo Gas Co., Ltd.
Tokyo
Japan

J. G. Korvink
Institute for Microsystem Technology
Albert Ludwig University Freiburg
Freiburg
Germany

C. Krantz-Rülcker
The Swedish Sensor Centre and
the Division of Applied Physics
Department of Physics and Measurement
Technology
Linköping University
Linköping
Sweden

I. Lundström
The Swedish Sensor Centre and
the Division of Applied Physics
Department of Physics and Measurement
Technology
Linköping University
Linköping
Sweden

C. R. Martin
Department of Chemistry and
Center for Research
at the Bio/Nano Interface
University of Florida
Gainesville, FL
USA

M. Mayer
ESEC SA
Cham
Switzerland

E. B. Rudnyi
Institute for Microsystem Technology
Albert Ludwig University Freiburg
Freiburg
Germany

Y. Shimizu
Tokyo Gas Co., Ltd.
Tokyo
Japan

T. Suzuki
Faculty of Engineering
Toyo University
Saitama
Japan

K. Takubo
Yamatake Corporation
Kawana, Fujisawa-shi
Kanagawa
Japan

D. Wilson
Department of Electrical Engineering
University of Washington
Seattle, WA
USA

F. Winquist
The Swedish Sensor Centre and
the Division of Applied Physics
Department of Physics and Measurement
Technology
Linköping University
Linköping
Sweden

M. Wirtz
Department of Chemistry and
Center for Research
at the Bio/Nano Interface
University of Florida
Gainesville, FL
USA

T. Yanada
Yamatake Corporation
Kawana, Fujisawa-shi
Kanagawa
Japan

PART 1

Sensor Technology

1.1 Review: Automatic Model Reduction for Transient Simulation of MEMS-based Devices

EVGENII B. RUDNYI and JAN G. KORVINK,
Institute for Microsystem Technology,
Albert Ludwig University Freiburg, Freiburg, Germany

Abstract

The rapid development of MEMS-based devices requires 3D time-dependent simulations for coupled physical domains (thermal, mechanical, electrical, etc.). This in turn requires the solution of high-dimensional ordinary differential equations (ODEs) that result from space discretization of the device. However, instead of a "brute force" approach to integrate a large system of ODEs, one can use modern mathematical methods to reduce the system's dimension. The goal of the present paper is to review them from an engineering perspective. It is shown that in many cases important for practice the order of ODEs can be reduced by several orders of magnitude almost without sacrificing precision.

Keywords: Large-scale dynamical system; model order reduction; control theory; moment matching; Krylov subspace; Arnoldi and Lanczos algorithms; proper orthogonal decomposition

Contents

1.1.1 Introduction

The goal of micro-electromechanical system (MEMS) computer-aided design and simulation is to represent accurately and efficiently the behavior of the system in question. This allows technologists to develop a better understanding of the system and, as a result, to choose an optimal design quickly. A hugely successful example of the application of computer-aided design (CAD) is in the simulation of electrical integrated circuits, for which the simulator's output is almost the same as that produced by a real circuit prototype. This drives MEMS designers to create similar techniques for MEMS simulations.

It so happens that electrical circuit and MEMS simulations are quite different in nature (see, for example, the discussion in Ref. [1]). A circuit is rather accurately described by lumped elements such as discrete resistors, capacitors, inductors, transistors and so on. The transient response of the circuit can be immediately written as a system of ordinary differential equations (ODEs) with the system's dimension of the order of the number of nodes connecting lumped elements in the circuit. On the other hand, the governing partial differential equations (PDEs) for MEMS devices do not always lend themselves to intuitive lumping as ODEs and hence are solved numerically by first spatially semi-discretizing them by means of finite element, boundary element and similar methods. This also leads to a system of ODEs, but its dimension depends on the quality of discretization and it could routinely lead to ODE system sizes of between 10^4 and 10^6 equations, especially in the case of 3D simulations. The relationships between differential equations, meshes and models are shown in Figure 1.1.1.

The arrows represent translations between descriptions: (1) lumping is done manually, either as a circuit equivalent or as an algebraic expression; (2) adaptive meshing determines the size of the subsequent model; (3) circuit equivalents or algebraic expressions are turned into a suitable set of ODEs; interconnecting many of these again leads to large systems; (4) semidiscretization of the PDEs on a mesh results in a set of ODEs; (5) an algebraic model reducer takes a large

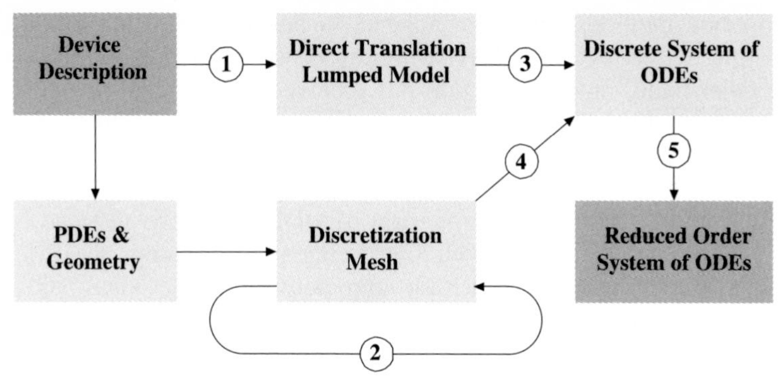

Figure 1.1.1. Some of the routes leading from a device description to a reduced order system of ODEs. The arrows represent translations between descriptions: 1) Lumping is done by hand, either as a circuit equivalent, or as an algebraic expression; 2) adaptive meshing determines the size of the subsequent model; 3) circuit equivalents or algebraic expressions are turned into a suitable set of ODEs; interconnecting many of these again leads to large systems; 4) semidiscretization of the PDEs on a mesh results in a set of ODEs; 5) an algebraic model reducer takes a large system of ODEs and produces a smaller (and hence reduced order) yet equivalent system of ODEs.

system of ODEs and produces a smaller (and hence reduced order) yet equivalent system of ODEs.

Recent advances in computer power allow us to solve very large systems of ODEs by brute force, one of the most striking examples being car crash simulations (see, for example, Ref. [2]). Nevertheless, this typically requires parallel computations (see the benchmark report in Ref. [3]) which increases the cost of simulation drastically and, as a result, limits simulation applicability considerably.

In order to facilitate computations, engineers often simplify the original rigorously derived governing equations or, instead, use simple empirical models: we use the term "quick-and-dirty" (QAD) calculations. Another approach, the topic of this chapter, is to perform model reduction, that is, to reduce formally the dimension of a system of ODEs derived from a rigorous approach before integrating it in time.

To this end, taken from current mechanical engineering practice, there are two popular methods and both are incorporated in some commercial software simulation tools: modal reduction [4] and dynamic condensation [5]. The idea behind modal reduction is to approximate a dynamic system response through a linear combination of several, often low-frequency, natural eigenmodes of the system. The second approach is based on the Guyan method [6] and is just an intuitive engineering extension of the Shur complement method from a stationary to a time-dependent formulation.

The main problem with all of the above order reduction methods is that their success depends primarily on engineering intuition, since they are not based on a solid mathematical background. Hence they could be referred to as nonautomatic model reduction methods and there appears to be no way to improve this situation. Certainly, without experience and intuition, we do not recommend their use.

On the other hand, model reduction has received a great deal of attention from mathematicians, who have developed a number of methods with which to approximate large-scale dynamics systems (for a mathematical review, see Ref. [7]) and which will be referred to as automatic model reduction. There are some spectacular examples where the dimension of ODEs could be reduced by several orders of magnitude, almost without sacrificing precision (eg, see [8] and [9]). However, there still remains a certain gap between these ideas and common MEMS engineering practice and the aim of this chapter is to start to fill this gap. Our review complements Ref. [7] (where automatic model reduction is considered mathematically) at the engineering level.

Nevertheless, the classification in this review is made on the basis of a mathematical perspective and therefore follows the structure of Ref. [7]. What we have found is that, even though different engineering communities are facing very different challenges, many solution techniques are related. At first glance, the simulation of groundwater flow in discretely fractured porous media has nothing to do with MEMS devices. It is therefore not surprising that these two engineering communities do not follow each other's work. However, the model reduction problem they are trying to solve is absolutely identical if we consider it from a mathematical viewpoint.

In principle, a system of ODEs can also be solved faster if it is possible to increase the efficiency of the time integrator. Recently, there have been some promising results in this direction based on matrix exponential approximations [10], but so far there are no engineering examples and hence this will be outside of the scope of this review.

We start our review with a statement of the mathematical problem for model reduction, where we introduce terms and give them the equivalents used in the MEMS community. Then we consider low-dimensional linear systems of ODEs. It is safe to state that, for this case, the problem of automatic model reduction is almost completely solved. It appears that almost all modern model reduction methods for large-scale systems are based, in one way or another, on Krylov subspace methods [11] and therefore a short introduction to Krylov subspace methods is given. After that, we switch to large-scale linear systems of ODEs. The challenge faced here is that the computational time required for a model reduction of a linear system of ODEs depends on the problem dimension (the number of equations in the system) to the cubic power. Computationally speaking, the algorithms for model reduction appropriate for small linear systems do not scale to large systems. Here one can say that, in principle, the answer to automatic model reduction is known but the challenge remains of how to compute it in reasonable time. Finally, we take a look at nonlinear systems of ODEs. Here success depends on a particular problem and there are almost no general results. Some algorithms for model reduction exist but, in contrast to linear systems, unfortunately, it seems that human intervention is inevitable.

It should be noted that we have not tried to reflect the priority of research groups in this field. In many cases, our literature citations should actually be read as "see, **for example**, Ref. [1.1]."

1.1.2 Mathematical Statement for Model Reduction

In the present review we limit our consideration to a system of first-order ODEs, written in the form

$$E \cdot \frac{d\boldsymbol{x}}{dt} = F \cdot \boldsymbol{x} + \boldsymbol{f} \tag{1.1.1}$$

where the unknown vector $\boldsymbol{x}(t) \in \Re^n$ contains unknowns functions in time, $E \in \Re^n \times \Re^n$ and $F \in \Re^n \times \Re^n$ are system matrices, typically sparse and often symmetric and the vector $\boldsymbol{f} \in \Re^n$ describes the system load. If the matrices contain constant coefficients then the system of ODEs is linear and otherwise we will call it nonlinear. (Strictly this is not correct. There is an intermediate case when coefficients depend on time explicitly, in which case it is termed a linear time-varying system [12].) Mechanical systems in motion, as well as general electrical circuits, are usually described by systems of ODEs of second order in time. It is a simple matter to convert them to the form of Equation (1.1.1) by increasing the number of unknowns and equations by a factor of two, eg, by treating the first derivatives in time as unknowns. Thus,

$$M \cdot \frac{d^2\boldsymbol{y}}{dt^2} + C \cdot \frac{d\boldsymbol{y}}{dt} + K \cdot \boldsymbol{y} = \boldsymbol{f} \tag{1.1.2}$$

together with the new variables $\boldsymbol{z} = d\boldsymbol{y}/dt$, becomes

$$\begin{bmatrix} M & 0 \\ 0 & I \end{bmatrix} \cdot \frac{d}{dt} \begin{bmatrix} \boldsymbol{z} \\ \boldsymbol{y} \end{bmatrix} = - \begin{bmatrix} C & K \\ -I & 0 \end{bmatrix} \cdot \begin{bmatrix} \boldsymbol{z} \\ \boldsymbol{y} \end{bmatrix} + \begin{bmatrix} \boldsymbol{f} \\ 0 \end{bmatrix} \tag{1.1.3}$$

which is again in the form of Equation (1.1.1). In some cases the methods treated in the review can be generalized to second-order systems of ODE directly.

The naming of system matrices and the notation are completely different for different engineering disciplines, but we hope that this does not pose an insurmountable problem. In order to perform a model reduction step, we rewrite Equation (1.1.1) from an implicit to an explicit system of ODEs:

$$\frac{d\boldsymbol{x}}{dt} = A \cdot \boldsymbol{x} + \boldsymbol{b} \tag{1.1.4}$$

where

$$A = E^{-1} \cdot F\,,\; A \in \Re^n \times \Re^n \quad \text{and} \quad \boldsymbol{b} = E^{-1} \cdot \boldsymbol{f}\,,\; \boldsymbol{b} \in \Re^n \tag{1.1.5}$$

It is necessary to stress that Equation (1.1.5) should be read in a mathematical and not in a computational sense. Mathematically this implies that matrix E is not degenerate (ie, it is invertible) and that this transformation is possible in principle. If matrix E is degenerate, then we do not have a system of ODEs, but rather a system of algebraic differential equations (ADEs). From a computational point of view, the operations in Equation (1.1.5) are highly disadvantageous: first, they are prohibitively expensive for large-scale systems, and second, they destroy the sparsity of the original matrices. In other words, computationally it is necessary to work with the two original sparse matrices. The question on how to compute Equation (1.1.5) effectively for the case of Krylov subspace methods is discussed in Section 1.1.4.2.

The main problem with Equation (1.1.4) is the high dimensionality of the vector $\boldsymbol{x}$, which is typically equal to the product of the number of unknowns in a system of PDEs to be solved by the number of nodes introduced during the discretization process. This in turn leads to the high dimension of system matrices and finally to the huge computational cost of solving the system's response.

In performing model reduction on Equation (1.1.4), the hope is that, for many systems of ODEs of practical importance, the behavior of vector $\boldsymbol{x}$ in time, $\boldsymbol{x}(t)$, is effectively described by some low-dimensional subspace as follows:

$$\boldsymbol{x} = X \cdot \boldsymbol{z} + \boldsymbol{\varepsilon}\,,\; \boldsymbol{z} \in \Re^k\,,\; \boldsymbol{k} \ll \boldsymbol{n} \tag{1.1.6}$$

Equation (1.1.6) states that, with the exception of a small error described by vector $\boldsymbol{\varepsilon} \in \Re^n$, the possible movement of the n-dimensional vector $\boldsymbol{x}$ belongs, for all times, to a k-dimensional subspace, with k much smaller than n, and is determined by an $n \times k$ transformation matrix X. The matrix X is composed from k n-dimensional vectors that form a basis for the reduced subspace, and the k-dimensional vector $\boldsymbol{z}$ represents a new low-order set of coordinates for the given basis.

The task of model reduction is to find such a subspace for which the error difference in Equation (1.1.6) is minimal according to some norm

$$\min \|\boldsymbol{\varepsilon}(t)\| = \min \|\boldsymbol{x}(t) - X \cdot \boldsymbol{z}(t)\| \tag{1.1.7}$$

Note that in Equation (1.1.7), we have functions in time, so that the norm in this case is represented by some integral over time [13]. When the subspace is found, Equation (1.1.4) should be projected on to it and this projection process produces a system of ODEs of reduced order k:

$$\frac{d\boldsymbol{z}}{dt} = \hat{A} \cdot \boldsymbol{z} + \hat{\boldsymbol{b}} \tag{1.1.8}$$

which can then be used later on, perhaps in another simulation package.

The physical background for model reduction so far is that the discretization grid used to solve the original PDEs is far from an optimal basis to represent the solution of the PDEs. From this point of view, the model reduction according to Equation (1.1.7) is, in a sense, similar to adaptive grid generation [14]. However, the opportunities of model reduction to minimize the problem dimensionality are much greater, because adaptive grid generation still deals with local shape functions (with local support) and the basis for the low-dimensional subspace in Equation (1.1.6) is formed from global domain functions, that is, each vector includes a contribution from the entire geometrical domain (much as eigenvectors do). From this point of view, model reduction complements adaptive grid generation or makes an alternative in a sense as will now be described.

An adaptive grid generation process starts with some initial grid and then the grid in different parts of the computational domain becomes refined or coarsened based on a priori or a posteriori local error estimators [15]. A model reduction strategy requires a fine initial grid, for which it produces an effective global low-dimensional basis, based on global error estimators. Then, in order to choose the best computational strategy, it is necessary to compare the time taken for model reduction of a system of ODEs built on the fine grid with the sum of times for adaptive grid generation and the subsequent model reduction of the refined grid system of ODEs.

We now take the next step and put the model reduction problem into a more general form. Often, engineers are not interested in the solution of Equation (1.1.4) over the entire computational domain, that is, for values at all nodes, but rather in only a few of their combinations. Control theorists [16] take this into account and convert Equation (1.1.4) to

$$\begin{cases} \dfrac{dx}{dt} = A \cdot \boldsymbol{x} + B \cdot \boldsymbol{u} \\ \boldsymbol{y} = C \cdot \boldsymbol{x} \end{cases} \tag{1.1.9}$$

Equation (1.1.9) treats the system as a "black box", which would be the case when a system's high-dimensional internal state vector $\boldsymbol{x}$, governed by ODEs, is not directly accessible to an external observer. The observer can influence the system state by some input functions, specified by the vector $\boldsymbol{u} \in \Re^m$, and which are distributed to the internal nodes in accordance with the *scatter* matrix $B \in \Re^n \times \Re^m$. The number of input signals $m \ll n$ is typically small and this means that matrix B has a small number of columns. On the other hand, the observer is interested in only a few outputs, specified by vector $\boldsymbol{y} \in \Re^p$ with the dimension $p \ll n$. The relationship between required outputs and the system state is given by the *gather* matrix $C \in \Re^p \times \Re^n$. As a result, we have a high-dimensional system of ODEs in relation to vector $\boldsymbol{x}$, the system state vector, which is governed by a small number of external inputs and, from the viewpoint of an external observer, contains a small number of relevant outputs. We will not describe here the well-known system-theoretical results of this equation, such as zero state and zero input, but refer the curious reader to the control theory literature [16].

Equation (1.1.9) is a generalization of Equation (1.1.4). If matrix B in Equation (1.1.9) represents a single vector, equal to vector $\boldsymbol{b}$ of Equation (1.1.1), then vector $\boldsymbol{u}$ will contain only one element, a single input, and we can equate it to a step function. Now let us say that matrix C is an identity matrix, that is, $\boldsymbol{y} = \boldsymbol{x}$, then we have a special case with respect to the original system of ODEs, which we call "single input – *complete* output" or SICO.

The multiple input case holds when matrix B has several columns corresponding to multi-load simulations or when the system is consecutively subjected to a variety of loads distributed to different nodes. In this case, each function in vector $\boldsymbol{u}$ has a "step" shape limited by the application time of the load (see Figure 1.1.2). Matrix C is usually formed by picking only those rows from the unit matrix which correspond to chosen nodes. In this case, vector $\boldsymbol{y}$ is just a small subset of the state vector $\boldsymbol{x}$.

The problem of model reduction in the case of Equation (1.1.9) consists in the reduction of the dimension of the state vector to order $k \ll n$ while retaining the same number of inputs and outputs

$$\begin{cases} \dfrac{dz}{dt} = \hat{A} \cdot z + \hat{B} \cdot \boldsymbol{u} \\ \hat{\boldsymbol{y}} = \hat{C} \cdot z \end{cases} \tag{1.1.10}$$

The input vector $\boldsymbol{u}$ in Equation (1.1.10) is exactly the same as in Equation (1.1.9), but the output vector $\hat{\boldsymbol{y}} \in \Re^k$ is just some approximation of the original vector $\boldsymbol{y} \in \Re^k$. This transformation is sketched in Figure 1.1.3.

The quality of the model reduction step of Equation (1.1.10) is determined by a norm

$$\min ||\boldsymbol{y}(t) - \hat{\boldsymbol{y}}(t)|| = \min ||\boldsymbol{y}(t) - \hat{C} \cdot z(t)|| \tag{1.1.11}$$

which ideally should hold for any input vector $\boldsymbol{u} \in \Re^m$. The difference between Equations (1.1.11) and (1.1.7) is that now we search for a reduced subspace given by Equation (1.1.6) to minimize the difference between given outputs only and not for the whole state vector. Certainly, if we have found a subspace that

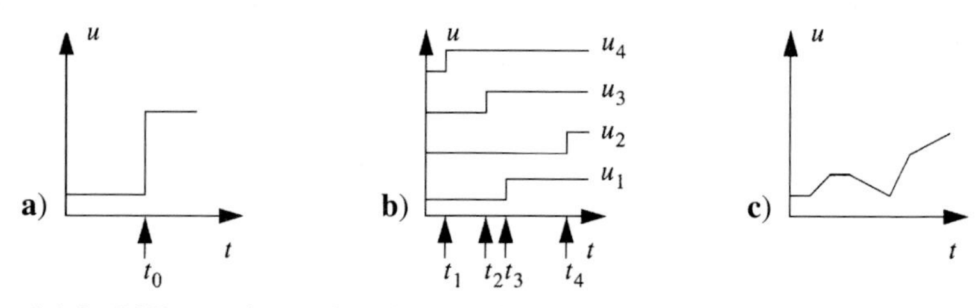

Figure 1.1.2. Different input functions, often provided in engineering simulation programs. (**a**) A step function activating at t_0. (**b**) A vector of step functions, each activating at a different time. (**c**) A piecewise linear function.

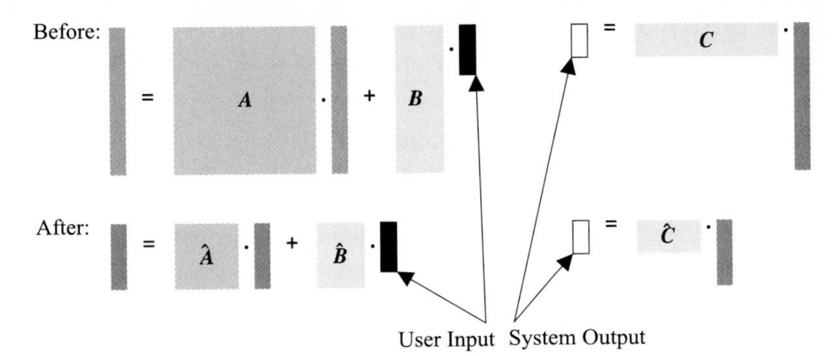

Figure 1.1.3. Sketch of the model reduction equations (1.1.9) before and (1.1.10) after the model reduction step. The dimensions of the system matrices A, B and C and the internal state vectors z and $\dot{z}$ are significantly smaller after model reduction. The input vector $\boldsymbol{u} \in \Re^m$ and output vector $\boldsymbol{y} \in \Re^k$ remain the same size.

minimizes Equation (1.1.7), then Equation (1.1.11) will be satisfied automatically. However, we expect that a subspace minimizing Equation (1.1.11) will have a much lower dimension than a subspace minimizing Equation (1.1.7).

If matrices B and C both consist of a single column and row, respectively, then the system is termed single input-single output (SISO), otherwise it is referred to as multiple input-multiple output (MIMO).

A dynamic system is often considered in the frequency domain, when the Laplace transform operator $L\{.\}$ is applied to the input and output vectors [13]

$$L\{\boldsymbol{y}(t)\} = \boldsymbol{Y}(s) \ , \ L\{\boldsymbol{u}(t)\} = \boldsymbol{U}(s) \tag{1.1.12}$$

and where the relationship between input and output is described by the transfer function

$$\boldsymbol{Y}(s) = G(s) \cdot \boldsymbol{U}(s) \tag{1.1.13}$$

Most of the results in model reduction obtained so far concern the case of a linear system of ODEs and where all the matrices of Equations (1.1.4) and (1.1.9) are composed of constant numbers. In this case, the transfer function is readily expressed via the system matrices as

$$G(s) = C \cdot (sI - A)^{-1} \cdot B \tag{1.1.14}$$

1.1.3 Small Linear Systems

Control theory has a very strong theoretical result for stable systems, ie, those systems for which the real parts of all the eigenvalues of the system matrix A in Equation (1.1.9) are negative. Each linear dynamic system (1.1.9) has n so-called Hankel singular values, σ_i (see Ref. [17] for mathematical details), which can be computed if one solves two Lyapunov equations

$$A \cdot P + P \cdot A^T = -B \cdot B^T \tag{1.1.15}$$

$$A^T \cdot Q + Q \cdot A = -C^T \cdot C \tag{1.1.16}$$

for the controllability grammian P and observability grammian Q. Then the Hankel singular values of the original dynamic system are equal to the square root of the eigenvalues of the product of the controllability and observability grammians

$$\sigma_i = \sqrt{\lambda_i(P \cdot Q)} \tag{1.1.17}$$

Once these values are known, there are a number of model reduction methods with guaranteed error bounds for the difference between the transfer function of an original k-dimensional system, as follows:

$$||G - \hat{G}||_\infty \leq 2(\sigma_{k+1} + \ldots + \sigma_n) \tag{1.1.18}$$

provided that the Hankel singular values have been sorted in descending order. Note that this equation is valid for arbitrary input functions. This means that model reduction based on these methods can be made fully automatic. A user just sets an error bound and then, by means of Equation (1.1.18), the algorithm finds the smallest possible dimension of the reduced system, k, which satisfies that bound. Alternatively, a user specifies the dimension of the reduced system and the algorithm estimates the error bound for the reduced system.

Another practical consequence of this result is that the success of model reduction depends only on the decay rate of the Hankel values. Figure 1.1.4 shows examples of the behavior of Hankel values for a few typical applications. If we can estimate this decay rate for a particular application, this would give us a complete answer as to the extent to which we could reduce the original system [18, 19].

The SLICOT library implements three methods, a balanced truncation approximation, a singular perturbation approximation and the Hankel-norm approximation, as well as including a special benchmark problem [20, 21]. All three methods and their variations are extensively used in control theory and there are numerous examples of their applications. However, they are outside the scope of the present review, since, owing to computational reasons, they are limited to relatively small systems.

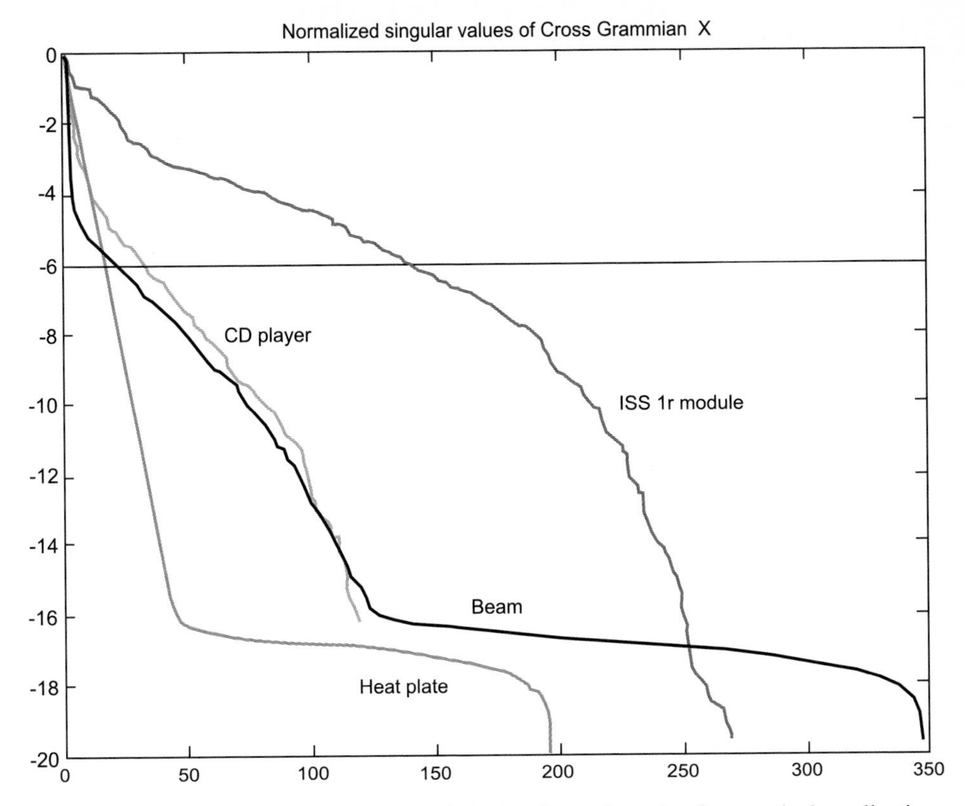

Figure 1.1.4. Decay of normalized Hankel singular values for four typical applications (from Ref. [7]). We expect these curves to also be typical for MEMS.

The time required to solve the Lyapunov equations and to perform a singular value decomposition grows as the cubic power in the number of equations or is $O(n^3)$. Hence, if the system order increases twofold, the time required to solve a new problem will increase about eightfold. In other words, even though the results described above are valid for all linear dynamic systems, in practice we can use them for small-order systems only.

The border between small and large systems depends on the computer power available and of course it steadily grows. According to Ref. [20], the model reduction of a randomly generated linear system of order 512 takes 76 s on a 400 MHz Pentium II processor PC. Since processors now promise a 2 GHz clock speed, this enables us to define current small systems as those with state vector dimensions in the range 1000–2000.

1.1.4 Introduction to Krylov Subspaces

It happens that, in many cases, very good candidates for the required low-order subspace of Equation (1.1.6) are Krylov subspaces and almost all modern model reduction methods for large-scale systems are based on them, one way or another. It should be noted that those iterative methods for solving a system of linear equations that are based on Krylov subspaces have been included in the list of the 10 top algorithms of the 20th century [11].

A Krylov subspace of k-th dimension of the matrix $A \in \Re^n \times \Re^n$ and vector $\boldsymbol{v} \in \Re^n$ is defined as a subspace spanned by the original vector $\boldsymbol{v}$ and the vectors produced by consecutive multiplication of the matrix A to this vector up to $k-1$ times, or

$$K_k^r(A, \boldsymbol{v}) = \text{span}\{\boldsymbol{v}, A \cdot \boldsymbol{v}, \ldots, A^{k-1} \cdot \boldsymbol{v}\} \tag{1.1.19}$$

The resulting vectors form a basis for k-dimensional subspace. However, if we compute them directly as written, then, because of rounding errors, they would become computationally linearly dependent even for relatively small k.

1.1.4.1 Arnoldi and Lanczos Algorithms to Build the Krylov Subspace

A numerically stable procedure for building a Krylov subspace (1.1.19) is an Arnoldi process [11, 22, 23]. It generates an orthonormal basis $X \in \Re^n \times \Re^k$ for the Krylov subspace and a Hessenberg matrix, $H_A \in \Re^k \times \Re^k$, related to the original matrix as follows:

$$X^* A X = H_A \tag{1.1.20}$$

The Hessenberg matrix for the Arnoldi process is made of an upper triangular matrix plus one diagonal below the main diagonal. It can be considered as an orthogonal projection of the matrix A on to the given Krylov subspace.

The main disadvantage of the Arnoldi method is that each new Arnoldi vector should be orthogonal to all previously generated vectors. This means that the computational cost grows disproportionately with the dimension of the subspace. The current alternative is to use a Lanczos algorithm, where the subspace (1.1.19) is considered as a right Krylov subspace. In addition to it and in parallel, the left Krylov subspace

$$K_k^l(A^*, \boldsymbol{w}) = \text{span}\{\boldsymbol{w}, A^* \cdot \boldsymbol{w}, \ldots, (A^*)^{k-1} \cdot \boldsymbol{w}\} \tag{1.1.21}$$

is also generated, where the vector $\boldsymbol{w}$ can be equal or not to vector $\boldsymbol{v}$ depending on the applications, and A^* is the conjugate transpose of the matrix A.

The Lanczos algorithm produces a pair of biorthogonal bases for subspaces (1.1.19) and (1.1.21) contained in the matrices X and Y such that

$$Y^* \cdot X = I \tag{1.1.22}$$

and a Hessenberg matrix H_L that is in tridiagonal form. This means that, for any iteration of the algorithm, it is necessary to deal with just two previously generated vectors. The Lanczos Hessenberg matrix is related to the original matrix as

$$Y^* \cdot A \cdot X = H_L \tag{1.1.23}$$

and can be considered to be an oblique projection of A on to the subspace (1.1.19) while remaining perpendicular to subspace (1.1.21). Figure 1.1.5 illustrates the orthogonal and oblique projections of a vector. Because the Lanczos algorithm is based on three-term recurrences, it is faster for large k. However, it is computationally less stable than the Arnoldi process: a typical trade-off of accuracy vs efficiency. The Lanczos and Arnoldi algorithms are mathematically equivalent if the matrix A is symmetric and the starting vectors $\boldsymbol{v}$ and $\boldsymbol{w}$ are the same, in other words, when the Krylov subspaces (1.1.19) and (1.1.21) are equivalent.

Instead of just one starting vector $\boldsymbol{v}$, one can take a number of starting vectors expressed by the matrix B. This leads to a generalization of the Arnoldi and Lanczos algorithms to the so-called block-Arnoldi and block-Lanczos algorithms [24, 25]. Here we define the appropriate right and left Krylov subspaces as

$$K_k^r(A, B) = \text{span}\{B, A \cdot B, \ldots, A^q \cdot B\} \tag{1.1.24}$$

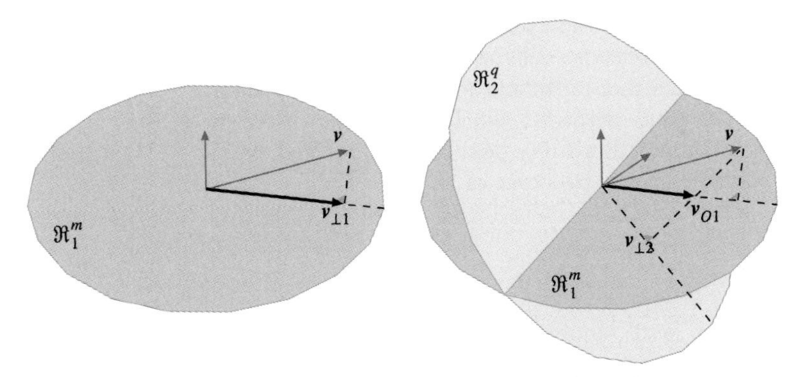

Figure 1.1.5. Example of an orthogonal and an oblique projection of a vector. The disks represent subspaces $\mathfrak{R}_1^k \subset \mathfrak{R}^n$ of the real linear space $\mathfrak{R}^n$ of a model reduction problem. A general vector $\boldsymbol{v} \in \mathfrak{R}^n$ is projected on to a subspace. The left figure illustrates an orthogonal projection $\boldsymbol{v}_{\perp 1} \in \mathfrak{R}_1^k$. The figure on the right demonstrates an oblique projection $\boldsymbol{v}_{O1} \in \mathfrak{R}_1^k$ determined by the "shadow" cast by an orthogonal projection of $\boldsymbol{v}$ on to a second subspace $\boldsymbol{v}_{\perp 2} \in \mathfrak{R}_2^k$.

$$K_k^l(A^*, C) = \text{span}\{C, A^* \cdot C, \ldots, (A^*)^q \cdot C\} \tag{1.1.25}$$

One difficulty with the block-Krylov subspaces is that it is more difficult to predict the number of multiplications in Equations (1.1.24) and (1.1.25) in order to generate a k-dimensional subspace. Typically, q is equal to the ratio of k by the number of columns of the matrices B or C, but the exact answer depends on the existence of linearly dependent vectors in Equations (1.1.24) and (1.1.25).

1.1.4.2 Computing the Inverse of the System Matrix

One computational advantage of all Krylov subspace methods is in their iterative nature, ie, to perform them a user only needs to provide consecutive matrix by vector multiplications. This allows us to exploit the sparse form of the matrices and to create fast application-specific implementations for the required $A \cdot \boldsymbol{v}$ product. The driver algorithms do not have to know the details of how matrix A is stored in the computer system.

For model reduction problems, the Krylov subspace (1.1.19) is actually based on the inverse of the system matrix A. Recalling Equation (1.1.5), this means that for both the Arnoldi and Lanczos processes it is necessary to compute the product

$$F^{-1} \cdot E \cdot \boldsymbol{v} \tag{1.1.26}$$

We now discuss, using this example, the advantage of the iterative structure of the Krylov subspace methods.

F and E are large-dimensional sparse matrices, but the product $F^{-1} \cdot E$ might be a dense matrix and the computational cost of computing this product is very high owing to the presence of the matrix inverse. Hence its direct computation is unwise. It is much more efficient to compute the product $F^{-1} \cdot E \cdot \boldsymbol{v}$. First, before the procedure, one performs an LU decomposition of F (or, equivalently, a Cholesky decomposition for a positive definite matrix, see [23]), which can take into account the sparse structure of F:

$$F = L \cdot U \tag{1.1.27}$$

where L and U are lower and upper triangular matrices, respectively. This is costly, but we will require L and U many times. Then, each multiplication $F^{-1} \cdot E \cdot \boldsymbol{v}$ is performed in three steps:

1. E is multiplied by $\boldsymbol{v}$, $\boldsymbol{a} = E \cdot \boldsymbol{v}$. E is sparse and so this is potentially a fast operation.
2. (a) The linear equations $L \cdot \boldsymbol{b} = \boldsymbol{a}$ are forward solved, so that $\boldsymbol{b} = L^{-1} \cdot E \cdot \boldsymbol{v}$. Since L is lower triangular, this is again a fast operation.

(b) The backward solution of the linear equations $U \cdot \boldsymbol{c} = \boldsymbol{b}$ then gives us the desired product $U^{-1} \cdot L^{-1} \cdot E \cdot \boldsymbol{v}$. Again, since U is upper triangular, this is a fast operation.

Once again, the above speedup is possible only because higher level algorithms do not need to have access to the full matrix A^{-1}; otherwise we would have no option but to compute it.

When the dimension of A grows large enough, LU decomposition is no longer useful because it takes too much time. Hence, the second step above changes to

3. The linear equations $F \cdot \boldsymbol{b} = \boldsymbol{a}$ are solved by an iterative method, $\boldsymbol{b} = F^{-1} \cdot E \cdot \boldsymbol{v}$. If one is lucky, an iterative method can be reasonably fast for a particular F matrix.

Iterative methods for the solution of a system of linear equations are also based on Krylov subspaces and it is important not to confuse them with those reviewed in this chapter. The modified step above implies that, for any computation of the subsequent Krylov vector, it is necessary to use second-level iterations to solve the linear system of equations. In addition to two books [22, 23], an excellent object-oriented template-oriented review of Krylov-based methods for the solution of linear systems can be found in [26]. It should be noted that the success of iterative Krylov methods for a linear solve step depends on the structure of the matrix and, for the general case, their effective use requires finding a preconditioner, another matrix P, which transforms the original linear system to an equivalent $P \cdot T \cdot \boldsymbol{b} = P \cdot \boldsymbol{a}$, but which has superior convergence properties. For a discussion on the importance of preconditioning for solving linear systems that are generated through the discretization of PDEs by the finite element method, see Ref. [27].

1.1.5 Large Linear Systems

As was already mentioned, algorithm time complexity limitations do not allow us to employ control theory algorithms directly for large-scale systems. As a result, most of the practical work in model reduction of large linear dynamic systems have been tied to Padé approximants of the transfer function (1.1.14) and we start the present section with these. These methods are computationally feasible but, on the other hand, they do not provide a global error estimate. Recently, there have been considerable efforts to find computationally effective strategies in order to apply methods based on Hankel singular values to large-scale systems and we briefly review them in the second part of this section.

1.1.5.1 Approximating a Transfer Function by Padé and Padé-type Approximants

For the case of single input-single output (SISO) systems, when matrices B and C are both composed of a single column and row, respectively, the transfer matrix (1.1.14) is a scalar rational function which can always be expressed in the factored form as

$$G(s) = \frac{a(s - z_1)\dots(s - z_{n-1})}{(s - p_1)\dots(s - p_n)} \tag{1.1.28}$$

where z_i and p_i are zeros and poles of the transfer function and a is a constant. In the multiple input–multiple output (MIMO) case, a transfer matrix is of dimension p by m, each element of which is a function of the form of Equation (1.1.28).

The idea of Padé [28] and Padé-type [29] approximants is to find a rational function of smaller dimension k, $\hat{G}(s)$, which retains the essential behavior of the large-dimensional original rational function. This is formulated in terms of moment matching in the expansion of the transfer functions around some given number s_o (in most applications $s_o = 0$),

$$G(s) = \sum_{i=0}^{\infty} m_i (s - s_o)^i \tag{1.1.29}$$

that is,

$$m_i = \hat{m}_i \quad \text{for} \quad i = 0, \dots, q \tag{1.1.30}$$

Padé approximants match the maximum number of moments, $q = 2k$, while Padé-type approximants match first $q < 2k$ moments. This is easily generalized to the MIMO system, where all moments will be $p \times m$ matrices.

It happens that the Arnoldi process for the right Krylov subspace

$$K_k^r\{(A - s_o I)^{-1},\ (A - s_o I)^{-1} \cdot B\} \tag{1.1.31}$$

produces such matrices H_A and X such that the reduced system

$$\hat{A} = H_A^{-1} \cdot (I + s_o H_A)\ ,\ \hat{B} = H_A^{-1} \cdot X^* \cdot (A - s_o I)^{-1} \cdot B\ ,\ \hat{C} = C \cdot X \tag{1.1.32}$$

implicitly matches the first k moments in Equation (1.1.29), that is, the Arnoldi process implicitly produces a Padé-type approximant of the original transfer function (1.1.14). On the other hand, if one performs Lanczos algorithms for the right (1.1.31) and left Krylov subspaces

$$K_k^l\{[(A - s_o I)^{-1}]^*,\ [(A - s_o I)^{-1}]^* \cdot C^*\} \tag{1.1.33}$$

matrices H_L, X and Y produce the reduce system

$$\hat{A} = H_L^{-1} \cdot (I + s_o H_L) \ , \ \hat{B} = H_L^{-1} \cdot Y^* \cdot (A - s_o I)^{-1} \cdot B \ , \ \hat{C} = C \cdot X \tag{1.1.34}$$

which will match $2k$ moments of the original dynamic system [24, 25, 30, 31]. Note that Equations (1.1.32) and (1.1.34) greatly simplify for the case of an expansion about $s_o = 0$ and that there are computationally more effective equations for producing the reduced matrices $\hat{B}$ and $\hat{C}$.

The Lanczos algorithm produces a reduced system closer to the original one, because the number of moments matched here is twice that of the Arnoldi process. This has a simple explanation. Model reduction by the Arnoldi process does not take into account matrix C at all, while model reduction by means of the Lanczos algorithm is made by an oblique projection on the right Krylov subspace (1.1.31) that takes into account the left Krylov subspace (1.1.33).

Still, both approaches are based on moment matching and they are by nature local, in the sense that, in general, they make a good approximation of the transfer function (1.1.14) near the expansion point s_o only. This can be improved by multi-point expansion, ie, expanding the transfer function about several points s_i and requiring the reduced transfer function to match the first moments at all expansion points. This idea was implemented in the so-called Rational Krylov method [31, 32], where the Arnoldi or Lanczos algorithms were applied to the union of the Krylov subspaces (1.1.31) and (1.1.33) for different values of s_i. The main methodological challenge here is to decide how to choose the expansion points and to determine how many are needed. Computationally this adds an additional load. If one uses LU decomposition for the inverse of the system matrices (see Section 1.1.4.2), then in this case it is necessary to perform an LU decomposition for each value of s_i.

The original dynamic system can be stable, that is, when time goes to infinity the values of $\boldsymbol{x}$ remain finite (bounded) and passive, which is to say, the system does not generate energy. If so, then it is important, especially in electrical circuit simulations, that the reduced system also possesses these properties. Unfortunately, both the "out-of-the-box" Arnoldi and Lanczos algorithms do not guarantee this and special attention should be paid to preserving the properties of the original dynamic system. It happens that the Arnoldi process is mathematically simpler than Lanczos algorithm (this is stressed by their names, process and algorithms, respectively). Probably for this reason, engineers often choose the Arnoldi process: the coordinate transformed Arnoldi [33] for stable model reduction and the provably passive model reduction method "block Arnoldi plus congruent transform" (or PRIMA) of Ref. [34]. On the other hand, mathematicians still prefer the Lanczos algorithms [24, 25, 35, 36] because, as was mentioned above, it takes into account the observability matrix C and it matches twice the number of moments of the Arnoldi process. They seem to prefer, while preserving the properties of the original dynamic system, to match as many moments as possible so as to obtain the most accurate representation for the same dimension k of the reduced model. It is also worth noting that, even though when some algorithm provably produces a passive reduced model, this does not mean that its computer implementation will really produce a passive model in practice, mainly because of the inevitable numerical rounding errors [24].

Now let us return to the original case of model reduction for systems (1.1.4)–(1.1.8). From a control theory viewpoint we term it single input-complete output-convection partial differential equation, then the Krylov subspace (1.1.31) with $s_o = 0$ is a very good choice for the lower dimensional subspace in Equation (1.1.6) [8, 9, 37–43]. In this case, the model reduction step (1.1.8) can be viewed as an approximate solution of the original system (1.1.4), because it is possible to recover the solution for all of the original unknowns by means of Equation (1.1.6). This work has been superseded by the use of a Krylov subspace (1.1.31) to approximate the matrix exponential [10, 23], but mathematically this is identical with a Padé-type approximant (1.1.32) when the matrix C is just discarded.

We next list examples of the papers in which Padé and Padé-type approximants via Krylov subspace methods have been used for the model reduction of a linear system (1.1.9). The papers come from several distinct communities:

- The largest community comes from electrical engineering where model reduction is mostly employed to deal with the so-called microchip interconnect problem [44–46]: mixed surface volume for 3D interconnect [47], lossy multiconductor transmission lines [48], 3D interconnect and packaging based on an alternative partial element equivalent circuit (PEEC) formulation [49], coupled lossy transmission lines [50], magnetoquasistatic analysis for packaging parasitics with skin effect [51], PEEC model of an electromagnetic problem [52], electromagnetic devices modeled by linearized Maxwell equations [53], full-wave electromagnetic analyses [54] and electromagnetic wave propagation by the finite element method (FEM) [55]. The actual number of publications on model reduction here is much greater.
- The ideas from electrical engineers have been used for the model reduction of wave propagation-like problems: the Helmholtz equation for exterior structural acoustics by FEM [56], neutron noise for nuclear reactor by the finite difference method [57] and aeroelastic analyses of turbomachines [58].
- Another community solves the advection-diffusion PDE, which arises in a variety of engineering disciplines. They mostly deal with the SICO case discussed above. Here, model reduction is at the beginning stage if we compare the number of papers in which model reduction is used to the total number of papers on the solution of advection-diffusion PDEs: advection dispersion equation for groundwater flow [37, 38], mass transport in hydrogeological environments [39], photon diffusion (optical tomography) problem [40], radionuclide decay-chain transport in porous media [41], groundwater flow in dual-porosity media [42], radionuclide decay chain transport in dual-porosity media [43], groundwater flow in discretely fractured porous media [8] and diffusion and convection dominated flow [9].
- Finally, we have the MEMS community which has just recently started to exploit the modern opportunities of model reduction: electrostatic gap-closing

actuator [59], linearized model for a micromirror [60] and the comb-drive resonator [61]. It is interesting that the MEMS community appears to have learned about model reduction from the electrical engineers and is not aware of work on the model reduction of advection-diffusion PDEs, even though this body of work is much closer to typical MEMS simulations.

1.1.5.2 Approximating Lyapunov Equations

Unfortunately, Padé and Padé-type approximants do not have global error estimates, similar to Equation (1.1.18), and this drives mathematicians to develop computationally effective strategies for large-dimensional systems based on the methods described in Section 1.1.3. In [7] these approaches are referred to as SVD-Krylov and in [62] there is a good overview of existing strategies.

The optimal minimal reduction methods for linear systems comprise two computationally expensive steps: solution of Lyapunov Equations (1.1.15) and (1.1.16) for the controllability and observability grammians and then eigenvalue-type decomposition of the product of two grammians, Equation (1.1.17). The computational time for both steps, even using the most advanced computational methods [63], grows as the cube of the system dimension.

A general idea for decreasing the computational time is to change the exact grammians to their low-rank approximations. It happens that this is possible if the numbers of inputs and outputs are much less than the dimension of the state vector, $m \ll n$ and $p \ll n$, and this is the case for the most important practical applications. As a result, it is possible to solve Lyapunov equations for low-rank grammian approximations much faster than for exact grammians [64–67]. For the case of a dense matrix A, the computational time is already proportional to the square of the system dimension n and it may be linearly proportional to n for the case of a sparse matrix A. Also, the advantage of these methods is that they can be formulated in terms of matrix-vector products only, as for the Krylov subspace methods. The second step, balancing, with the use of low-rank grammians, is also much faster because there are special algorithms that can take this into account [62, 67].

A very simple case of model reduction arises when the inputs are the same as the outputs and matrix A is symmetric. Note that if matrices E and F in Equation (1.1.1) are symmetric and E is positive definite, then by an appropriate coordinate transformation one can obtain Equation (1.1.4) with a symmetric matrix A [33]. In this case, the grammians are equal to each other because Equations (1.1.15) and (1.1.16) become the same: then it is necessary to solve just a single Lyapunov equation and there is no need to perform balancing. Another approach is to use, instead of the two Lyapunov Equations (1.1.15) and (1.1.16), the Sylvester equation [68]:

$$A \cdot R + R \cdot A = -B \cdot C \tag{1.1.35}$$

to find the so-called cross-grammian R. It happens that in the case of a linear dynamic system with a symmetric transfer function, the Hankel singular values are equal to the eigenvalues of the cross-grammian and here there is also no need for balancing. This is always true for any SISO system, because in this case the transfer function is a scalar. In the MIMO case, one can use a transformation described in Ref. [68] in order to convert any linear dynamic system to one with a symmetric transfer function.

Some methods for model reduction based on solving large-dimensional Lyapunov equations are implemented in the library LYAPACK [69] (this requires MATLAB). As mentioned in Ref. [69], Lyapunov equations of order more than 12000 were solved by LYAPACK within a few hours on a regular workstation.

1.1.6 Nonlinear Systems

Now let us allow the elements of the system matrices to depend on the state vector $\boldsymbol{x}$ and on time. If they depend explicitly on time only, then we have a special case of a time-varying system and there are examples of extending Krylov subspace model reduction methods to this case [54, 70].

Note that, even when system matrices depend on $\boldsymbol{x}$, Equation (1.1.4) is a special case of a general nonlinear system:

$$\frac{d\boldsymbol{x}}{dt} = f[\boldsymbol{x}(t), \boldsymbol{u}(t)] \tag{1.1.36}$$

An evident solution for model reduction is to split the whole system into nonlinear and linear parts and then to apply the model reduction to the linear subparts [24], thus reducing the total number of unknowns in the state vector. Another popular alternative is to linearize the nonlinear system around an operating point and then to apply model reduction for the resulting linear system. The answer as to whether this is possible depends on the application in question. There is an interesting example in Refs [71] and [72] where, in order to improve the precision of the linearization process, the authors included quadratic terms in the expansion.

There are some methods for model reduction of nonlinear systems applicable to small-dimensional problems [73, 74] and some special cases where it is possible to find particular approaches which allow us to use ideas from the previous section [59, 75, 76]. Nevertheless, to our knowledge, for the general case of model reduction of large nonlinear systems, there appears to be one approach only, which we consider in the next section.

1.1.6.1 Proper Orthogonal Decomposition

For systems with strong nonlinear effects, linearization is impossible because a linearized system cannot capture the complexity of the original phenomena. We remind ourselves that nonlinear systems may show instabilities such as snap-through and bifurcations and ultimately the onset of chaotic behavior, all of which should be represented in the reduced system. In this case, in order to find an appropriate low-dimensional subspace (1.1.6), one can use results of the full order simulation of the original dynamic system (1.1.4) and this is implemented within the proper orthogonal decomposition (POD; another popular name is Karhunen-Loève decomposition) [77]. This is the main difference with respect to linear systems, where the model reduction process can be based on the system matrices without performing a full order simulation.

Let us consider a slightly simplified procedure for a finite-dimensional system. The first step is to perform one or more simulations and to collect a series of so-called "snapshots":

$$W = \{\boldsymbol{x}_i\} \ , \ W \in \Re^n \times \Re^s \tag{1.1.37}$$

where matrix W is composed from s state vectors $\boldsymbol{x}_i$, corresponding to different times of simulations of Equation (1.1.4). This is the most crucial step during POD because the reduced basis will be obtained from matrix W only and if it does not give a good representation of the whole ensemble of possible values of $\boldsymbol{x}$, then the generated low-dimensional basis will lead to a poor quality of approximation. If, for linear systems, it was possible to perform model reduction for any input functions, for nonlinear systems it is necessary to choose the most typical input functions and to perform simulations with them. Unfortunately, there exist almost no formal rules as to how to choose the number "snapshots" to collect and at what times they should be taken. Hence, POD is more of an "art" and typically, for any new nonlinear system, it is necessary to make a special investigation in this respect.

Nevertheless, the following POD steps are completely formal. For a given "snapshot" matrix W, it is formally possible to find a low-rank approximation within a given error margin by means of a singular value decomposition (SVD) [7, 23]:

$$W = U \cdot \Sigma \cdot V^T = \sum_{i=1}^{s} \sigma_i (\boldsymbol{u}_i \cdot \boldsymbol{v}_i^T) \ , \ \Sigma \in \Re^n \times \Re^s \ , \ U \in \Re^n \times \Re^n \ , \ S \in \Re^s \times \Re^s \tag{1.1.38}$$

where $\Sigma = \text{diag}\{\sigma_i\}$ is a diagonal matrix of singular values, $U = \{\boldsymbol{u}_i\}$ is a matrix of left singular vectors, and $V = \{\boldsymbol{v}_i\}$ is a matrix of right singular vectors. Provided that the singular values of W rapidly decay, we can take only a small number singular vectors, $k \ll s$, corresponding to the largest singular values and this gives us a low-rank approximation of matrix W of the form

$$\hat{W} = \hat{U} \cdot \hat{\Sigma} \cdot \hat{V}^T = \sum_{i=1}^{k} \sigma_i (\boldsymbol{u}_i \cdot \boldsymbol{v}_i^T) \tag{1.1.39}$$

where the reduced matrices are formed from the full matrix by leaving only k dominant vectors. Equation (1.1.39) shows that all observations are effectively described by a small number of vectors $\boldsymbol{u}_i$, which gives a reduced basis on which to project the original differential equation:

$$X = \hat{U} \tag{1.1.40}$$

The transition from Equation (1.1.37) to (1.1.39) can be made completely automatic because according to SVD theory there is an error estimate based on singular values with the norm

$$||W - \hat{W}|| \tag{1.1.41}$$

and Equation (1.1.39) actually reduces this norm to a minimum. The problem is that it is difficult to predict, a priori, whether this error estimate can be used for the transition from Equation (1.1.4) to (1.1.8), because this already strongly depends on the quality of the generated "snapshots", that is, whether they are representative or not.

The final step is to project original nonlinear equation on to the low-dimensional basis. For Equation (1.1.4), when the elements of A and $\boldsymbol{b}$ depend on $\boldsymbol{x}$, we can write

$$\hat{A} = X^* \cdot A \cdot X \quad \text{and} \quad \hat{\boldsymbol{b}} = X^* \cdot \boldsymbol{b} \tag{1.1.42}$$

and, for the general case of Equation (1.1.36), the reduced model becomes

$$\frac{dz}{dt} = X^* \cdot \boldsymbol{f}[X \cdot z(t) \; , \; \boldsymbol{u}(t)] \tag{1.1.43}$$

There is a hidden computational problem with Equation (1.1.42), momentarily ignoring Equation (1.1.43), namely how to compute the reduced system matrices. Matrix A contains some functions of $\boldsymbol{x}$ and hence Equation (1.1.42) should be computed by means of symbolic manipulations. This is practically unfeasible. In the general case one may only compute the right sides in Equation (1.1.42) for each time step during the simulation of the reduced model and this then constitutes the main computational cost. For example, in Ref. [78], the size of the state vector has been reduced from 21540 to 15 (about 1500 times) but, for the above reason, the time of the simulation was reduced only by a factor of six.

POD has been used extensively in fluid dynamics in order to model turbulence [77]. Recently, it has been employed in a variety of disciplines tied with nonlinear dynamics: rapid thermal processing systems [79], control of a solid fuel ignition [80], chemical vapor deposition [81, 82] and other distributed reacting systems [83–86], cascading failures in power systems [87], feedback control of systems governed by a nonlinear PDE [88–90] and various mechanical engineering problems [78, 91–93]. The MEMS community has also started to employ this technique [71, 94, 95].

The SVD decomposition of a matrix is a computationally demanding method: the time grows as the cube of the matrix dimension. This means that when the dimension of matrix W grows we might not have enough computational resources in order to make the decomposition (1.1.38). It happens that again iterative methods based on the Krylov subspaces can help to find the dominant singular vectors without performing the full SVD decomposition [96, 97] and thus keeping computational time within reasonable limits.

The original POD procedure does not take into account the information about required system inputs and outputs and this limits its applicability in system simulation. Recently, the method has been generalized [98, 99] in order to take into account ideas from the linear control theory. The generalization is based on the introduction of "empirical grammians" which are computed based on "empirical snapshots". This opens up new perspectives for applications of POD to nonlinear model reduction and hopefully in the future we will see further development of these ideas.

1.1.7 Conclusion

Let us summarize the current status of automatic model order reduction. The situation is reasonably good for large-scale linear dynamic time-invariant systems. The moment matching methods for model reduction based on the Arnoldi and Lanczos algorithms are in a mature state. They scale well with the size of the system, their behavior is fairly predictable and they are easily implemented in almost any computational environment. As already mentioned, the Arnoldi process is more computationally stable and one can implement it much more easily than the Lanczos algorithm. On the other hand, the latter can match more moments and thus provides a better approximation of the original system. As a result, the Arnoldi process is the best choice for those who would like to implement moment matching methods fast and from scratch and it is better to obtain the implementation of the Lanczos algorithm from professional sources.

A typical question with moment matching techniques is when to stop model reduction. A good strategy is provided in Ref. [52], where a local error estimate is suggested for model reduction based on the Lanczos algorithm. First, it is necessary to estimate a range of frequencies in which the approximation of the transfer function is required. It is possible to set s_o to an expansion point in the middle of this range and then to use the local error estimate on the border of this range as a monitor as to when to stop the model reduction process, because the approximation error typically increases faster the further one is from the expansion point. This procedure still does not give a global error estimate as the balanced truncation approximation does, but for most engineering purposes this should be good enough.

Another problem is that Padé and Padé-type approximants are local by their nature and they might be not optimal if one would like to obtain a good approximation of the transfer function over a wide range of s, that is, the dimension of the reduced model might then be too large. Here one can think of a Rational Krylov approximation or to employ the two-step strategy suggested in Ref. [51]. First one computes a medium-order model by means of moment matching techniques around a chosen s_o (for many cases outside electrical engineering $s_o = 0$ seems to be a satisfactory choice) and then to employ a truncated balanced approximation to reduce the intermediate model as much as possible.

The development of model reduction based on the solution of Lyapunov equations is the next logical step for a large linear system. It is evident that in the few next years we will see more practical examples in this area and as the experience of mathematicians grows one can expect more practical outcomes for engineers. This will bring us truly automatic model reduction, just as we have for the case of small linear systems right now, provided that the minimum over the norm (1.1.18) is sufficient for the application. Let us stress this with an example from Ref. [100]. The norm (1.1.18) measures the absolute error over the whole frequency range and if the transfer function changes by many orders of magnitude, then the balanced truncation approximation could describe the transfer function behavior fairly well if we consider it from the viewpoint of the absolute error, but not that well if we consider the relative error.

The situation with nonlinear systems is completely different and human intervention in some form appears to be inevitable here. First, it is necessary to see if a problem in question can be handled by

- linearization;
- splitting to linear and nonlinear subparts;
- some special effective case for a particular nonlinear dynamic system.

If not, then the choice is clearly POD, where the main questions are how many "snapshots" should be generated and how often. Alternatives here are to follow the example of a similar nonlinear system or to make a special investigation in order to establish the special behavior and requirements of the system. Nevertheless, the POD suggests an appropriate framework for general nonlinear model reduction because it is possible to state that human intervention here is limited to decision making. After a researcher has decided on how to obtain matrix (1.1.37), all other POD steps can be made fully automatic. POD is especially attractive for those applications where it is possible to obtain a reduced system matrix (1.1.42) in a closed form, that is, when the governing equations can be directly projected to the reduced basis.

1.1.8 On-line Resources

The advent of the Internet has made accessible a wide variety of information resources. There are good slide shows on model reduction with illustrations and examples [101, 102]. Below are the home pages of some of the scientists involved in model reduction, where one can find additional resources:

- A. C. Antoulas http://www-ece.rice.edu/~aca/
- P. Benner http://www.math.tu-berlin.de/~benner/
- D. Boley http://www-users.cs.umn.edu/~boley/
- R. W. Freund http://cm.bell-labs.com/who/freund/
- B. B. King http://www.math.vt.edu/people/bbking/
- J. E. Marsden http://www.cds.caltech.edu/~marsden/
- S. Lall http://element.stanford.edu/~lall/
- T. Penzl http://www.mathematik.tu-chemnitz.de/in_memoriam/penzl/
- Y. Saad http://www-users.cs.umn.edu/~saad/
- P. Van Dooren http://www.auto.ucl.ac.be/~vdooren/
- A. Varga http://www.robotic.dlr.de/control/num/modred.html
- NICONET http://www.win.tue.nl/niconet/NIC2/NICtask2.html

Also, there are a number of theses, available on the Internet, which provide a good introduction in a particular field: adaptive meshing [103], Krylov subspaces [104], control theory [105], moment-matching model reduction [106, 107], SVD-Krylov model reduction [108] and POD [109, 110].

1.1.9 Acknowledgments

The research described in this chapter was partly made possible by Grant IST-1999-29047 from the European Commission Community Research, Information Society Technologies Program and partly by Grant DFG 1883/3-1 from the Deutsche Forschungsgemeinschaft and partly by an operating grant from the University of Freiburg. Part of the manuscript was completed while J. G. Korvink was on a sabbatical at Ritsumeikan University in Kusatsu, Japan. The authors thank Tamara Bechtold for invaluable help during the writing of the review and Prof. Dr. Boris Lohmann for critically reading the manuscript.

1.1.10 References

[1] Mukherjee, T., Fedder, G.K., Ramaswany, D., White, J., *IEEE Trans. Comput.-Aided Des. Integr. Circuits Syst.* **19** (2000) 1572–1589.
[2] The NCAC Animation Archives of Impact Simulations, http://www.ncac.gwu.edu/archives/animation/.
[3] LS-DYNA performance for car-to-car and car-to-barrier benchmarks on IBM SP and Compaq AlphaServer SC, http://www.epm.ornl.gov/evaluation/CRASH/index.html, September 2000.
[4] Emmenegger, M., Korvink, J.G., Bächtold, M., von Arx, M., Paul, O., Baltes, H., *Sens. Mater.* **10** (1998) 405–412.
[5] Bouhaddi, N., Fillod, R., *Comput. Struct.* **60** (1996) 403–409.
[6] Guyan, J., *AIAA J.* **3** (1965) 380.
[7] Antoulas, A.C., Sorensen, D.C., *Approximation of Large-scale Dynamical Systems: an Overview, Technical Report*, Rice University, Houston, TX, 2001, http://www-ece.rice.edu/ ~aca/mtns00.pdf.
[8] Woodbury, A., Zhang, K.N., *Adv. Water Resour.* **24** (2001) 621–630.
[9] Pini, G., Gambolati, G., *Int. J. Numer. Methods Fluids* **35** (2001) 25–38.
[10] Hochbruck, M., Lubich, C., Selhofer, H., *SIAM J. Sci. Comput.* **19** (1998) 1552–1574.
[11] van der Vorst, H.A., *Comput. Sci. Eng.* **2** (2000) 32–37.
[12] Hairer, E., Norsett, S.P., Wanner, G., *Solving Ordinary Differential Equations. I. Nonstiff Problem*; Berlin: Springer, 1986.
[13] Antoulas, A.C., *Frequency Domain Representation and Singular Value Decomposition, UNESCO EOLSS (Encyclopedia for the Life Sciences)*, Contribution 6.43.13.4, June 2001.
[14] Zienkiewicz, O.C., Taylor, R.L., *The Finite Element Method*; New York: McGraw-Hill, 1989.
[15] Müller, J., Korvink, J.G., *Proc. SPIE* **4175** (2000) 82–93.
[16] Polderman, J.W., Willems, J.C., *Introduction to Mathematical Systems and Control: a Behavioral Approach*, Applied Mathematics, Vol. 26; Berlin: Springer, 1998.
[17] Antoulas, A.C., in: *Wiley Encyclopedia of Electrical and Electronics Engineering*, Webster, J.G. (ed.); New York: Wiley, 1999, Vol. 11, pp. 403–422.
[18] Penzl, T., *Syst. Control Lett.* **40** (2000) 139–144.
[19] Antoulas, A.C., Sorensen, D.C., Zhou, Y., *On the Decay Rate of the Hankel Singular Values and Related Issues, Technical Report*, Rice University, Houston, TX, 2001, http://www- ece.rice.edu/~aca/decay_fin.ps.
[20] Varga, A., in: *Applied and Computational Control, Signals and Circuits*, Datta, B.N. (ed.); Boston: Kluwer, 2001, Vol. 2, pp. 239–282.
[21] Varga, A., *Model Reduction Routines for SLICOT, NICONET Report 1999-8*, NICONET, 1999, ftp:// wgs.esat.kuleuven.ac.be/pub/WGS/REPORTS/NIC1999-8.ps.Z.
[22] Saad, Y., *Iterative Methods for Sparse Linear Systems*; Boston: PWS, 1996.
[23] Golub, G.H., Van Loan, C.F., *Matrix Computations*; Baltimore: Johns Hopkins University Press, 3rd edn., 1996.
[24] Freund, R.W., in: *Applied and Computational Control Signals and Circuits*, Datta, B.N. (ed.); Boston: Birkhauser, 1999, pp. 435–498.
[25] Freund, R.W., *J. Comput. Appl. Math.* **123** (2000) 395–421.

[26] Barrett, R., Berry, M., Chan, Y., Demmel, J., Donato, J., Dongarra, J., Eijkhout, V., Poso, R., Romine, C., van der Vorst, H., *Templates for the Solution of Linear Systems: Building Blocks for Iterative Methods*; Philadelphia: Society for Industrial and Applied Mathematics, 1994, p. 112; see also http://www.netlib.org/templates/templates.ps.
[27] Ferencz, R.M., Hughes, T.J.R., in: *Handbook of Numerical Analysis*, Ciarlet, P.G., Lions, J.L. (eds.); Amsterdam: Elsevier Science, 1998, Vol. 6, pp. 3–174.
[28] Baker, G.A., Graves-Morris, P.R., in: *Encyclopedia of Mathematics and Its Applications*; Cambridge: Cambridge University Press, 2nd ed., 1996, Vol. 59.
[29] Brezinski, C., *Padé-type Approximation and General Orthogonal Polynomials, International Series of Numerical Mathematics*, Vol. 50; Basle: Birkhäuser, 1980, p. 250.
[30] Grimme, E.J., Sorensen, D.C., Van Dooren, P., *Numer. Algorithms* **12** (1996) 1–31.
[31] Gallivan, K., Grimme, E., Van Dooren, P., *Numer. Algorithms* **12** (1996) 33–63.
[32] Ruhe, A., Skoogh, D., in: *Applied Parallel Computing, Lecture Notes in Computer Science*, Vol. 1541; Berlin: Springer, 1998, pp. 491–502.
[33] Silveira, L.M., Kamon, M., Elfadel, I., White, J., *Comput. Methods Appl. Mech. Eng.* **169** (1999) 377–389.
[34] Odabasioglu, A., Celik, M., Pileggi, L.T., *IEEE Trans. Comput.-Aided Des. Integr. Circuits Syst.* **17** (1998) 645–654.
[35] Bai, Z.J., Freund, R.W., *Linear Algebra Appl.* **332** (2001) 139–164.
[36] Gallivan, K., Grimme, E., Sorensen, D., Van Dooren, P., *Math. Res. Ser.* **7** (1996) 87–116.
[37] Li, H., Woodbury, A., Aitchison, P., *Int. J. Numer. Methods Eng.* **43** (1998) 221–239.
[38] Li, H.N., Aitchison, P., Woodbury, A., *Int. J. Numer. Methods Eng.* **42** (1998) 389–408.
[39] Farrell, D.A., Woodbury, A.D., Sudicky, E.A., *Adv. Water Resour.* **21** (1998) 217–235.
[40] Su, Q., Syrmos, V.L., Yun, D.Y.Y., *Circuits Syst. Signal Process.* **18** (1999) 291–314.
[41] Li, H.N., Woodbury, A., Aitchison, P., *Int. J. Numer. Methods Eng.* **44** (1999) 355–372.
[42] Zhang, K., Woodbury, A.D., Dunbar, W.S., *Adv. Water Resour.* **23** (2000) 579–589.
[43] Zhang, K., Woodbury, A.D., *J. Contam. Hydrol.* **44** (2000) 387–416.
[44] Cheng, C.-K., Lillis, J., Lin, S., Chang, N., *Interconnect Analysis and Synthesis*; New York: Wiley, 2000.
[45] Achar, R., Nakhla, M.S., *Proc. IEEE* **89** (2001) 693–728.
[46] Ruehli, A.E., Cangellaris, A.C., *Proc. IEEE* **89** (2001) 740–771.
[47] Chou, M., White, J. K., *IEEE Trans. Comput.-Aided Des. Integr. Circuits Syst.* **16** (1997) 1454–1476.
[48] Celik, M., Cangellaris, A.C., *IEEE Trans. Comput.-Aided Des. Integr. Circuits Syst.* **16** (1997) 485–496.
[49] Kamon, M., Marques, A., Silveira, L.M., White, J., *IEEE Trans. Compon. Packag. Manuf. Technol. Part B Adv. Packag.* **21** (1998) 225–240.
[50] Knockaert, L., De Zutter, D., *AEU-Int. J. Electron. Commun.* **53** (1999) 254–260.
[51] Kamon, M., Wang, F., White, J., *IEEE Trans. Circuits Syst. II Analog Digit. Signal Process.* **47** (2000) 239–248.
[52] Bai, Z.J., Slone, R.D., Smith, W.T., *IEEE Trans. Comput.-Aided Des. Integr. Circuits Syst.* **18** (1999) 133–141.

[53] Bracken, J.E., Sun, D.K., Cendes, Z., *Comput. Methods Appl. Mech. Eng.* **169** (1999) 311–330.
[54] Balk, I., *IEEE Trans. Adv. Packag.* **24** (2001) 304–308.
[55] Slone, R.D., Lee, R., *Radio Sci.* **35** (2000) 331–340.
[56] Malhotra, M., Pinsky, P.M., *J. Comput. Acoust.* **8** (2000) 223–240.
[57] Kuang, Z.F., Pazsit, I., Ruhe, A., *Ann. Nucl. Energy* **28** (2001) 1595–1611.
[58] Willcox, K., Peraire, J., White, J., *Comput. Fluids* **31** (2001) 369–389.
[59] Bai, Z., Bindel, D., Clark, J., Demmel, J., Pister, K.S.J., Zhou, N., in: *Technical Proceedings of the Fourth International Conference on Modeling and Simulation of Microsystems*; Cambridge: Computational Publications, 2001, pp. 31–34.
[60] Ramaswamy, D., White, J., in: *Technical Proceedings of the Fourth International Conference on Modeling and Simulation of Microsystems*; Cambridge: Computational Publications, 2001, pp. 27–30.
[61] Srinivasan, V., Jog, A., Fair, R.B., in: *Technical Proceedings of the Fourth International Conference on Modeling and Simulation of Microsystems*; Cambridge: Computational Publications, 2001, pp. 72–75.
[62] Penzl, T., *Algorithms for Model Reduction of Large Dynamical Systems, Preprint SFB393/99-40*, 1999, http://www.mathematik.tu-chemnitz.de/preprint/quellen/1999/SFB393_40.ps.gz.
[63] Penzl, T., *Adv. Comput. Math.* **8** (1998) 33–48.
[64] Hodel, A., Poolla, K., Tenison, B., *Linear Algebra Appl.* **236** (1996) 205–230.
[65] Penzl, T., *SIAM J. Sci. Comput.* **21** (2000) 1401–1418.
[66] Antoulas, A.C., Sorensen, D.C., Gugercin, S., *A Modified Low-rank Smith Method for Large-scale Lyapunov Equations, Technical Report*, Rice University, Houston, TX, 2001, http://www.caam.rice.edu/caam/trs/2001/TR01–10.ps.
[67] Li, J.R., White, J., *SIAM J. Matrix Anal. Appl.* **24** (2002) 260–280.
[68] Sorensen, D.C., Antoulas, A. C., *The Sylvester Equation and Approximate Balanced Reduction, Technical Report*, Rice University, Houston, TX, 2002, http://www-ece.rice.edu/~aca/crossgram.pdf.
[69] Penzl, T., *LYAPACK: a MATLAB Toolbox for Large Lyapunov and Riccati Equations, Model Reduction Problems and Linear-Quadratic Optimal Control Problems*, Version 1, 2000, http://www.netlib.org/lyapack/.
[70] Roychowdhury, J., *IEEE Trans. Circuits Syst. II Analog Digit. Signal Process.* **46** (1999) 1273–1288.
[71] Chen, J., Kang, S.-M., in: *Technical Proceedings of Third International Conference on Modeling and Simulation of Microsystems*; Cambridge: Computational Publications, 2000, pp. 213–216.
[72] Chen, Y., White, J., in: *Technical Procedings of the Third International Conference on Modeling and Simulation of Microsystems*; Cambridge: Computational Publications, 2000, pp. 477–480.
[73] Lohmann, B., *IEEE Trans. Control Syst. Tech.* **3** (1995) 102–109.
[74] Lohmann, B., *Math. Mod. Syst.* **1** (1995) 77–90.
[75] Phillips, J.R., in: *Proceedings of DAC*; New York: ACM, 2000, pp 184–189.
[76] Gunupudi, P.K., Nakhla, M.S., *IEEE Trans. Adv. Packag.* **24** (2001) 317–325.
[77] Holmes, P., Lumley, J.L., Berkooz, G., *Turbulence, Coherent Structures, Dynamical Systems and Symmetry*; Cambridge: Cambridge University Press, 1996.
[78] Krysl, P., Lall, S., Marsden, J.E., *Int. J. Numer. Methods Eng.* **51** (2001) 479–504.
[79] Banerjee, S., Cole, J.V., Jensen, K.F., *IEEE Trans. Semicond. Manuf.* **11** (1998) 266–275.

[80] Hinze, M., Kauffmann, A., *Reduced Order Modeling and Suboptimal Control of a Solid Fuel Ignition Model, Report 636*, Technische Universität, Berlin, 1999, http://www.math.Tu-Dresden.de/~hinze/Psfiles/SFIM.ps.gz.
[81] Kepler, G.M., Tran, H.T., Banks, H.T., *Optim. Control Appl. Methods* **21** (2000) 143–160.
[82] Kepler, G.M., Tran, H.T., Banks, H.T., *IEEE Trans. Semicond. Manuf.* **14** (2001) 231–241.
[83] Shvartsman, S.Y., Kevrekidis, I.G., *Phys. Rev. E* **58** (1998) 361–368.
[84] Shvartsman, S.Y., Theodoropoulos, C., Rico-Martinez, R., Kevrekidis, I.G., Titi, E.S., Mountziaris, T.J., *J. Process Control* **10** (2000) 177–184.
[85] Bendersky, E., Christofides, P.D., *Chem. Eng. Sci.* **55** (2000) 4349–4366.
[86] Raimondeau, S., Vlachos, D.G., *J. Comput. Phys.* **160** (2000) 564–576.
[87] Parrilo, P.A., Lall, S., Paganini, F., Verghese, G.C., Lesieutre, B. C., Marsden, J.E., in: *Proceedings of the American Control Conference*; Piscataway: IEEE **6** (1999) 4208–4212.
[88] Baker, J., Christofides, P.D., *Int. J. Control* **73** (2000) 439–456.
[89] Banks, H.T., del Rosario, R.C.H., Smith, R.C., *IEEE Trans. Autom. Control* **45** (2000) 1312–1324.
[90] Atwell, J.A., King, B.B., *Math. Comput. Model.* **33** (2001) 1–19.
[91] Georgiou, I.T., Sansour, J., in: *Computational Mechanics, New Trends and Applications*, Idelsohn, S., Onate, E., Dvorkin, E. (eds.); CIMNE, Barcelona, 1998.
[92] Georgiou, I.T., Schwartz, I.B., *SIAM J. Appl. Math.* **59** (1999) 1178–1207.
[93] Steindl, A., Troger, H., *Int. J. Solids Struct.* **38** (2001) 2131–2147.
[94] Hung, E.S., Senturia, S.D., *J. Microelectromech. Syst.* **8** (1999) 280–289.
[95] Liang, Y.C., Lin, W.Z., Lee, H.P., Lim, S.P., Lee, K.H., Feng, D. P., *J. Micromech. Microeng.* **11** (2001) 226–233.
[96] Fahl, M., *Math. Comput. Model.* **34** (2001) 91–107.
[97] Chahlaoui, Y., Gallivan, K., Van Dooren, P., *SIAM J. Matrix Anal. Appl.* in press.
[98] Lall, S., Marsden, J.E., Glavaski, S., in: *Proceedings of the IFAC World Congress*; Kidlington: Elsevier Science **6** (1999) 473–478.
[99] Lall, S., Marsden, J.E., Glavaski, S., *Int. J. Robust Nonlin. Control* **12** (2002) 519–535.
[100] Gallivan, K., Grimme, E., Van Dooren, P., in: *Error Control and Adaptivity in Scientific Computing*, Zenger, B. (ed.); Kluwer, 1999, pp. 177–190.
[101] Antoulas, A.C., Van Dooren, P., *Model Reduction of Large-scale Dynamical Systems*, Course presented at the SIAM Annual Meeting, 2000, http://www.auto.ucl.ac.be/~vdooren/ siam.html.
[102] White, J., *Techniques for Model Order Reduction*, http://rle-vlsi.mit.edu/people/users_krylov/white/public_html/pubs/Mor.ppt.
[103] Müller, J., *Accurate FE-Simulation of Three-dimensional Microstructures*, Thesis, Fakultät für Angewandte Wissenschaften, Albert-Ludwigs Universität Freiburg im Breisgau, 2001, http://www.freidok.uni-freiburg.de/volltexte/251/pdf/thesis.pdf.
[104] Ernst, O.G., *Minimal and Orthogonal Residual Methods and their Generalizations for Solving Linear Operator Equations*, Thesis, Fakultät für Mathematik und Informatik, TU Bergakademie Freiberg, 2000, http://www.mathe.tu-freiberg.de/~ernst/publ/habil.pdf.
[105] Wortelboer, P.M.R., *Frequency-weighted Balanced Reduction of Closed-loop Mechanical Servo-systems: Theory and Tools*, Thesis, Delft University of Technology, 1994, http://www.ocp.tudelft.nl/sr/Downloads/thesis/wortelboer/wortelb.zip.

[106] Grimme, E.J., *Krylov Projection Methods For Model Reduction*, Thesis, Graduate College, University of Illinois, Urbana-Champaign, IL, 1997, http://www.cs.fsu.edu/~gallivan/theses/grimmephd.ps.gz.
[107] Skoogh, D., *Krylov Subspace Methods for Linear Systems, Eigenvalues and Model Reduction*, Thesis, Department of Mathematics, University of Göteborg, 1998, http://www.md.chalmers.se/~skoogh/skooghPHD.html.
[108] Li, J.-R., *Model Reduction of Large Linear Systems via Low Rank System Gramians*, Thesis, Department of Mathematics, MIT, Boston, 2000, http://www.cims.nyu.edu/~jingli/thesis.ps.
[109] Newman, A.J., *Modeling and Reduction with Applications to Semiconductor Processing*, Thesis, Department of Electrical and Computer Engineering, University of Maryland, College Park, MD, 1999, http://www.isr.umd.edu:80/TechReports/ISR/1999/PhD_99-5/PhD_99-5.phtml.
[110] Atwell, J.A., *Proper Orthogonal Decomposition for Reduced Order Control of Partial Differential Equations*, Thesis, Virginia Polytechnic Institute, 2000, http://www.math.vt.edu/people/bbking/papers/atwell.pdf.

List of Symbols and Abbreviations

Symbol	Designation
A	system matrix
B	scatter matrix
C	gather matrix
E	system matrix
$\boldsymbol{f}$	system load
F	system matrix
H	Hessenberg matrix
k	dimension of reduced system
L	Laplace transform operator
L	lower triangular matrix
m	number of inputs
n	dimension of original system
p	number of outputs
P	controllability grammian
q	maximum number of moments
Q	observability grammian
t	time
$\boldsymbol{u}$	vector
U	upper triangular matrix
$\boldsymbol{v}$	vector
$\boldsymbol{w}$	vector
$\boldsymbol{x}$	vector
X	orthonormal matrix

Symbol	Designation
X	transformation matrix
$\boldsymbol{y}$	vector
$\boldsymbol{Y}$	transfer function
ε	error
σ_i	Hankel singular value

Abbreviation	Explanation
CAD	computer-aided design
FEM	finite element method
MEMS	micro-electromechanical system
MIMO	multiple input-multiple output
ODE	ordinary differential equation
PDE	partial differential equation
POD	proper orthogonal decomposition
QAD	"quick-and-dirty"
SICO	single input-complete output
SISO	single input-single output
SVD	singular value decomposition

1.2 Nanotube Membrane Sensors: Resistive Sensing and Ion Channel Mimetics

M. Wirtz and C. R. Martin
Department of Chemistry and Center for Chemical Research at the Bio/Nano Interface, University of Florida, Gainesville, USA

Abstract

Nanotubule membranes are utilized for sensing applications and ion channel mimetics. The nanotubule membranes are composed of either gold or alumina. The gold nanotubule membranes are prepared via electroless deposition of Au on to the pore walls of a polycarbonate membrane, ie, the pores act as templates for the nanotubes. These membranes are a new class of molecular sieves and can be used to separate small molecules on the basis of molecular size. In addition, the use of these membranes in new approaches to electrochemical sensing is discussed. In this case, a current is forced through the nanotubes, and analyte molecules present in a contacting solution phase modulate the value of this transmembrane current. We further discuss synthetic micropore and nanotube membranes that mimic the function of a ligand-gated ion channel, ie, these membranes can be switched from an 'off' state (no or low ion current through the membrane) to an 'on' state (higher ion current) in response to the presence of a chemical stimulus, eg, drug or surfactant. Ion channel mimics are based on both modified Au nanotube and microporous alumina membranes.

Keywords: Nanotubule membranes; resistive sensing; ion channel mimetics; electroless deposition; molecular sieving; ion current; ion channel pores

Contents

1.2.1 Introduction

We have been exploring the transport and electrochemical properties of nanotube membranes prepared by the template method [1–3], a general approach for preparing nanomaterials. This method entails the synthesis or deposition of the desired material within the cylindrical and monodisperse pores of a nanopore membrane or other solid. We have used polycarbonate filters, prepared via the track-etch method [4], and nanopore aluminas, prepared electrochemically from Al foil [5], as our template materials. Cylindrical nanostructures with monodisperse diameters and lengths are obtained and, depending on the membrane and synthetic method used, these may be solid nanowires or hollow nanotubes. We, and others, have used this method to prepare nanowires and tubes composed of metals [5–15], polymers [16–19], semiconductors [20, 21], carbons [22–24], and Li^+ intercalation materials [25–27]. It is also possible to prepare composite nanostructures, both concentric tubular composites, where an outer tube of one material surrounds an inner tube of another [28, 29], and segmented composite nanowires [30].

One of our earliest applications of the template method was to prepare ensembles of microscopic [31, 32] and nanoscopic [33, 34] electrodes. Such electrodes are prepared by depositing noble metals within the pores of the polycarbonate filtration membranes. Initially, we deposited the metal in the pores using electrochemical plating methods [31], but we ultimately discovered that electroless plating allowed for more uniform metal deposition [33]. In the electroless method, metal deposition begins at the pore walls, creating, at short deposition time, hollow metal nanotubes within the pores [8–12, 35, 36].

Coincidentally, there is also a long-standing interest in our research group in the area of membrane-based chemical separations [37–39]. This interest led us to undertake a series of fundamental investigations of the transport properties of gold nanotube membranes. We discovered that by controlling the deposition time, we could prepare Au nanotubes that had effective inside diameters of molecular dimensions (<1 nm) [9]. This suggested that these membranes might be useful as molecular sieves. In addition, because these membranes are composed of an electronically conductive material, it occurred to us that excess charge could be applied to the tubes by electrochemical charging in an electrolyte solution. We reasoned that it might be possible to use this excess charge to regulate ion transport across these membranes [8, 35]. Furthermore, because the tubes are composed of gold, it seemed possible that we could use well-known Au–thiol chemistry to change the chemical environment within the tubes and, via this route, introduce chemical transport selectivity into these membranes [10–12, 36].

Another application for these nanotube membranes is in electroanalytical chemistry where the membrane is used to sense analyte species [40, 41]. In that work, membranes containing gold nanotubes with inside diameters that approached molecular dimensions (1–4 nm) were used [40]. The Au nanotube membrane was placed between two salt solutions and a constant transmembrane potential was applied. The resulting transmembrane current, associated with migration of ions through the nanotubes, was measured. When an analyte molecule whose diameter was comparable to the inside diameter of the nanotubes was added to one salt solution, this molecule partitioned into the nanotubes and partially occluded the pathway for ion transport. This resulted in a decrease in the transmembrane ion current, and the magnitude of the drop in current was found to be proportional to the concentration of the analyte [40].

In the experiment discussed above, a baseline transmembrane ion current was established, and the analyte molecule, in essence, turned off this current. It occurred to us that there might be an advantage in doing the opposite, ie, starting with an ideally zero current situation and having the analyte molecule switch on the ion current. That is, we would like to make a synthetic membrane that mimics the function of a ligand-gated ion channel. An example is the acetylcholine-gated ion channel [42], which is closed ('off' state) in the absence of acetylcholine but opens (and supports an ion current, 'on' state) when acetylcholine binds to the channel. In order to accomplish this, the 'off' state was obtained by making gold and alumina membranes hydrophobic, and the 'on' state was obtained by introducing ions and electrolyte into the membrane [43]. Ions were introduced by either partitioning a hydrophobic ionic species (eg, a drug or a surfactant) into the membrane or by deprotonation of a surface-bound hydrophobic carboxylic acid.

In this chapter we discuss the use of the nanotube membranes as molecular sieves, chemical sensors, and synthetic ion channel mimics. We begin by briefly reviewing some of the experimental details such as the electroless plating method and the gas flux method used to determine nanotube inside diameter of gold nanotubule membranes. We then discuss investigations of molecular size-based transport selectivity in this type of membrane. This is followed by a discussion of an extremely sensitive chemical sensing method utilizing the Au nanotube membranes. Lastly, we describe alumina and gold nanotube membranes that mimic ligand-gated ion channels. In these modified membranes the current can be switched on by surfactants or drugs.

1.2.2 Membrane Preparation and Analysis

1.2.2.1 Template Membranes and Electroless Plating

Commercially available 'track-etched' polycarbonate filters are used as the templates to prepare the Au nanotubes. The track-etch process [4] entails bombarding a solid material (in this case a ca. 10 μm thick polycarbonate film) with a collimated beam of high-energy nuclear fission fragments to create parallel damage tracks in the film. The damage tracks are then etched into monodisperse cylindrical pores by exposing the film to a concentrated solution of aqueous base. The diameter of the pores is determined by the etch time and the etch solution temperature. The density of pores is determined by exposure time to the fission-fragment beam. Membranes with pore diameters ranging from as small as 10 nm to as large as ca. 10 μm are available commercially.

The membranes used for these studies had nominal pore diameters of 30 nm and contained 6×10^8 pores per cm^2 of membrane surface area. The nominal pore diameter (supplied by the manufacturer) was obtained from scanning electron microscopic images of the film surface. Microscopic investigations of template-synthesized nanostructures prepared within the pores of such membranes have shown that the diameter of the pore in the center of the membrane is larger than the diameter at the membrane surface, ie, that cigar-shaped pores are obtained [14, 17]. It has been suggested that this pore geometry arises because the fission fragment that creates the damage track also generates secondary electrons, which contribute to the damage along the track [14]. The number of secondary electrons generated at the faces of the membrane is less than in the central region of the membrane, and this is why 'bottleneck' pores are obtained.

The electroless plating method used to deposit the Au nanotubes [11, 33] within the pores of these membranes (Figure 1.2.1). Briefly, the template membrane is first 'sensitized' by immersion in $SnCl_2$ solution, which results in deposition of Sn(II) on all of the membrane's surfaces (pore walls and membrane faces). The sensitized membrane is then immersed in $AgNO_3$ solution, and a surface redox reaction occurs (Equation (1.2.1)) which yields nanoscopic metallic Ag particles on the membrane surfaces:

$$\mathrm{Sn(II)_{surf}} + 2\mathrm{Ag(I)_{aq}} \rightarrow \mathrm{Sn(IV)_{surf}} + 2\mathrm{Ag(0)_{surf}} \tag{1.2.1}$$

where the subscripts surf and aq denote species adsorbed on the membrane surfaces and species dissolved in solution, respectively. The membrane is then immersed in a commercial gold plating solution and a second surface redox reaction occurs, to yield Au nanoparticles on the surfaces:

$$\mathrm{Au(I)_{aq}} + \mathrm{Ag(0)_{surf}} \rightarrow \mathrm{Au(0)_{surf}} + \mathrm{Ag(I)_{aq}} \tag{1.2.2}$$

These surface-bound Au nanoparticles are good autocatalysts for the reduction of Au(I) to Au(0) using formaldehyde as the reducing agent. As a result, Au deposition begins at the pore walls, and Au tubes are obtained within the pores (Figure 1.2.2) [8–12, 35, 36]. In addition, the faces of the membrane become coated with thin Au films. These surface films do not, however, block the mouths of the nanotubes, and there are open nanoscopic channels running from one face of the membrane to the other. By controlling the electroless plating time, the inside diameter of these nanotubes can be controlled at will, down to molecular dimensions.

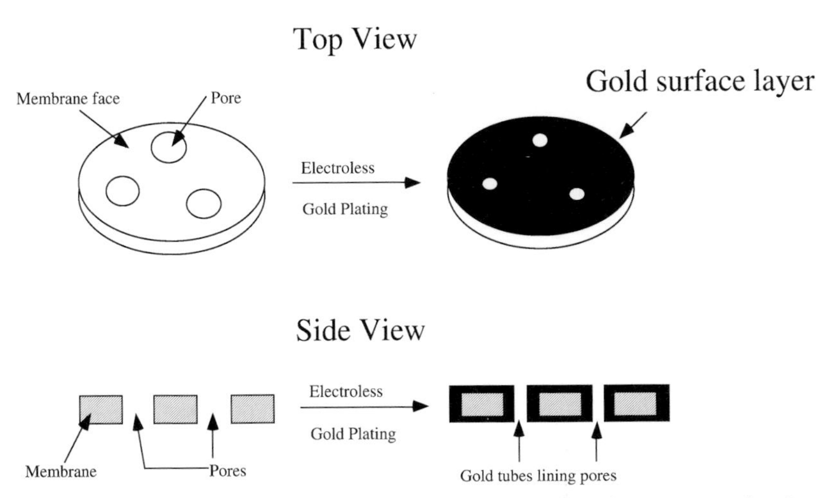

Figure 1.2.1. Schematic of the electroless plating process used to prepare the Au nanotube membranes.

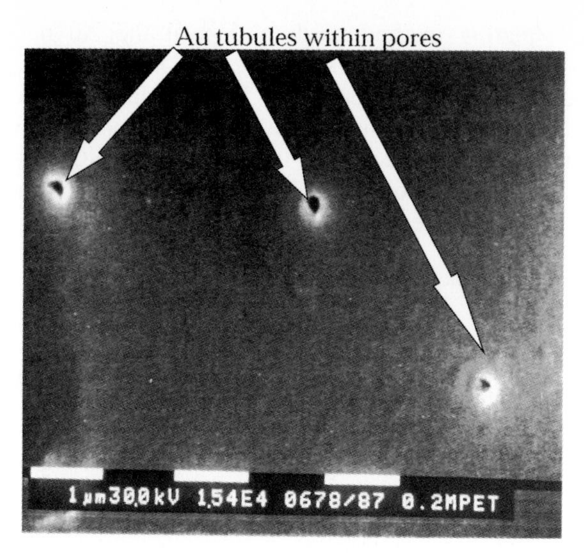

Figure 1.2.2. Scanning electron micrograph of the surface of a polycarbonate template membrane showing three Au nanotubes deposited within the pores. To visualize the tubes by scanning electron microscopy, the membrane had larger pores than those used to prepare the nanotube membranes discussed here.

1.2.2.2 Estimation of the Nanotube Inside Diameter

We use a gas-transport method to obtain an estimate of the inside diameter (id) of the template-synthesized Au nanotubes [11]. Briefly, the tube-containing membrane is placed in a gas-permeation cell, and the upper and lower half-cells are evacuated. The upper half-cell is then pressurized, typically to 20 psi with H_2, and the pressure–time transient associated with leakage of H_2 through the nanotubes is measured using a pressure transducer in the lower half-cell. The pressure–time transient is converted to gas flux (Q, mol s^{-1}) which is related to the radius of the nanotubes (*r*, cm) via [11, 44]

$$Q = 4/3(2\pi/MRT)^{1/2}(nr^3\Delta P/l) \tag{1.2.3}$$

where ΔP is the pressure difference across the membrane (dyn cm^{-2}), M is the molecular weight of the gas, R is the gas constant (erg K^{-1} mol^{-1}), n is the number of nanotubes in the membrane sample, l is the membrane thickness (cm), and T is the temperature (K). At long plating times, membranes containing nanotubes with ids of molecular dimensions are obtained (Figure 1.2.3).

In using Equation (1.2.3) we assume (1) that we know the number of nanotubes (n) in the membrane sample, (2) that the nanotubes have a constant inside diameter down their entire length, and (3) that the mechanism of gas transport through the membrane is Knudsen diffusion in the nanotubes. We have discussed the validity of each of these assumptions in detail in a recent review [45].

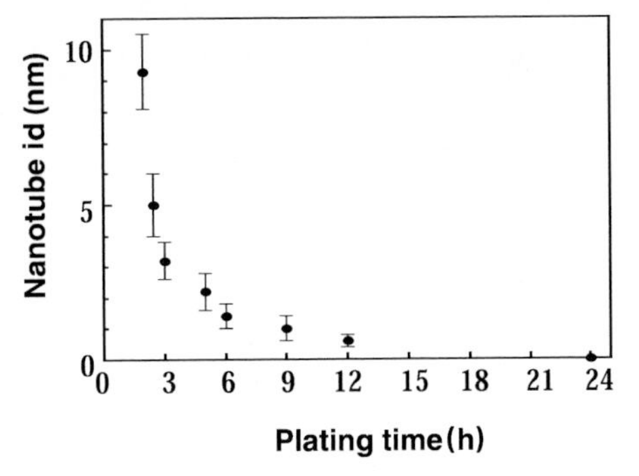

Figure 1.2.3. Variation of the nanotube effective inside diameter with plating time.

1.2.3 Molecular Sieving and Filtration in Au Nanotube Membranes

Molecular sieving experiments were conducted using a simple U-tube permeation cell, where the membrane to be studied separates the 'feed' and 'permeate' half-cells. The feed half-cell is an aqueous solution containing the molecule or molecules whose transport properties are to be evaluated. The permeate half-cell initially contains only water or a salt solution. Passive diffusion drives the permeate molecule from the feed half-cell through the membrane and into the permeate half-cell. The time course of the transport process is followed by periodically assaying the permeate half-cell for the permeate molecule(s). The transport data are plotted as moles of permeate molecules transported against permeation time [9–12]. Straight-line plots are typically obtained, and the flux of the permeate molecule can be calculated from the slope. We will call such plots 'flux plots' in this chapter.

1.2.3.1 Molecular Sieving in Single-molecule Permeation Experiments

In these experiments, the flux plot for a particular molecule is determined with only that molecule present in the feed solution. The feed solution is then replaced with a solution of a second molecule and the flux plot for this molecule is obtained for the same membrane. A membrane-transport selectivity coefficient (α) can then be obtained by ratioing the fluxes for the two permeate molecules. Since molecular size-based selectivity is of interest here, one of the permeate molecules used was large, the tris-bipyridyl complex of Ru(II), $Ru(bpy)_3^{2+}$, and the other was smaller, methyl viologen, MV^{2+} (Figure 1.2.4).

The ratio of the diffusion coefficients for MV^{2+} and $Ru(bpy)_3^{2+}$ in free aqueous solution is 1.5 [40, 45, 46]. For this reason, if a simple solution-like diffusion process were operative in the nanotubes, a selectivity coefficient of $\alpha = 1.5$ would be anticipated. In contrast, even for the largest id nanotubes investigated

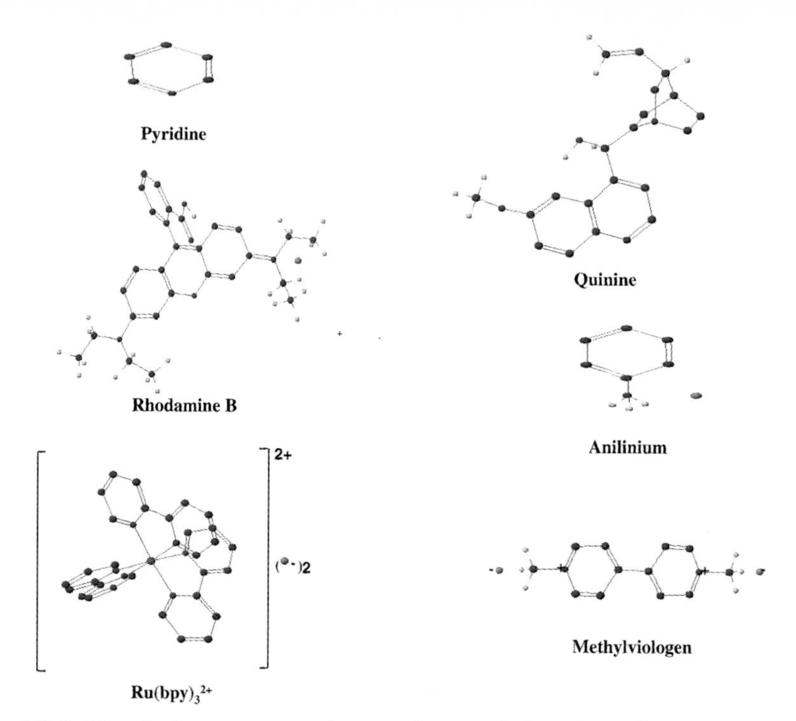

Figure 1.2.4. Chemical structures and approximate relative sizes of the three big molecule–small molecule pairs used in the molecular filtration experiments. Quinine, MV^{+2}, and $Ru(bpy)_3^{+2}$ were also used as analytes in the sensor work.

(5.5 nm), the selectivity coefficient was substantially greater, $\alpha = 50$ (Figure 1.2.5 a) [9]. These data show that size-based molecular sieving occurs in these large-id (larger than molecular dimensions) nanotubes.

Molecular sieving is a result of hindered diffusion of the molecules in the Au nanotubes [47]. The simplest way to understand hindered diffusion is to consider first the Stokes–Einstein equation that relates the diffusion coefficient (D_s) to the molecular radius (r_m) for diffusion in free solution:

$$D_s = kT/6\pi\eta p r_m \qquad (1.2.4)$$

where k is the Boltzmann constant, T is the absolute temperature, and η is the viscosity. The denominator $6\pi\eta p r_m$ can be thought of as a molecular-friction coefficient that determines the resistance to diffusion in the solution. As would be expected, this molecular-friction term increases with increasing size of the molecule and increasing viscosity of the solution.

In the Au nanotube membranes, this molecular-friction coefficient is larger than in free solution because collisions with the nanotube wall increase the frictional drag on the molecule [44]. In addition, the rate of diffusive mass transport in the nanotube is decreased, relative to a contacting solution phase, because of

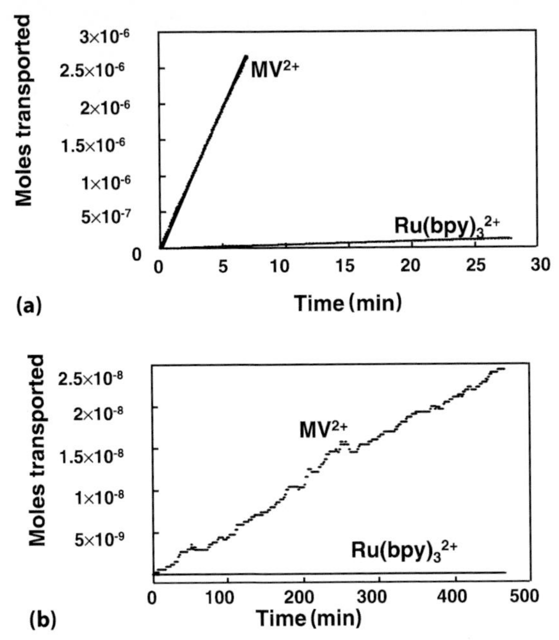

Figure 1.2.5. Single-molecule permeation experiments showing moles of MV^{2+} and $Ru(bpy)_3^{2+}$ transported versus time. Membranes contained nanotubes with id (**a**) 5.5 and (**b**) <0.6 nm. Only MV^{2+} was transported through this membrane.

steric reasons [47]. Consider a molecule of radius r_{mol} diffusing within a nanotube of comparable radius r_{tube}. The extent to which the diffusion coefficient for a molecule in the nanotube (D_{tube}) is decreased relative to its value in free solution (D_{sol}) is related to the parameter λ, which is the ratio of the radius of the diffusing molecule to the radius of the nanotube [47]:

$$\lambda = r_{mol}/r_{tube} \tag{1.2.5}$$

A large number of theoretical expressions have been derived that predict how the ratio D_{tube}/D_{sol} varies with λ [44, 47, 48]. The extremes are easy to define: when $\lambda=0$ ($r_{mol} << r_{tube}$) $D_{tube}/D_{sol}=1$, and when λ approaches unity (tube and molecule are the same size) D_{tube}/D_{sol} must approach zero. The Renkin equation:

$$D_{tube}/D_{sol} = 1 - 2.104\lambda + 2.09\lambda^3 - 0.95\lambda^5 \tag{1.2.6}$$

is an often-used example of the relationship between D_{tube}/D_{sol} and λ [48]. Plots of this equation and various other expressions for the relationship between D_{tube}/D_{sol} and λ can be found in the literature [44, 47, 48].

Equations (1.2.5) and (1.2.6) show that for any nanotube id, diffusivity in the nanotube membrane will be lower for the larger $Ru(bpy)_3^{2+}$ than for the smaller MV^{2+}. This is reflected in the transport data, where the flux of the larger

$Ru(bpy)_3^{2+}$ is decreased more than the flux for the smaller MV^{2+}. As a result, $\alpha=50$ is obtained [9]. Equations (1.2.5) and (1.2.6) predict that as the nanotube id is made smaller, the α value should become even larger, which is also reflected in the transport data. Values for the 5.5, 3.2 and 2.0 nm id nanotube membranes are $\alpha=50$, 88, and 172, respectively [9].

1.2.3.2 Molecular Filtration in Two-molecule Permeation Experiments

The smallest id nanotube membrane investigated (id $\approx$ 0.6 nm) provides a measurable flux for MV^{2+}, but the larger $Ru(bpy)_3^{2+}$ could not be detected in the permeate solution, even after a 2 week permeation experiment (Figure 1.2.5 b). These data suggest that clean separation (molecular filtration) of these two species should be possible with this nanotube membrane. This was proved by doing two-molecule permeation experiments, where both the larger and smaller molecules (Figure 1.2.4) were present in the feed half-cell together. A simple U-tube cell was used, and the permeate solution was periodically assayed, using UV–vis absorption or fluorescence, for both molecules. For all three of the large molecule–small molecule pairs shown in Figure 1.2.4, the small molecule could be easily detected in the permeate solution but the large molecule was undetectable [9].

These data show that within the limits of the measurement, the Au nanotube membrane can cleanly separate large molecules from small molecules. However, one could argue that the large molecule is, indeed, present in the permeate solution but at a concentration just below the detection limit of the analytical method employed. This argument allows us to define a minimum transport selectivity coefficient (α_{min}) for each small molecule–large molecule pair investigated, where α_{min} is defined as the measured concentration of the small molecule in the permeate solution divided by the detection limit for the large molecule. The α_{min} values obtained are extraordinary (Table 1.2.1). It is important to stress again that, in all three cases, the larger molecule was undetectable in the permeate solution.

Table 1.2.1. Minimal membrane transport selectivity coefficients

Permeate pair	α_{min}
Pyridine–quinine	15000
Anilinium–rhodamine B	130000
MV^{2+}–$Ru(bpy)_3^{2+}$	1500

1.2.4 Chemical Sensing with Au Nanotube Membranes

In addition to the above possible applications in size-based separations, these Au nanotube membranes have been used as sensors for the determination of ultratrace concentrations of ions and molecules [40, 41, 46]. In this case, the nanotube membrane was allowed to separate two salt solutions, a constant transmembrane potential was applied, and the resulting transmembrane current was measured. When an analyte with dimensions comparable to the inside diameter of the nanotubes was added to one of the salt solutions, a decrease in transmembrane current was observed. The magnitude of this drop in transmembrane current (Δi) is proportional to the analyte concentration.

1.2.4.1 Calibration Curves and Detection Limits

As in the transport experiments, a U-tube cell was assembled with the nanotube membrane separating the two halves of the cell. The two half-cells were filled with the desired electrolyte and an electrode was placed in each half-cell. Three different sets of electrodes and electrolytes were used. The first set consisted of two Pt plate electrodes, and the electrolyte used in both half-cells was 0.1 M KF. The second set consisted of two Ag/AgCl wires, and the electrolyte used in both half-cells was 0.1 M KCl. The third set consisted of two Ag/AgI wires immersed in 0.1 M KI.

As noted above, the experimental protocol used with these cells was to immerse the electrodes into the appropriate electrolyte and apply a constant potential between the electrodes. The resulting transmembrane current was measured and recorded on an *X–t* recorder. After obtaining this baseline current, the anode half-cell was spiked with a known quantity of the desired analyte (Figure 1.2.4). This resulted in a change in the transmembrane current, Δi (Figure 1.2.6). A potentiostat was used to apply the potential between the electrodes and measure the transmembrane current. The transmembrane potential used was on the order of 0.5 V [40, 46].

Plots of log Δi versus log [analyte] for the analytes $Ru(bpy)_3^{2+}$, MV^{2+}, and quinine (Figure 1.2.4) were obtained using Ag/AgCl electrodes and 0.1 M KCl as the electrolyte in both half-cells (Figure 1.2.7). For these experiments, a membrane with 2.8 nm id Au nanotubes was used. A log–log format is used for these ‘calibration curves’ because of the large dynamic range (spanning as much as five orders of magnitude in analyte concentration) obtained with this cell. Analogous calibration curves were obtained for the other electrode/electrolyte systems investigated. The detection limits [40] obtained are shown in Table 1.2.2. For the divalent cationic electrolytes, the detection limits were lowest (best) in the Ag/AgI/KI cell and worst in the Pt/KF cell. The detection limit for quinine was the same in both the Ag/AgI/KI and Ag/AgCl/KCl cells. In general, the detection limit decreases as the size of the analyte molecule increases (see Figure 1.2.4).

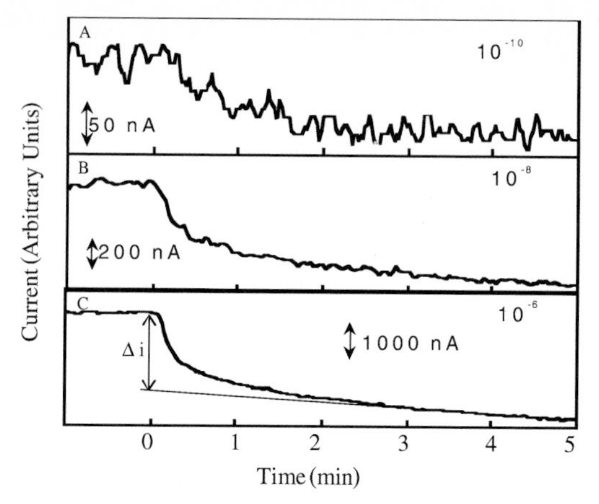

Figure 1.2.6. Nanotube membrane sensor current–time transients associated with spiking the anode half-cell with the indicated concentrations of $Ru(bpy)_3^{2+}$. Tube id=2.8 nm; Ag/AgCl/KCl cell; Δi determined as shown in (C).

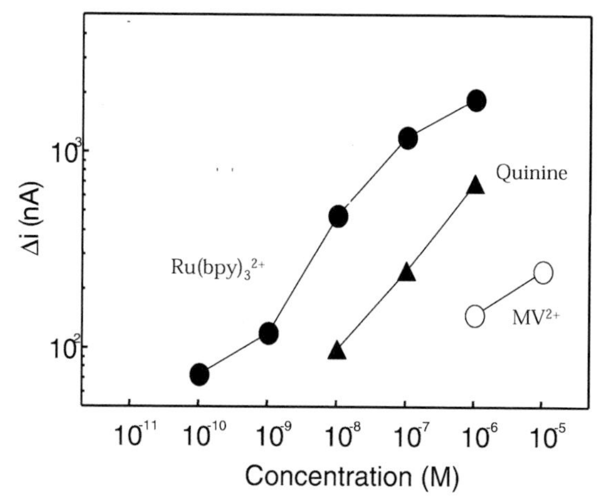

Figure 1.2.7. Calibrations curves for the indicated analytes. Membrane and cell as described in Figure 1.2.6.

Finally, the detection limits obtained (down to 10^{-11} M) are extraordinary and compete with even the most sensitive of modern analytical methods.

The majority of the quinine in both the KCl and KI solutions is present as the monoprotonated (monocationic) form. Perhaps the reason why the detection limits for $Ru(bpy)_3^{2+}$ and MV^{2+} are lower in the Ag/AgI/KI cell whereas the detection limit for quinine is the same in both this cell and the Ag/AgCl/KCl cell has to do with the difference in charge of these analytes (predominantly monocationic versus dicationic). To explore this point, the detection limits for a neutral

Table 1.2.2. Detection limits obtained for the three different electrode–electrolyte systems studied (nanotubule id=2.8 nm)

Cell	Analyte	Detection limit (M)
Pt/KF	$Ru(bpy)_3^{2+}$	10^{-9}
Ag/AgCl/KCl	$Ru(bpy)_3^{2+}$	10^{-10}
	Quinine	10^{-8}
	MV^{2+}	10^{-6}
	2-Naphthol	10^{-6}
Ag/AgI/KI	$Ru(bpy)_3^{2+}$	10^{-11}
	Quinine	10^{-8}
	MV^{2+}	10^{-7}
	2-Naphthol	10^{-6}

analyte, 2-naphthol, were obtained in both the Ag/AgI/KI and Ag/AgCl/KCl cells. Like quinine, the detection limit for this neutral analyte was the same in both cells (10^{-6} M, Table 1.2.2).

In the membrane transport studies it was shown that $Ru(bpy)_3^{2+}$ and MV^{2+} come across such membranes as the ion multiples $Ru(bpy)_3^{2+}(X^-)_2$ and $MV^{2+}(X^-)_2$ (X^-=anion) [9]. In the KI cell, the ion multiple contains two larger (relative to chloride) iodide anions. Perhaps the larger size of the iodide ion multiple accounts for the lower detection limit in the KI-containing cell. If this is true then the difference between the quinine cation paired with one I^- versus this cation paired with one Cl^- is not great enough to cause the detection limit for this predominantly monovalent analyte to be significantly different in the Ag/AgI/KI versus the Ag/AgCl/KCl cells (Table 1.2.2).

The final variable to be investigated is the effect of nanotube inside diameter on detection limit. To explore this parameter, membranes with nanotube inside diameters of approximately of 3.8, 2.8, 2.2, 1.8, and 1.4 nm were prepared and used in the Ag/AgI/KI cell [40]. Calibration curves for the analytes $Ru(bpy)_3^{2+}$, MV^{2+}, and quinine were generated as before, and detection limits were obtained from these calibration curves. Figure 1.2.8 shows plots of detection limits for these three different analytes versus the nanotube inside diameter in the membrane used. A minimum in this plot is observed for each of the three analytes.

The nanotube membrane that produces the minimum (best) detection limit depends on the size of the analyte. These molecules decrease in size in the order $Ru(bpy)_3^{2+}$ > quinine > MV^{2+}. The nanotube membrane that yields the lowest detection limit follows this size order, that is, the nanotube diameters that produce the lowest detection limit for $Ru(bpy)_3^{2+}$, quinine, and MV^{2+} are 2.8, 2.2, and 1.8 nm, respectively. For the roughly spherical analytes, the optimum tube diameter is slightly over twice the diameter of the molecule.

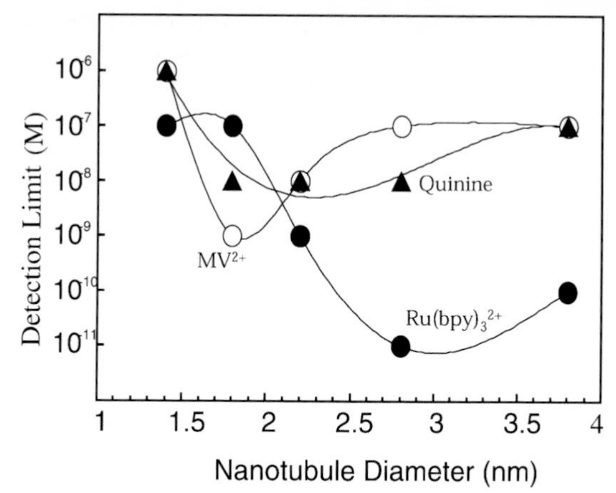

Figure 1.2.8. Detection limits for MV^{2+}, quinine, and $Ru(bpy)_3^{2+}$ versus id of the nanotubes used in the sensor.

1.2.4.2 Molecular Size based Selectivity

The data presented above show a strong correlation between detection limit and the relative sizes of the nanotube and the analyte molecule (Figure 1.2.8). This indicates that this device should show molecular size-based selectivity. This is not surprising given the transport studies discussed previously. To explore size-based selectivity, a series of solutions were prepared containing decreasing concentrations of the analyte species, but containing a constant (higher) concentration of an interfering species. The interfering species was smaller than the analyte species. The response of the nanotube membrane (nanotube diameter= 2.8 nm) to these solutions was then measured starting from lowest to highest concentration of the analyte species.

The small pyridine molecule was used as the first interfering species. When present at a concentration of 10^{-4} M, pyridine offered very little interference for any of the analytes $Ru(bpy)_3^{2+}$, MV^{2+}, or quinine. The detection limits in the presence of 10^{-4} M pyridine were 10^{-10} M for $Ru(bpy)_3^{2+}$, 10^{-6} M for MV^{2+}, and 10^{-7} M for quinine, within an order of magnitude of the detection limit with no added interfering species (Table 1.2.2). Put another way, this nanotube membrane sensor can detect 10^{-10} M $Ru(bpy)_3^{2+}$ in the presence of six orders of magnitude higher pyridine concentration.

A second set of experiments was done using the larger MV^{2+} as the interfering species. Now at low concentrations of analyte, there is a region where the device produces a constant response due to the constant concentration (10^{-4} M) of this interfering species, ie, the much higher concentration of the MV^{2+} swamps the response of the device. However, as the concentration of $Ru(bpy)_3^{2+}$ increases, there is a concentration range where the device responds to this analyte species without interference from the MV^{2+}. This concentration range begins at concentrations of $Ru(bpy)_3^{2+}$ above 10^{-8} M. That is, the size-based selectivity is such that the larger analyte species, $Ru(bpy)_3^{2+}$, can be detected down to 10^{-8} M in the presence of four orders of magnitude higher concentration of the smaller interfering species, MV^{2+}.

These data can be quantified by defining the selectivity coefficient $K_{bpy/MV}$ as the slope of the calibration curve for the analyte, $Ru(bpy)_3^{2+}$, divided by the slope for the interfering species, MV^{2+}. This analysis is problematic because the calibration curves are nonlinear and because the device is not very sensitive to MV^{2+} [40]. However, taking the data from the central part of the $Ru(bpy)_3^{2+}$ calibration curve gives a slope of ~ 400 A M^{-1}; dividing by the slope for the MV^{2+} data gives $K_{bpy/MV} = 4000$. These experiments show that, in agreement with the transport studies, the nanotube membrane-based sensor can show excellent size-based selectivity.

1.2.5 Synthetic Ion Channel Pores

We have conducted experiments that provide proof of the basic concept that an analyte molecule can switch on an ion current in a synthetic membrane-based ion channel mimic [43]. The membrane used for most experiments was a commercially available microporous alumina filter. The pores in this membrane were made hydrophobic by reaction with an 18-carbon (C_{18}) alkylsilane. When placed between two salt solutions, the pores in this C_{18}-derivatized membrane are not wetted by water, yielding the 'off' state of the membrane. When exposed to a solution containing a sufficiently high concentration of a long-chain ionic surfactant (the analyte), the surfactant molecules partition into the hydrophobic membrane, and ultimately cause the pores to flood with water and electrolyte. As a result, the membrane will now support an ion current, and the ion channel-mimetic membrane is switched to its 'on' state. Cationic drug molecules can also switch this membrane from the off to the on state. Furthermore, when a hydrophobic –COOH-containing silane is used, the off/on transition can be induced by controlling the pH of the contacting solution phases.

1.2.5.1 Membrane Preparation

The alumina membranes were Anopore® (Whatman, Clifton, NJ, USA) that had nominally 200 nm diameter pores and were 60 μm thick (Figure 1.2.9). The alumina membranes were modified with octadecyltrimethoxysilane [43]. Alumina membranes were also derivatized with a silane that terminated with the 20-carbon carboxylic acid $-(CH_2)_{20}-COOH$. For the latter modification, a two-step method based on a literature procedure was used [43, 49, 50].

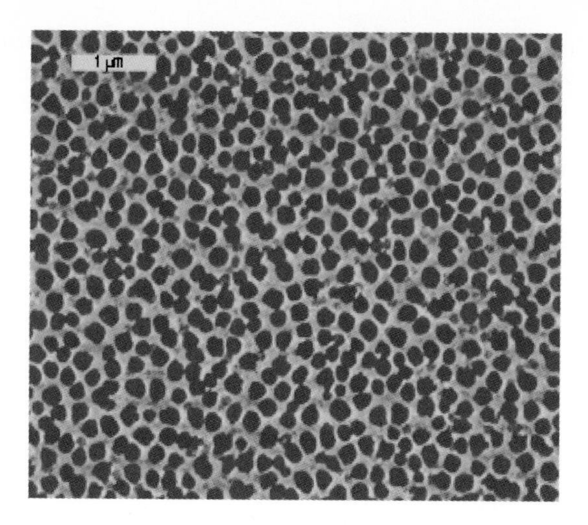

Figure 1.2.9. Scanning electron micrograph of the surface of an Anopore® alumina membrane.

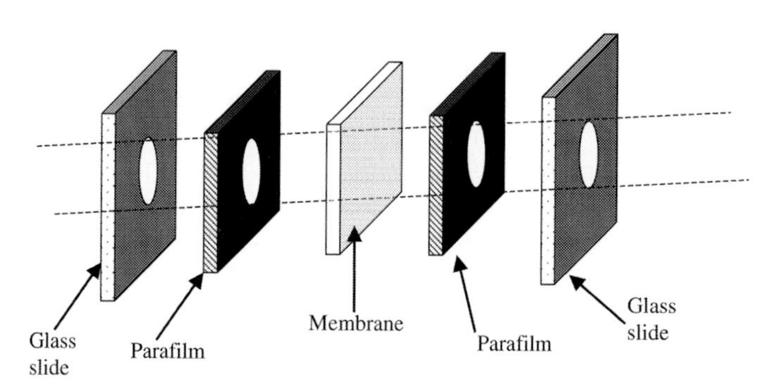

Figure 1.2.10. Membrane assembly.

Following attachment of the desired silane, the membrane was mounted into the assembly device shown in Figure 1.2.10. After assembling the various components, the assembly was heated for 3 min at 150 °C in an oven to melt the Parafilm, which acted as a glue to hold the various pieces together. This assembly exposes 0.079 cm^2 of membrane area to the contacting electrolyte solutions. We also conducted experiments with gold nanotube membranes containing nanotubes with inside diameters of ~ 2 nm. The Au nanotubule membranes were rendered hydrophobic by chemisorbing octadecanethiol to the nanotube walls [9].

1.2.5.2 AC Impedance Experiments with 1-Dodecanesulfonic Acid (DBS) Analyte

The membrane assembly was mounted between the halves of a U-tube permeation cell, and both half-cells were filled with ~20 mL of 0.1 M KCl. An Ag/AgCl working electrode was immersed in one half-cell solution, and a Pt counter electrode and an Ag/AgCl reference electrode were placed in the other half-cell [51–53].

AC Impedance measurements proved to be a useful way to demonstrate the analyte-induced switching of the membrane between the off and on states. The uppermost curve in Figure 1.2.11 is the Nyquist plot for a C_{18}-modified alumina membrane with 0.1 M KCl solutions, and no analyte (DBS), on either side of the membrane. As in prior investigations of ion channel and ion channel mimetic membranes [54, 55], the impedance data were interpreted in terms of the equivalent circuit shown in the inset in Figure 1.2.11, where R_s is the solution resistance, R_m is the membrane resistance, and C is the membrane capacitance. The dashed curve is the best fit to the experimental data, from which the R_m (Figure 1.2.12) and C values were obtained. Also shown in Figure 1.2.11 are impedance data after spiking the half-cell electrolyte solutions to the indicated concentrations with the analyte (DBS).

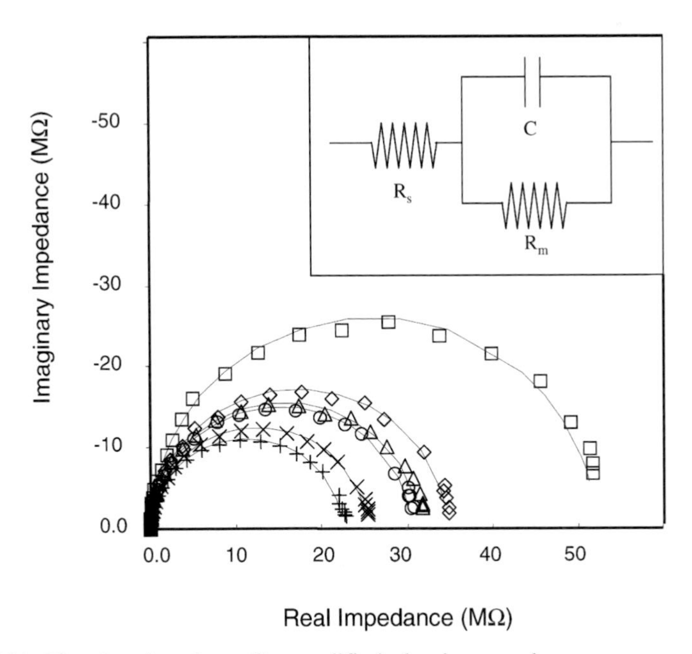

Figure 1.2.11. Nyquist plots for a C_{18}-modified alumina membrane upon exposure to increasing concentrations of DBS in 0.1 M KCl. The points are the experimental data. The lines are calculated data obtained using the equivalent circuit shown in the inset. Concentrations of DBS were as follows: □, 0; ◆, 1.4; △, 3; ○, 10; ×, 40; +, 100 nM.

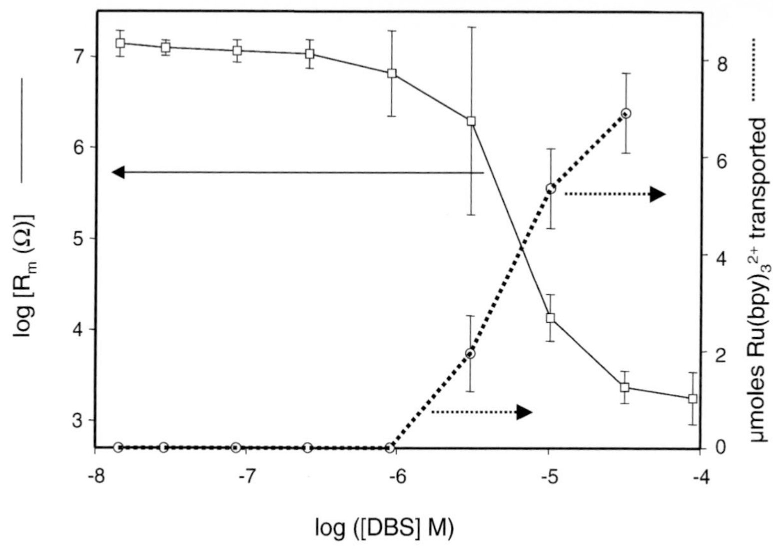

Figure 1.2.12. Plots of log (membrane resistance) (left *y*-axis) and μmoles of $Ru(bpy)_3^{2+}$ transported across the membrane (right *y*-axis) versus log[DBS] for a C_{18}-modified alumina membrane. The error bars represent the standard deviation of three separate experiments.

In the absence of DBS, the membrane resistance is very large, >50 MΩ as opposed to ~5 Ω for the alumina membrane before modification with the C_{18} silane. Transport experiments (see below) show that this is because the very hydrophobic C_{18}-modified pores are not wetted by water. This is supported by contact angle measurements on the membrane surface, where a water contact angle of 130 (±8)° was obtained for the C_{18}-treated alumina membrane as opposed to ~8 (±1)° for the untreated membrane. It should be noted that our C_{18}-treated alumina water contact angle is slightly higher than literature values, which are around 110° [56, 57]. This is presumably due to the increased roughness of the alumina surface.

While over the concentration range 10^{-9}–10^{-7} M there is some drop in membrane resistance with increasing DBS concentration (Figure 1.2.13), R_m remains very large (>20 MΩ). However, over the DBS concentration range between 10^{-6} and 10^{-5} M there is a precipitous, four orders of magnitude, drop in R_m (Figure 1.2.12). This drop signals the analyte-induced switching of the membrane from the off to the on states. The capacitance data also show the effect of this off/on transition. At concentrations of DBS below the transition, the membrane capacitance is extremely low (eg, 0.46 nF per cm^2 of membrane surface area at a DBS concentration of 10^{-6} M), but jumps by two orders of magnitude when the membrane is switched to the on state (22 nF per cm^2 at [DBS] = 10^{-5} M). We note that the switching concentration of DBS is much lower than the critical micelle concentration (CMC), which is 1.1 mM [58].

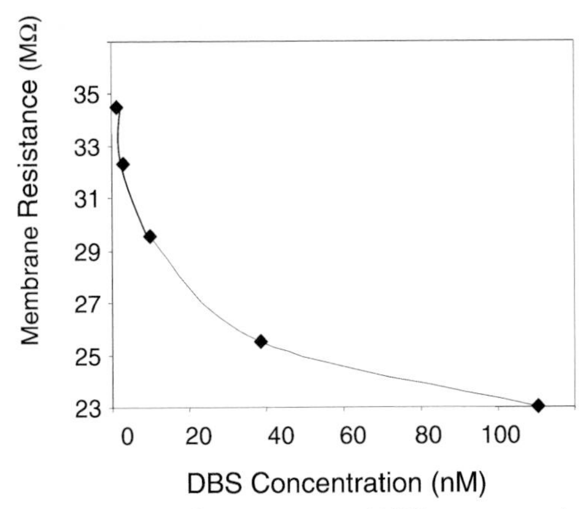

Figure 1.2.13. Plot of membrane resistance versus DBS concentration.

1.2.5.3 Transport Experiments

These were conducted by mounting the membrane between the two halves of a U-tube permeation cell and adding 0.1 M KCl to each half-cell. The feed half-cell was also 50 μM in either $Ru(bpy)_3^{2+}$ or naphthalene disulfonate (NDS^{2-}), the permeate ions. These permeate ions were chosen because transport from the feed solution through the membrane and into the permeate solution can be easily monitored by measuring the UV absorbance of the permeate solution (both permeate ions were detected at 286 nm). Transport occurred by diffusion of the permeate ion down the concentration gradient across the membrane; both the feed and permeate half-cells were vigorously stirred during the permeation experiments.

The experimental protocol used was as follows. An increment of the analyte surfactant (for these experiments DBS) was added to both the feed and permeate half-cells and permeation was allowed to occur for 24 h. After this time, the permeate half-cell was sampled and the UV absorbance was used to determine the moles of the permeate ion transported. The permeate solution was then returned to the permeate half-cell and a second increment of DBS was added. Permeation was again allowed to occur for 24 h and the amount of permeate ion transport was again determined. This process was repeated for various DBS concentrations over the range from 10^{-8} to 10^{-4} M.

The data obtained for $Ru(bpy)_3^{2+}$ transport are shown in Figure 1.2.12. At DBS concentrations below 10^{-6} M there is no detectable $Ru(bpy)_3^{2+}$ in the permeate solution. It is important to emphasize that each permeation data point in Figure 1.2.12 corresponds to an additional 24 h of permeation time. Hence, by the time the DBS concentration was increased to 9×10^{-7} M, the total permeation time was 5 days. The inability to detect $Ru(bpy)_3^{2+}$ in the permeate solution after 5 days of permeation, shows that over the DBS concentration range 0 to $\sim 10^{-6}$ M, the pores in the C_{18} membrane are not wetted by water, making the rate of

$Ru(bpy)_3^{2+}$ transport immeasurable small. These data, again, show that at DBS concentrations below 10^{-6} M, the membrane is in the off state.

At DBS concentrations above 10^{-6} M, $Ru(bpy)_3^{2+}$ transport is switched on, and flux increases with concentration of DBS for concentrations above this value. The impedance and transport data tell a consistent story about the effect of DBS on the C_{18}-derivatized membrane (Figure 1.2.12). At low DBS concentrations ($<10^{-6}$ M) where the membrane resistance is in the $10^7\ \Omega$ range, $Ru(bpy)_3^{2+}$ is not transported. The sudden drop in R_m at DBS concentrations above $\sim 10^{-6}$ M is seen in the transport experiments as an abrupt switching on of $Ru(bpy)_3^{2+}$ transport across the membrane.

In order to understand the nature of this abrupt switch to the 'on' state, we compared $Ru(bpy)_3^{2+}$ and NDS^{2-} fluxes across bare (no C_{18}) alumina membranes with fluxes across C_{18}-modified membranes that had been exposed to 10^{-4} M DBS. The $Ru(bpy)_3^{2+}$ fluxes for the bare alumina membrane and for the C_{18}-modified membrane that had been exposed to 10^{-4} M DBS are the same. In the case of the bare membrane, the pores are flooded with water and transport occurs by diffusion through these water-filled pores. The equivalence of the flux for the C_{18}-modified membrane that had been exposed to 10^{-4} M DBS clearly shows that the pores in this membrane are also flooded with water. The same results were obtained for the flux of the anionic permeate ion NDS^{2-}. An anion was studied to insure that transport of the cationic $Ru(bpy)_3^{2+}$ was not being facilitated in some way by the anionic surfactant incorporated within the pores. Again, these data show that the transition from the off state to the on state occurs because at DBS concentrations above 10^{-6} M, sufficient DBS has partitioned into the membrane that the pores spontaneously flood with water and electrolyte.

1.2.5.4 X-ray Photoelectron Spectroscopy (XPS)

XPS was used to show that the prototypical analyte dodecylbenzene sulfonate (DBS) is present on the C_{18}-modified alumina surface after exposure of the membrane to DBS solution. However, the XPS cross-section for S from the DBS proved too weak to obtain unambiguous evidence; furthermore, O, C, and Na^+ (the counterion for the DBS) are ubiquitous, and therefore not useful as probes to prove that DBS is present on the surface. For this reason, we used a surface ion-exchange reaction to replace Na^+ with Cs^+ as the counterion for the surface-bound DBS. We then used XPS to look for the presence of Cs^+ on the C_{18}-modified surface that had been treated with DBS, using an identical surface that was exposed to the Cs^+ solution but not to DBS as the control.

Figure 1.2.14 shows XPS data for a C_{18}-modified membrane that had been exposed to an aqueous 2.0 mM solution of Na^+-DBS, rinsed, exposed to a 100 mM aqueous solution of $CsNO_3$, and then rinsed extensively again. The Cs 3d peaks at 724 and 738 eV are clearly evident [59]. This may be contrasted with the control surface, a C_{18}-modified alumina membrane that was exposed to the Cs^+ solution but not to DBS, where no Cs signal is seen (Figure 1.2.14). These data

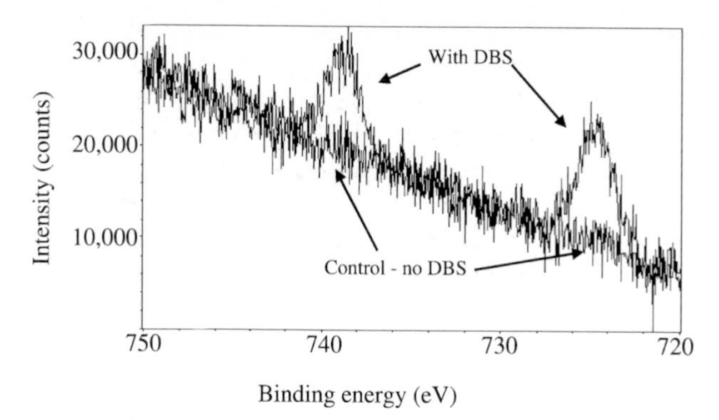

Figure 1.2.14. XPS data for a C_{18}-modified alumina membrane surface that was exposed to a DBS solution and then to a Cs^+ solution and for an identical surface that was exposed to Cs^+ but not to DBS.

show that exposure of the membrane to DBS results in partitioning of this analyte species on to the C_{18}-modified surface.

1.2.5.5 Measurements of Ion Current

While the transport experiments show that the analyte DBS can switch on ion (eg, $Ru(bpy)_3^{2+}$ and NDS^{2-}) transport across the membrane, we also wanted to obtain a direct measure of the ion current. To do this, a constant transmembrane potential of 1.5 V was applied and the resulting transmembrane ion current was measured. The current was monitored for 30 min and then the half-cell solutions were spiked with DBS to a total concentration of 10^{-9} M. The current was again measured for 30 min and the half-cells were spiked again with DBS. This process was repeated for various DBS concentrations over the range from 10^{-9} to $10^{-3.5}$ M.

Figure 1.2.15 shows the measured ion current versus time data; at the indicated times the electrolyte solutions were spiked to the indicated concentrations with DBS. The ion current data show the same general trend as both the impedance and transport data: at concentrations below $\sim 10^{-6}$ M the ion current is at a very low baseline value and at concentrations above $\sim 10^{-6}$ M the ion current abruptly switches on. In addition, the $10^{-5.5}$ M datum shows that the transition from the low-current to the high-current state occurs very abruptly.

Both the impedance and ion current data show that when the membrane is in the off state, some small baseline current does flow across the membrane. It is important to note that the resistance value for the off state obtained by the impedance and ion current measurements are essentially identical. As shown in Figure 1.2.12, the impedance measurement yields a value of $\sim 10^7\ \Omega$. The ion current in the off state is $\sim 1.5 \times 10^{-7}$ A, which for a 1.5 V transmembrane potential yields a membrane resistance of $\sim 10^7\ \Omega$. The issue left to resolve, however, is

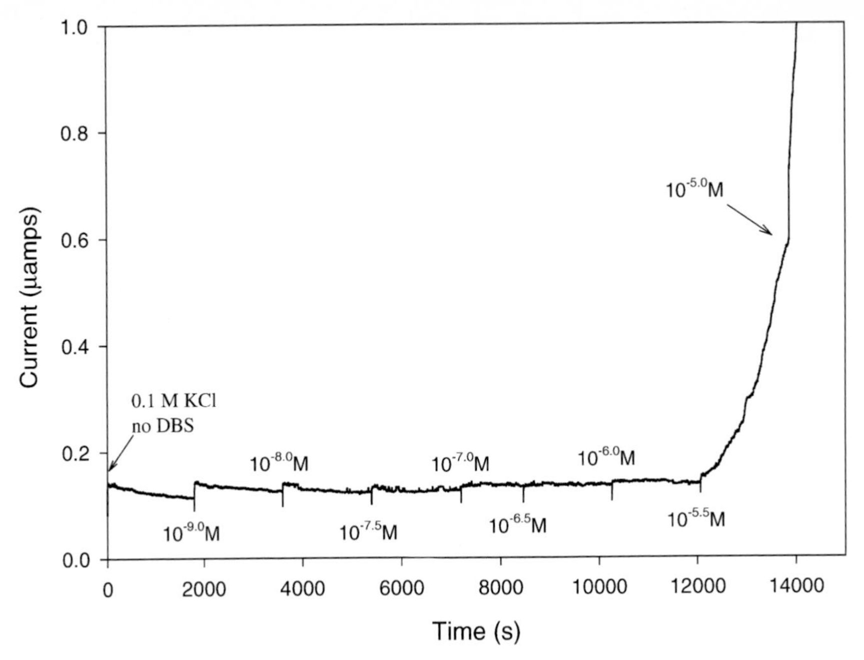

Figure 1.2.15. Ion current through a C_{18}-modified alumina membrane versus time. The contacting solution phases were spiked with the indicated concentrations of DBS at the indicated times. The electrolyte was 0.1 M KCl. A constant transmembrane potential of 1.5 V was applied.

what is supporting this baseline ion current when the membrane is in the off state? At this point we cannot say other than to suggest that this current results from some surface conduction process that occurs along the pore walls when the pores are devoid of water. In the absence of DBS, this surface conduction process may involve residual surface hydroxyl sites. The impedance data (Figure 1.2.12) indicate that in the presence of DBS, the surfactant itself is involved in the conduction process.

1.2.5.6 Effect of Alkyl Chain Length and Nature of the Surfactant Head Group

The above data show that partitioning of the analyte DBS into the membrane is responsible for the transition to the on state. It seems likely that this partitioning process is driven by the hydrophobic effect. To prove this point, C_{18}-modified alumina membranes were exposed to 0.1 wt% solutions of various alkyl sulfonate, alkyl benzene sulfonate, and trimethyl alkyl ammonium surfactants. This 0.1 wt% corresponds to a change in concentration from 8.1 mM for the smallest surfactant to 2.8 mM for the largest. The membranes were then rinsed with water

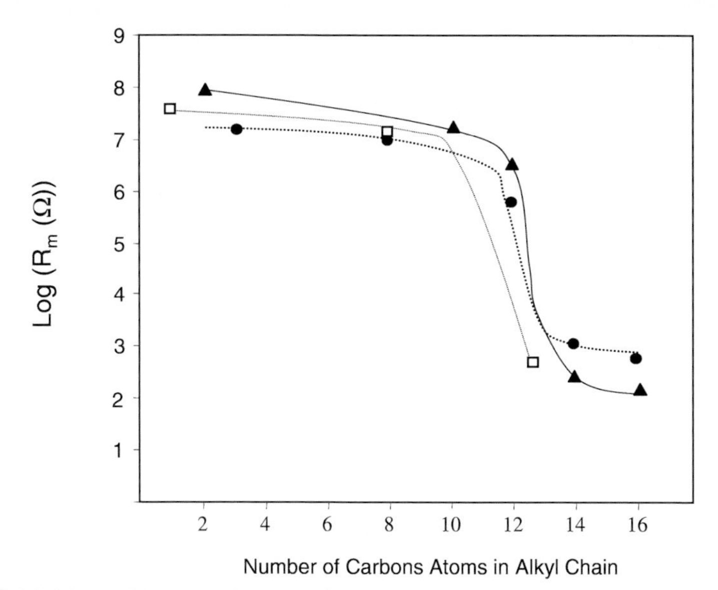

Figure 1.2.16. Plots of log(membrane resistance) for C_{18}-modified alumina membranes as a function of the carbon chain length of the surfactant exposed to the membrane. Prior to measurement, the membranes were soaked in 0.1 wt% surfactant solutions. □, Benzenesulfonate; ●, sulfonate; ▲, trimethylammonium surfactant families.

and the impedances determined in 0.1 M KCl that was devoid of surfactant. Figure 1.2.16 shows plots of log (membrane resistance) versus number of carbon atoms in the alkyl side chain for all of the surfactants investigated. These data show that in order for a surfactant to switch the membrane to the on state, the alkyl chain must be above some minimum length. For the alkyl benzenesulfonates, a 12-carbon chain is required. For the alkyl sulfonates and alkylammoniums, the alkyl group must be at least 14 carbons long (Figure 1.2.16).

These data clearly show that the hydrophobic effect is responsible for driving the surfactant into the C_{18}-modified membrane. The difference between the alkyl benzenesulfonates and the other two classes of surfactants reflects the added contribution of the benzene group to the hydrophobicity. With the added benzene group, a 12-carbon alkyl chain can switch the membrane to the on state; without the benzene group a 14-carbon chain is needed (Figure 1.2.16). The data in Figure 1.2.16 also show that for the long-chain surfactants the adsorption of the surfactant to the membrane is essentially irreversible if the membrane is exposed to aqueous solution. However, we have found that the surfactant can be removed from the membrane by rinsing with ethanol. After the ethanol rinse the membrane resistance returns to values equivalent to the off state obtained prior to exposure to surfactant.

1.2.5.7 Detection of Drug Molecules

To explore further the role of the hydrophobic effect in driving the analyte species into the C_{18}-derivatized alumina membrane, we investigated the effect of hydrophobic cationic drug molecules on the membrane resistance. The molecules and their molecular weights are amiodarone (645 g mol^{-1}), amitriptyline (278 g mol^{-1}), and bupivacaine (288 g mol^{-1}) (Figure 1.2.17). Because its molecular weight is more than double those of the other drugs and because it contains very hydrophobic iodo substituents, amiodarone is by far the most hydrophobic of these molecules. If the hydrophobic effect is responsible for driving molecules into the C_{18}-derivatized membrane, then the transition from the off to the on state would occur at lowest concentrations for amiodarone, and this is what is observed experimentally (Figure 1.2.17). There is only a 3% difference in the molecular weights of amitriptyline and bupivacaine; however, bupivacaine presents two additional opportunities for hydrogen bonding with water: the lone pairs on the carbonyl group and the lone pair of the nonprotonated nitrogen. For this reason, bupivacaine is much more hydrophilic, and it would be expected to be the mostly poorly detected of the three drugs; Figure 1.2.17 shows that this is also observed experimentally.

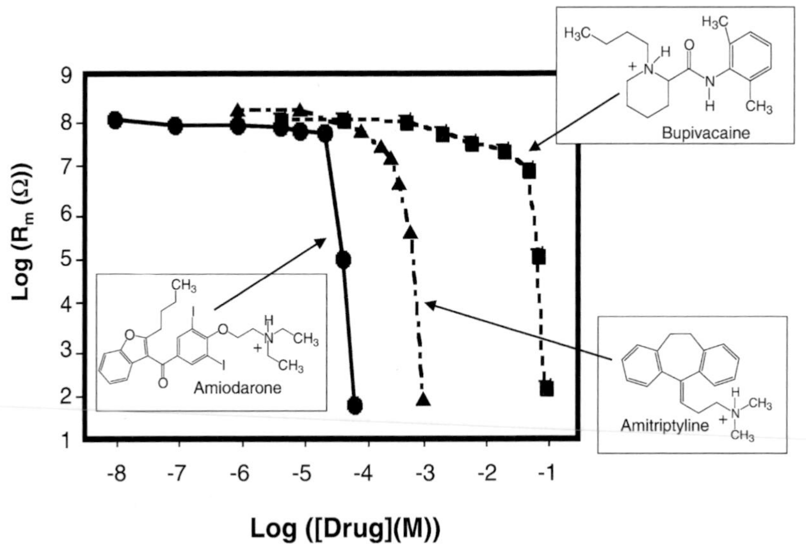

Figure 1.2.17. Plots of log(membrane resistance) versus log[drug] for the indicated drugs and a C_{18}-modified alumina membrane.

1.2.5.8 Switching the Membrane in Response to Solution pH

We have also investigated the effect of solution pH on the resistance of a membrane that was derivatized with the hydrophobic carboxylated silane –Si–$(CH_2)_3NHCO–(CH_2)_{20}–COOH$. At low pH values, the membrane resistance is high ($>10^5$ Ω), signifying the off state (Figure 1.2.18). That the resistance of the off state for this membrane is lower than that for the C_{18}-derivatized membrane is not surprising because the 18-carbon alkyl chain is certainly more hydrophobic than the $–(CH_2)_3NHCO–(CH_2)_{20}–COOH$ chain. A small but measurable decrease in membrane resistance with increasing pH is observed over the pH range 5.5–8.0. Above pH 8 there is a large (more than four orders of magnitude) drop in membrane resistance, signifying the transition to the on state. The membrane resistance after this pH is, in fact, indistinguishable from that of the bare alumina membrane, of the order of 5 Ω.

The transition to the on state is associated with the deprotonation of the carboxylic acid group. This is a reversible effect; the fully deprotonated membrane obtained after exposure to pH 10 buffer can be rinsed and dried and then re-exposed to pH 5.5 buffer to regenerate the off state of the membrane. Finally, if it is assumed that the surface-bound long chain carboxylic acid has a typical aqueous solution phase K_a ($\sim 10^{-5}$), then the transition to the on state occurs when 99.9% of the –COOH groups have been deprotonated. This is, however, only an approximation because the very hydrophobic environment within the silane-modified pores may substantially change the K_a relative to the solution value. In fact, when Kane and Mulvaney [60] and Hu and Bard [61] examined the pK_as of aliphatic carboxylic acids in well-formed monolayers, they found an increase of 2–3 units in the pK_a values. This may account for the on state of our membrane occurring at pH 8.

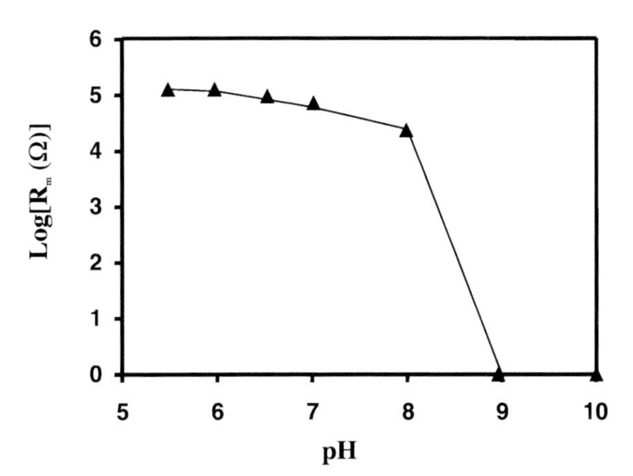

Figure 1.2.18. Plots of log(membrane resistance) versus pH for a $–(CH_2)_3NHCO–(CH_2)_{20}–COOH$-modified alumina membrane.

1.2.5.9 Effect of Pore Density and Pore Diameter on Analyte Detection

Au nanotube membranes that were rendered hydrophobic by chemisorbing octadecanethiol [11] were used to explore these issues. These membranes contained $\sim$ 2 nm id Au nanotubes and had a total porosity of only $2\times10^{-3}\%$. This may be contrasted with the alumina membranes which contain pores that are two orders of magnitude larger in diameter and are $\sim$ 30% porous. In a study analogous to Figure 1.2.12, we examined the resistance of these membranes with increasing concentration of DBS. An abrupt transition from a very high-resistance off state to a four orders of magnitude lower resistance on state is observed over essentially the same concentration range as is observed for the C_{18}-derivatized alumina membranes (Figure 1.2.19).

These results show that for membranes containing either C_{18}-modified alumina micropores or C_{18}-modified Au nanotubes, the concentration of DBS that induces the transition from the off to the on state does not depend on porosity or pore/tube diameter. Although at first glance this may seem surprising, it is important to point out that changing the porosity or pore/nanotube diameter in reality changes only the surface area of C_{18} groups available for the adsorption of analyte, ie, the low-porosity Au nanotube membranes have a lower surface area available for analyte adsorption than the alumina membranes. This situation can be modeled by assuming that the analyte is adsorbed on the C_{18}-modified surfaces via a Langmuir isotherm and that a certain critical fraction (Θ_c) of the surface must be covered with analyte in order to flood the pores with electrolyte. Θ_c is related to the equilibrium constant for the adsorption reaction (β) and the activity of DBS in the solution phase (a_{DBS}) via [62]

$$\Theta_c = (\beta a_{DBS})/(1 + \beta a_{DBS}) \tag{1.2.7}$$

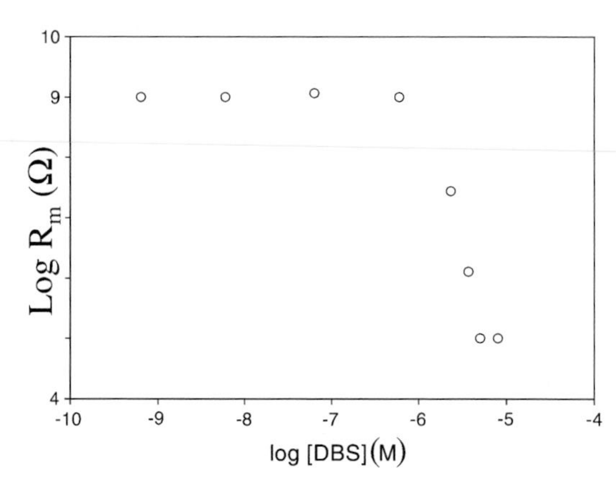

Figure 1.2.19. Plots of log(membrane resistance) versus log[DBS] for a C_{18}-modified Au nanotube membrane.

Equation (1.2.7) shows that Θ_c depends only on the value of the equilibrium constant (the same for both membranes, since both contain C_{18} alkyl groups) and the activity of surfactant in solution. Because surface area does not enter into the equation, the concentration of DBS required to switch the membrane to the on state should be independent of membrane porosity and pore diameter, and this is what is observed experimentally.

1.2.6 Conclusion

In this review, we have discussed the use of Au nanotube membranes to separate molecules on the basis of size. Furthermore, we have described a highly sensitive method of electroanalysis based on these membranes. Besides showing size-based selectivity, previous studies have demonstrated that these Au nanotube membranes can show ionic charge-based transport selectivity and that the membranes can be electrochemically switched between anion-transporting and cation-transporting states [45]. Hence these membranes can be viewed as universal ion exchangers. Furthermore, chemical transport selectivity can be introduced into these membranes by chemisorbing thiols to the inside tube walls [10, 11]. In this case, the chemisorbed thiol changes the chemical environment within the nanotubes and this, in turn, changes the transport properties of the membrane. For example, membranes modified with hydrophobic thiols selectively transport hydrophobic molecules [11]. Hence these nanotube membranes can utilize all of the selectivity paradigms (sterics, electrostatics, and chemical interactions) that Nature uses in the design of exquisitely selective molecular recognition schemes (eg, protein [63] and ion channels [42]). Nanotube membranes can be viewed as model systems for these naturally-occurring nanotubes. Indeed, we have shown that synthetic micropore and nanotube membranes can mimic the function of ligand-gated ion channels, ie, they can be switched from an off state to an on state in response to the presence of a chemical stimulus. This concept of ion channel mimetic sensing, as originally proposed by Umezawa's group [64], has been of considerable interest in analytical chemistry [65–67]. There is also considerable appeal in using naturally occurring and genetically engineered protein channels as sensors (see [45] and references therein). Such research at the bio/nano interface is of great current interest in our group.

1.2.7 Acknowledgement

Aspects of this work were supported by the Office of Naval Research and the National Science Foundation.

1.2.8 References

[1] Martin, C.R., *Science* **266** (1994) 1961–1966.
[2] Hulteen, J.C., Martin, C.R., *J. Mater. Chem.*, **7** (1997) 1075–1087.
[3] Martin, C.R., Mitchell, D.T., *Anal. Chem.*, **70** (1998) 322A–327A.
[4] Fleischer, R.L., Price, P.B., Walker, R.M., *Nuclear Tracks in Solids*; Berkeley, CA: University of California Press, 1975.
[5] Hornyak, G.L., Patrissi, C.J., Martin, C.R., *J. Phys. Chem. B.*, **101** (1997) 1548–1555.
[6] Possin, G.E., *Rev. Sci. Instrum.*, **41** (1970) 772.
[7] Williams, W.D., Giordano, N., *Rev. Sci. Instrum.*, **55** (1984) 410.
[8] Nishizawa, M., Menon, V.P., Martin, C.R., *Science*, **268** (1995) 700–702.
[9] Jirage, K.B., Hulteen, J.C., Martin, C.R., *Science,* **278** (1997) 655–658.
[10] Hulteen, J.C., Jirage, K.B., Martin, C.R., *J. Am. Chem. Soc.,* **120** (1998) 6603–6604.
[11] Jirage, K.B., Hulteen, J.C., Martin, C.R., *Anal. Chem.,* **71** (1999) 4913–4918.
[12] Hou, Z., Abbott, N.L., Stroeve, P., *Langmuir,* **16** (2000) 2401–2404.
[13] Tourillon, G., Pontinnier, L., Levy, J.P., Langlais, V., *Electrochem. Solid-State Lett.,* **3** (2000) 20–23.
[14] Schonenberger, C., van der Zande, B.M.I., Fokkink, L.G.J., Henry, M., Schmid, C., Kruger, M., Bachtold, A., Huber, R., Birk, H., Staufer, U., *J. Phys. Chem. B,* **101** (1997) 5497–5505.
[15] Preston, C.K., Moskovits, M.J. *J. Phys. Chem.,* **97** (1993) 8405.
[16] Martin, C.R., in: *Handbook of Conducting Polymers*, 2nd edn., Reynolds, J.R., Skotheim, T., Elsebaumer, R. (eds.); New York: Marcel Dekker, 1997, Chap. 16, pp. 409–421.
[17] Duchet, J., Legras, R., Demoustier-Champagne, S., *Synth. Met.,* **98** (1998) 113–122.
[18] Demoustier-Champagne, S., Stavaux, P.-Y., *Chem. Mater.,* **11** (1999) 829–834.
[19] Sukeerthi, S., Contractor, Q., *Anal. Chem.,* **71** (1999) 2231–2236.
[20] Lakshmi, B.B., Patrissi, C.J., Martin, C.R., *Chem. Mater.*, **9** (1997) 2544–2550.
[21] Lakshmi, B.B., Dorhout, P.K., Martin, C.R., *Chem. Mater.,* **9** (1997) 857–862.
[22] Che, G., Lakshmi, B.B., Martin, C.R., Fisher, E.R., *Langmuir,* **15** (1999) 750–758.
[23] Che, G., Fisher, E.R., Martin, C.R., *Nature*, **393** (1998) 346–349.
[24] Kyotani, T., Tsai, L.F., Tomita, A., *Chem. Commun.* (1997) 701.
[25] Patrissi, C.J., Martin, C.R., *J. Electrochem. Soc.*, **146** (1999) 3176–3180.
[26] Li, N., Patrissi, C.J., Martin, C.R., *J. Electrochem. Soc.,* **147** (2000) 2044–2049.
[27] Che, G., Jirage, K.B., Fisher, E.R., Martin, C.R., Yoneyama, H., *J. Electrochem. Soc.,* **144** (1997) 4296–4302.
[28] Cepak, V.M., Hulteen, J.C., Che, G., Jirage, K.B., Lakshmi, B.B., Fisher, E.R., Martin, C.R., *J. Mater. Res.*, **13** (1998) 3070–3080.
[29] Cepak, V.M., Hulteen, J.C., Che, G., Jirage, K.B., Lakshmi, B.B., Fisher, E.R., Martin, C.R., *Chem. Mater.,* **9** (1997) 1065–1067.
[30] Martin, B.R., Dermody, D.J., Reiss, B.D., Fang, M., Lyon, L.A., Natan, M.J., Mallouk, T.E., *Adv. Mater.,* **11** (1999) 1021–1025.
[31] Penner, R.M., Martin, C.R., *Anal. Chem.,* **59** (1987) 2625–2630.
[32] Cheng, I.F., Martin, C.R., *Anal. Chem.*, **60** (1988) 2163–2165.
[33] Menon, V.P., Martin, C.R., *Anal. Chem.,* **67** (1995) 1920–1928.

[34] Hulteen, J.C., Menon, V.P., Martin, C.R., *J. Chem. Soc., Faraday Trans. 1,* **92** (1996) 4029–4032.
[35] Kang, M.S., Martin, C.R., *Langmuir.* **17** (2001) 2753–2759.
[36] Lee, S.B., Martin, C.R., *Anal. Chem.*, **73** (2001) 768–775.
[37] Parthasarathy, R.V., Menon, V.P., Martin, C.R., *Chem. Mater.*, **9** (1997) 560–566.
[38] Chen, W.-J., Martin, C.R., *J. Membr. Sci.,* **104** (1995) 101–108.
[39] Liu, C., Martin, C.R., *Nature,* **352** (1991) 50–52.
[40] Kobayashi, Y., Martin, C.R., *Anal. Chem.*, **71** (1999) 3665–3672.
[41] Kobayashi, Y., Martin, C.R., *J. Electroanal. Chem.*, **431** (1997) 29–33.
[42] Voet, D., Voet, J.G., *Biochemistry,* 2nd edn.; New York: Wiley, 1995, pp. 1297–1298.
[43] Steinle, E.D., Mitchell, D.T., Wirtz, M., Lee, S.B., Young, V.Y., Martin, C.R., *Anal. Chem.*, in press.
[44] Kathawalla, I.A., Anderson, J.L., Lindsey, J.S., *Macromolecules,* **22** (1989) 1215–1219.
[45] Martin, C.R., Nishizawa, M., Jirage, K.B., Kang, M.S., *J. Phys. Chem. B,* **105** (2001) 1925–1934.
[46] Bayley, H., Martin, C.R., *Chem. Rev.,* **100** (2000) 2575–2594.
[47] Deen, V.M., *AICHE J.,* **33** (1987) 1409–1425.
[48] Nitsche, J.M., Balgi, G., *Ind. Eng. Chem. Res.,* **33** (1994) 2242–2247.
[49] Archibald, D.D., Qadri, S.B., Gaber, B.P., *Langmuir,* **12** (1996) 538–546.
[50] Cohen, Y., Levi, S., Rubin, S., Willner, I., *J. Electroanal. Chem.,* **417** (1996) 65–75.
[51] Armstrong, R.D., Covington, A.K., Evans, G.P., *J. Electroanal. Chem.,* **159** (1983) 33–40.
[52] Xie, S.-L., Cammann, K., *J. Electroanal. Chem.,* **229** (1987) 249–263.
[53] Zhang, W., Spichiger, U.E., *Electrochim. Acta,* **45** (2000) 2259–2266.
[54] Ding, L., Li, J., Dong, S., Wang, E., *J. Electroanal. Chem.,* **416** (1996) 105, **112.**
[55] Ikematsu, M., Iseki, M., Sugiyama, S., Mizukami, A., *Biosystems,* **35** (1995) 123–128.
[56] Calistri-Yeh, M., Kramer, E.J., Sharma, R., Zhao, W., Rafailovich, M.H., Sokolov, J., Brock, J.D., *Langmuir,* **12** (1996) 2747–2755.
[57] Pursch, M., Vanderhart, D.L., Sander, L.C., Gu, X., Nguyen, T., Wise, S.A., Gajewski, D. A., *J. Am. Chem. Soc.,* **122** (2000) 6997–7011.
[58] Ohki, K., Tokiwa, F., *J. Chem. Soc. Jpn. Chem. Ind. Chem.,* **91** (1970) 534–539.
[59] Muilenberg, G.E. (ed.), *Handbook of X-Ray Photoelectron Spectroscopy*; Eden Prairie: Perkin-Elmer, 1978.
[60] Kane, V., Mulvaney, P., *Langmuir,* **14** (1998) 3303–3311.
[61] Hu, K., Bard, A.J., *Langmuir,* **13** (1997) 5114–5119.
[62] Bard, A.J., Faulkner, L.R., *Electrochemical Methods*; New York: Wiley, 1980.
[63] Bayley, H., *Sci. Am.*, September (1997) 62.
[64] Sugawara, M., Kojima, K., Sazawa, H., Umezawa, Y., *Anal. Chem.,* **59** (1987) 2842–2846.
[65] Xiao, K.P., Bühlmann, P., Umezawa, Y., *Anal. Chem.,* **71** (1999) 1183–1187.
[66] Wu, Z., Tang, J., Cheng, Z., Yang, X., Wang, E,. *Anal. Chem.,* **72** (2000) 6030–6033.
[67] Katayama, Y., Ohuchi, Y., Yang, X., Wang, E., *Anal. Chem.*, **72** (2000) 4671–4674.

List of Symbols and Abbreviations

Symbol	Designation
a_{DBS}	activity of DBS in solution phase
C	membrane capacitance
D	diffusion coefficient
i	transmembrane current
I	membrane thickness
k	Boltzmann constant
M	molecular weight
n	number of nanotubes in membrane
P	pressure
Q	gas flux
r	radius of nanotubes
r_{m}	molecular radius
r_{tube}	radius of nanotube
R	resistance
T	absolute temperature
α	selectivity coefficient
β	equilibrium constant for adsorption reaction
η	viscosity
λ	ratio of radius of diffusing molecule to radius of nanotube
Θc	critical fraction of surface

Abbreviation	Explanation
CMC	critical micelle concentration
DBS	1-dodecanesulfonic acid
XPS	x-ray photoelectron spectroscopy

1.3 Signal Processing Architectures for Chemical Sensing Microsystems

D. M. Wilson, University of Washington, Seattle, WA, USA
T. Roppel, Auburn University, Auburn, AL, USA

Abstract

This paper reviews several levels of signal processing and associated architectures for chemical sensing microsystems that use either arrays of optical or of physical sensors. Many chemical sensors, because of their interaction and vulnerability to the environment, have been eliminated from inclusion in sensing systems that require high precision and accuracy. This discussion evaluates parametric vs. nonparametric techniques and linear vs. nonlinear signal processing approaches for addressing chemical classification problems using imperfect sensing technologies. Hardware implementations of signal processing and biologically inspired signal processing are also reviewed. Future research into the development of more accurate chemical classification systems demands the customization of current approaches, so that underlying principles of chemical sensors and associated interfering influences do not overburden the computational space, thereby allowing higher accuracy rates.

Keywords: chemical sensors; gas sensors; electronic nose; neuromorphic engineering

Contents

1.3.1 Introduction

With the advancement of chromatographic and spectroscopic analysis microsystems, the role of physical sensors that directly interact with the chemical stimulus, such as the chemical field effect transistor (ChemFET), chemiresistor, surface acoustic wave (SAW) and related optical sensors, has been increasingly questioned in the crowded chemical sensor market. Fundamentally, a sensor whose coating interacts directly with the environment, with a minimum of additional transduction and processing at the chemistry stage, will continue to produce the fastest response time and lowest false negative rate. The inherent usefulness of these two system characteristics will continue to make physical chemical sensors a key contributing factor to many complete chemical sensing systems. Most compact chemical sensor technologies, however, are not ideally selective, and require participation in arrays as well as appropriate signal processing to be useful for solving chemical classification problems in real-world environments. In this chapter, the use of creative signal processing architectures for circumventing the limitations inherent in these technologies is reviewed and concluded with a positive outlook on the future research and commercialization of systems containing chemical sensors that are not perfectly selective.

An area of research and product focus that deserves particular attention for the use of nonselective chemical sensors arrays is portable systems. Portability necessitates a reduction of computing performed at the software level and an increase in computing performed at the hardware level. Portability constraints effectively reduce the computational space and overheads available to devise a solution for a chemical classification problem. Even in situations where a wireless link enables the portable sensing system to be in communication with a central computer, bit rates are severely limited compared with the almost unlimited data transfer rate assumed as part of many chemical discrimination systems demonstrated using a sensor array that is directly wired to a computer. Pattern recognition engines, especially linear statistical analysis techniques and artificial neural networks, have demonstrated significant robustness and discrimination capability using arrays of chemical sensors. Translation of these systems to general-purpose hardware has been completed in some research efforts. To improve system efficiency further, however, it is the task of the interface electronics to compress information provided by the sensor array in a way that conforms to the reduced bit rate of a portable system while still serving the needs of the pattern recognition engine to perform at a specified accuracy. Whether in a stand-alone system or in a wireless network of sensing nodes, the role of interface electronics for portable systems is greatly increased over the standard amplify and readout functions typical of many sensing systems on a chip. One should be forewarned, however, that implementation in hardware falls far behind what is possible in software for chemical discrimination; in this context, while hardware shows great promise in delivering high-performing handheld and redundancy systems in the future, development will not progress as quickly as that for software-based systems. The effective interaction between hardware and software in portable or miniaturized systems is very important to the success of the overall system. For this reason, hardware implementations of interface, processing electronics, and signal processing are also discussed in this paper in the context of the larger systems to which they contribute.

Following a discussion of hardware designs directed at chemical sensing systems, biologically inspired systems are discussed in the context with which inspiration from biology can provide effective approaches to processing larger arrays of nonselective chemical sensors. Inspiration from biology can provide a means to reduce the influence of interfering signals on the performance of chemical sensor arrays (eg, humidity, concentration) in conjunction with mainstream signal processing techniques. Mainstream signal processing techniques can be loosely classified as (a) linear or nonlinear and (b) parametric or nonparametric. Parametric techniques require some prior insight into the behavior of a chemical sensor array as a known relationship is fit with real data. Nonparametric techniques require far less understanding of basic underlying input–output relationships in the system. These pattern recognition techniques are frequently used in chemical sensing systems because the interactions with real-world sensing environments are complex, overlapping, and difficult to define in a parametric manner. All four classes of signal processing techniques are reviewed with particular attention to the most common approaches in each class of signal processing.

1.3.2 Hardware-based Signal Processing Architectures

For chemical sensing microsystems, hardware implementations of signal processing have been limited to applications where bandwidth, noise reduction, or portability are critical. In other types of sensing systems, limited bandwidth is a frequent justification for hardware-based signal processing, so that thousands or millions of signals can be transferred off the sensing plane to the central processing core. Chemical sensing microsystems, however, are limited to less than 100 sensors, with few exceptions. As expected from this scale of spatial resolution, limited effort has been committed to efficiently scanning chemical sensor signals off the sensing plane. Hardware-based signal processing directed at signal acquisition has often been catalyzed by limitations in existing data acquisition equipment rather than by fundamental limitations in the scanning process itself. For example, PCB modules are used in an array of 19 conducting polymer films to monitor temperature and humidity in the immediate sensing environment, convert the sensor signals to their digital equivalents, and multiplex these signals into the next stage of computation [1]. However, outside of the research environment, these PCB modules would require limited customization in the product development stage. Other similar efforts are directed at implementing pattern recognition algorithms in general purpose hardware especially digital signal processors [2]. With the advent of programmable systems on chip and other semi-programmable integrated circuits, it is expected that scanning and acquisition of chemical sensors will not, in general, require custom hardware for implementation, but will become a straightforward outcome of the product development process.

Unlike other sensing systems, chemical sensing systems exhibit a wide variety of noise components in their output signals which subsequently limit system resolution and sensing capability. Signal processing architectures and algorithms are often implemented in hardware for the purposes of reducing the contributions from a variety of noise sources and subsequently improving the resolution of analyte discrimination or concentration detection. Noise reduction in these systems is loosely defined as the removal of a portion of each sensor signal that is not directly correlated to the interaction between sensor and chemical environment. In this context, noise reduction can range across a wide variety of tasks such as:

- reduction of thermal noise in chemiresistors by optimizing electrode structure;
- reduction of shot noise in ChemFETs by using resonant measurement techniques;
- compensation for the influence of humidity through on-board humidity sensing and feedback;
- preprocessing of sensor signals to reduce the effects of widely varying baseline states;
- bandpass filtering to reduce the influence of transient chemical events and the impact of long-term drift.

Noise reduction techniques directed at chemical sensors, arrays, and systems have been evaluated in a limited number of efforts. Apsel et al. developed a continuously programmable, floating gate circuit that adapts to a broad range of composite polymer chemiresistor baseline values and produces a circuit output only when the time constant of an event is on the order of a chemical interaction of interest between the sensing environment and the sensor material [3]. Similar efforts using discretely programmable elements directed at compensating for variable baseline states in composite polymer chemiresistors, metal oxide chemiresistors, and ChemFETs have been demonstrated by McKennoch and Wilson [4, 5]. An example of baseline compensation and its impact on system performance is shown in Figure 1.3.1. In this example, three chemiresistors of identical polymer compositions exhibit baseline values that range from 25 000 to 300 000 ohms. With no signal processing, this variation can consume a large amount of resolution in the analog-to-digital converter that converts sensor outputs to their digital equivalents in the signal processing stream. If the maximum response of each sensor is only 10% of its baseline, the total dynamic range of the first, uncompensated situation is 25 000–330 000 ohms. With compensation, the baseline output voltages are all corrected to the same value (Figure 1.3.1 B), allowing each state of the subsequent analog converter to contain response-relevant information (as opposed to more superfluous baseline variation information). If a 12-bit analog-to-digital converter is used as the next stage in the signal processing stream, baseline compensation, in this example, results in a 68-fold improvement in resolution of the digital information provided to the primary signal processing center for concentration detection and analyte discrimination. These compensation circuits are designed such that compensation for variable baseline can be done at any time during sensor operation, thereby addressing initial fabrication variation as well as sensor drift.

Noise reduction can be implemented in a wide variety of different ways to improve the accuracy of chemical sensor signals in contributing to system level decision making. Whenever the noise is inherent to the sensor transduction mechanism or sensor interaction to the environment, it only makes sense to do the signal processing associated with the noise reduction in close proximity to the sensor itself. Local signal processing often demands the need for hardware implementations of these noise reduction techniques.

Hardware implementations of signal processing architectures and algorithms also become critically important when system portability is a central issue to the success of a chemical sensor system. Many pattern recognition algorithms can be directly converted to programmable digital microprocessor or microcontrollers without the need for significant restructuring or reduction of the signal processing algorithm. For example, Ortega et al. implemented an FFT of dynamic waveforms and resulting self-organizing map analysis of these waveforms directly on to an Analog Devices digital signal processor without loss or change of the signal processing algorithm. They demonstrate the use of the DSP-based hardware on tin oxide sensors excited by programmable temperature pulses to produce dynamic waveforms sufficient for discriminating methane from carbon monoxide [6]. In other situations, it may be necessary to restructure a signal processing al-

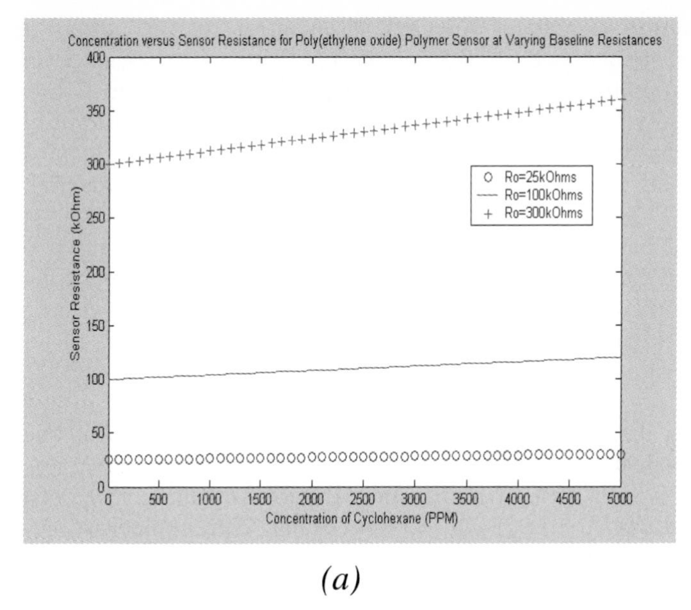

(a)

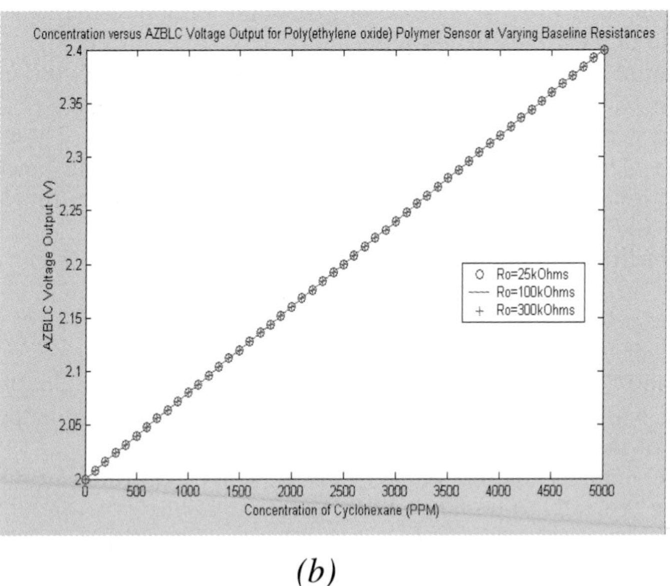

(b)

Figure 1.3.1. Baseline compensation as a form of noise reduction in composite polymer chemiresistors. (**a**) Fabrication variations cause typical variations in baseline resistance values of the chemiresistor to vary by an order of magnitude. Baseline variations consume resolution in subsequent analog-to-digital conversion that could be otherwise consumed by the sensor response. (**b**) Baseline compensation allows the full resolution of the analog-to-digital conversion process to be consumed by sensor response rather than more superfluous baseline information.

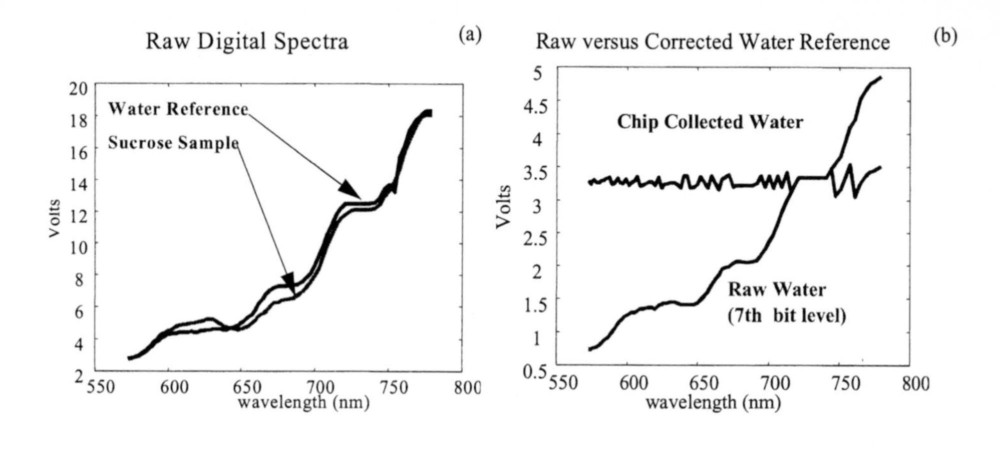

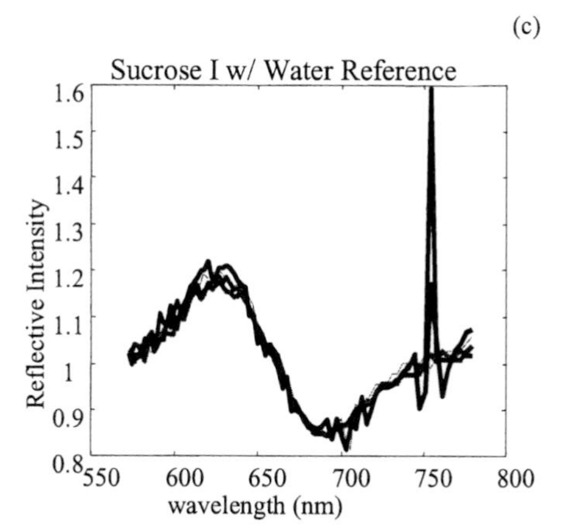

Figure 1.3.2. Portable SPR signal processing using integrated optical computing enables the (**a**) raw digital spectra to be (**b**) flat-field corrected (autocalibrated) so that (**c**) subsequent measurements are made independent of the background environment (reference medium). Auto-calibration not only provides resilience to a wide variety of reference media, it also facilitates quick detection of concentration (as evidenced by the minimum point in (**c**) whose location is directly proportional to concentration) [7].

gorithm in such a way that it is better suited to low power consumption in a hardware implementation. Wilson et al. demonstrated the restructuring of multivariate regression techniques into a programmable optical computation architecture implemented in a standard complementary metal–oxide semiconductor (CMOS) that effectively computes the concentration of a targeted analyte in a surface plasmon resonance-sensing architecture [7]. The restructuring allows compensation for variable optical characteristics of the background medium (solution) by adjusting photodetector integration times (in hardware) so that the output of the system in the absence of an analyte is a flat-line (Figure 1.3.2 b). Once compensated in this manner, the response of the SPR sensor in the presence of an analyte is more easily analyzed for concentration information (Figure 1.3.2c). In this case, the location of the minimum or the sum of the responses across wavelength can be extracted directly from the photodetector processing circuits and are directly correlated with concentration via a simple, low-memory consumption, look-up table obtained from probe calibration.

The nature of chemical analyte discrimination and concentration detection is often so complex in real-world environments that translation of an effective signal processing and pattern recognition algorithm to hardware requires a restructuring and streamlining of the algorithm to facilitate real-time operation. Often, features of the sensor space such as response rank relative to other sensors [8, 9], average in a like group of sensors, or outlier status [10] must first be extracted to reduce the computational burden on subsequent signal processing. A great deal of research and development remains to be invested in the development of this type of hardware to facilitate effective portable systems for field deployment, disposable use, or similar application.

1.3.3 Biologically Inspired Signal Processing Architectures

The term 'electronic nose' has typically been used to describe the use of arrays of broadly selective chemical sensors to discriminate among chemicals or mixtures of chemicals in the presence of other interferent chemicals. The functionality of the 'electronic nose' has typically focused on chemical diversity rather than on highly specific analyte detection, thereby distinguishing it from its laboratory-based counterparts, including gas chromatography and spectroscopic techniques. Similarities between most electronic noses and the biological olfactory system, however, have been limited to the use of broadly selective arrays of sensors and the use of artificial neural networks to process these array signals into an odor discrimination decision. In fact, many 'electronic noses' use pattern recognition and signal processing techniques that do not relate to the biological nose at all. To date, no 'electronic nose' system has been constructed that clearly models both the architecture and the functionality of the biological olfactory system.

A limited number of research groups are involved in the construction of biologically inspired chemical sensing systems (Keller at Batelle laboratories [11]; Pearce at the University of Leicester, UK [12], Goodman and Lewis at the California Institute of Technology [13], and Wilson at the University of Washington [14]). Complementing these efforts, Natale et al. and Lewis et al. have made direct comparisons between the performance of an 'electronic nose' and the olfactory system for detecting large numbers and classes of odors [15, 16].

Some of the research efforts directed at using biological inspiration to strengthen artificial chemical sensing systems design have been directed at neuromorphic (biologically inspired architecture as well as functionality) 'electronic noses'; these neuromorphic systems require that the array architecture, and signal processing stream as well as the central processing techniques (comparable to the brain's role in olfaction) be modelled after olfaction. Since the (perceived) ideal sensing situation where each sensor in a system is sensitive to one and only one single molecule component in the sensing environment is rarely available, biologically inspired architectures offer the promise of resolving largely overlapping specificities between sensors into robust chemical sensing systems. Many of the solid-state sensors have such overlapping specificities and are also compatible with large array sizes, making them ideal candidates for designing, simulating, and testing biologically inspired architectures on artificial systems.

Individual sensory receptors in the biological system possess many of the same types of variability as available artificial sensor technologies, a similarity that supports the use of biological inspiration in designing both preprocessing and mainstream signal processing architectures in chemical sensing systems. Keller has divided the biological process of olfaction and the signal processing of artificial olfaction into seven stages: (1) sniffing; (2) reception and binding; (3) stimulus; (4) transmission; (5) identification; (6) action; and (7) cleaning [17]. The olfactory system (Figure 1.3.3) in most higher-level animals contains a huge number of receptors from 2–20 million in the human olfactory system to 200 million in the canine olfactory system. The large number of signals generated by interactions between these receptors and the odor composition in the olfactory mucous layer are aggregated, compared, and contrasted before being presented to a content-addressable memory in the olfactory cortex. The cortex associates an

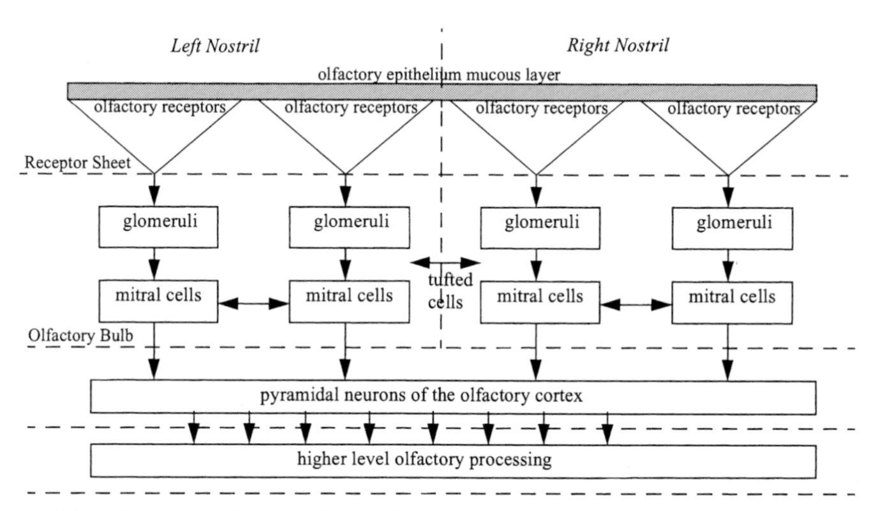

Figure 1.3.3. Basic signal processing pathways in the olfactory system.

odor with some odor or mixture of odors presented to the animal in its past history, either environmental or genetic.

The sniffing process (stage 1 of olfaction) produces airflow into the nostrils that is accompanied by extensive mixing through the production of turbulent air flow by the turbinate bones in order to produce uniform odorant concentration [17]. During stage 2 (reception and binding), sensory transduction occurs on the olfactory epithelium where the olfactory receptors are covered by a partially selective, chromatographic buffer layer of mucous. Transduction is mediated by semi-selective messenger proteins that bind with odorants in the vapor phase and deliver the odorant to the receptor sheet through the mucous buffer layer. The number of receptors on the epithelial sensing plane ranges from 1–2 million in the salamander to over 200 million in the dog. Excessive concentrations of odorants in the airflow are moderated by an increase in mucous concentration, a negative feedback loop that stabilizes the odorant concentration that is delivered to the olfactory receptor sheet. Binding between odorant and receptor is converted to a neuron signal during the stimulus stage of the signal stream.

During the fourth stage of olfaction (transmission), the millions of stimulus signals are combined into bundles called glomeruli and aggregated in the process of being transferred via glomeruli to the olfactory bulb. The reduction or aggregation of signals in travelling from olfactory sensing plane to olfactory bulb varies, but a 25000:1 receptor-to-bulb projection ratio is typical. It is unclear exactly how receptors are organized in this aggregation process. The two most popular theories are conflicting; one predicts that identical receptors project on to the same area of the olfactory bulb, producing a mapping for each unique odor; another theory predicts that the projection of information on to the olfactory bulb is distributed rather than localized by function. Future research is expected to provide more insight into the true function of the glomeruli to resolve these two conflicting theories. In spite of this conflict, it is clear that, during signal transmission to the olfactory bulb, aggregation before pattern recognition provides robustness to the system that is not available at the individual receptor level. Once in the olfactory bulb, data are compressed even further to facilitate real-time computation in the olfactory cortex and to improve further the robustness of olfactory decisions. As with aggregation, the exact features that are extracted to form this reduced data set are not well known. However, it is known that glomeruli signals are excited and inhibited across the olfactory bulb to enhance strong signals and inhibit weaker ones. Adaptation processes, communication between the two olfactory bulbs, cross-inhibition and excitation of glomeruli signals and other processes embedded in the architecture of the olfactory system all support the preprocessing of olfactory signals into a compressed, but complex olfactory map of the chemical composition in the ambient air. Olfactory systems, beyond a certain threshold, are indifferent to concentration and auto-adapt to high levels of odors so that a narrow dynamic range of inputs is presented to the olfactory receptors and glomeruli (despite a wide dynamic range of odor concentrations in the environment) [18–24].

After aggregation, compression, and cross-correlation of signals in the two olfactory bulbs during transmission of the olfactory map, the fifth stage of biological olfaction (identification) occurs [18–24]. Identification is performed in a manner compatible with the concept of a content-addressable memory. The closest relationships between the present odors and odors of memory are identified, and the map associated with the new odor is used to modify the connections in the content-addressable memory. During the last stage of olfaction, a breath of fresh air removes the odor composition from the receptor sheet, and the olfactory process being anew.

Research efforts directed at modeling the complex architecture of the olfactory system (from the olfactory mucus to the construction of an olfactory image in the olfactory bulb) have been limited but directed at efficient implementations in silicon or other microfabrication substrates for 'micronose' arrays. Goodman et al. at Cal Tech have directed their published efforts at this point to the auto-adaptation of chemical (olfactory) sensor signals so that baseline drift and background odors can be screened from subsequent pattern recognition and interpretation algorithms [13]. Comparable efforts by Wilson et al. at the University of Washington have also emphasized baseline compensation, where adaptation occurs at discrete rather than continuous time intervals, and distortion of sensor response is minimized by customizing the compensation and adaptation circuits to individual sensor technologies [4, 5]. Pearce et al. at the University of Leicester have focused their efforts on population coding the placement of sensors in an artificial olfactory system to enhance significantly the selectivity to analytes (odors) of interest without a corresponding degradation in dynamic range (total amount of analytes that can be sensed by the system) [12]. Further efforts by the University of Washington group have emphasized the use of local (linear and nonlinear) communication among sensors and feature extraction methods to reflect the aggregating properties of the glomeruli and early processing in the olfactory bulb [14, 25].

Additional efforts directed at using biological inspiration for the design of artificial chemical sensing systems exist, but have been directed at the implementation of neuromorphic architectures in the olfactory cortex where pattern recognition or odor discrimination takes place. Efforts to modify neural network architectures to be more biologically similar to the olfactory cortex have been modeled after the search for food in insects [26], the need for hardware models of chaotic dynamics in the olfactory cortex [27], and modular analog implementations of artificial neural networks geared toward olfactory processing [28]. Artificial neural networks, in general, lack the temporal dynamics that supplement large interconnection architectures in biological systems and are discussed as a separate class of signal processing in a subsequent section. Despite the fact that they generally lack temporal dynamics as do most artificial neural networks, self-organizing maps are especially relevant to biologically inspired systems. These maps organize odors according to similarities in sensor array responses and classify each incoming odor according to its closest relationship with existing odor patterns stored in the map (content-addressable memory format).

Methodologies for converting biological olfactory architectures to silicon or other substrates are not straightforward. Part of the difficulty in developing such design methodologies is due to the gaps in the spectrum of knowledge regarding the biological olfactory system. Complete models of the olfactory system have yet to be developed in a form and manner that are compatible with engineering implementation of these models into artificial systems. In addition, differences between artificial olfactory sensors such as metal oxide or polymer chemiresistor and FET-based devices and biological olfactory receptors are significant despite fundamental similarities in overlapping specificities and stability; these differences require many additional levels of research and modeling to achieve methodologies that can be effectively applied to the construction of arrays using each sensor technology. Initial results in this field of olfactory modeling, however, are promising and additional effort in this area is anticipated in the ongoing community effort to bring chemical microsystems to practical fruition. Several processes to be modelled and imitated in biological systems can be identified. *Adaptation* to variable baselines coincides with biology's ability to accommodate an olfactory receptor turnover on the average of 90 days. *Redundancy* provides resilience to broken, aging, or immature sensors by reducing the noise (variation) of each aggregate signal from a group of like sensors (homogeneous sub-array). *Aggregation* of and communication between sensor signals extracted from related but not identical sensors are based on interconnections between neurons in local maps in the olfactory bulb and strengthen the recognition and identification capability of the overall system. *Modulation* of concentration at the signal level, similar to the function provided by the olfactory mucous, provides improved mapping capability for odorant identification when performed in a way that coincides with the fundamental response (transfer) characteristic of the sensor or odor receptor.

1.3.3.1 Example: Redundancy of Sensor Signals (Composite Polymer Chemiresistors)

Homogeneous arrays, or arrays of identical sensors, can be used in chemical sensing systems, as in biology, to offset the detrimental effects of sensor poisoning, drift, and other erroneous behavior. Statistically, the variance of a homogeneous array decreases as the square root of the number of sensors in the array. The impact of reduction in variance on the chemical classification capability of the array varies with the type and number of sensors in the array as well as the analyte to be identified and its potential interferents. Regardless of these factors, reduction in variance of any single sensor will reduce the variance of the array response in multidimensional sensor space. An example of this improvement in sensor variance and in array performance is shown in Figure 1.3.4 for an array of tin oxide and composite polymer chemiresistors. In Figure 1.3.4a, the use of redundant sensors of a particular type results in a decrease in variance that approximates the square root of N (the degree of redundancy) for an array of tin oxide sensors up to 40-fold redundancy. Likewise, for a heterogeneous array of polymer chemiresistors consisting of homogeneous sub-arrays, performance of the array is improved (Figure 1.3.b) as the addition of redundancy increases the resolving power or discriminability of the array. Further details of this homogeneous array analysis can be found in [29] and [30].

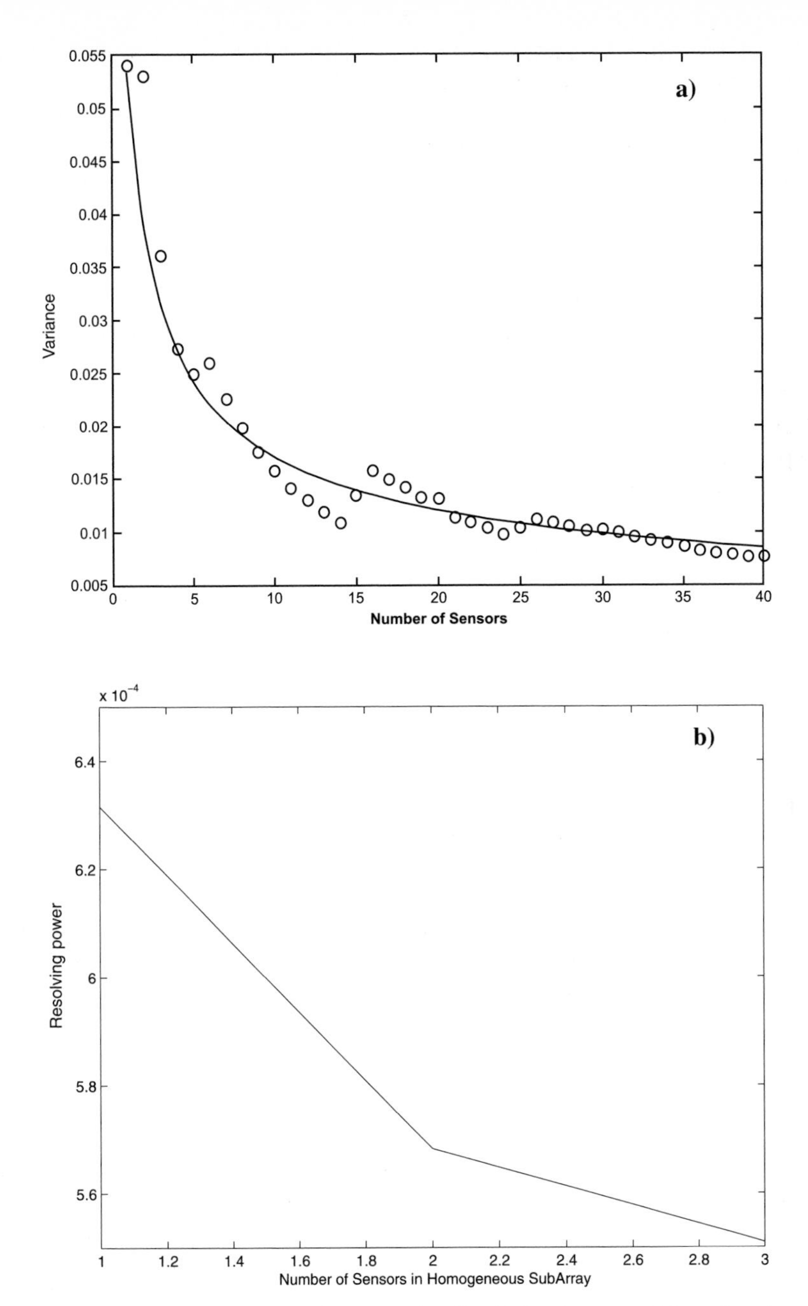

Figure 1.3.4. Use of redundancy to improve the discrimination capability of a chemiresistor array. Shown are the results of averaging multiple like (homogeneous) sensors (**a**) on the variation in the individual sensor signal in an array of tin oxide sensors and (**b**) on the variation in discriminability in an array of composite polymers.

1.3.3.2 Example: Modulation of Concentration (Surface Acoustic Wave Chemical Sensors)

The biological olfactory system modulates concentration at the olfactory receptor sheet, ensuring that a small dynamic range of odorant concentrations reaches the olfactory receptors themselves. Messenger protein activity and olfactory mucus thickness are dynamically modulated to maintain this limited dynamic range. In artificial chemical sensing systems, it is often difficult to modulate concentration prior to sensor reception because of the limited ability to control the sampling of airflow into the system. No intermediate liquid layer exists, for the most part, in vapor-phase systems between sensing environment and receptor. As a result, it is common practice to normalize sensor signals as a preprocessing step in an effort to reduce the influence of concentration on the distribution of sensors in multidimensional odorant space, prior to processing these signals in a pattern recognition engine.

Normalization is often a linear step, involving the division of each sensor output by a reference output or by the maximum output within a particular sensor technology or set of operating conditions. Linear normalization of signals whose response across concentration is nonlinear complicates the subsequent signal processing, inevitably requiring that the pattern recognition engine be nonlinear in making decision boundaries for odor discrimination. Linearization of the sensor response according to the fundamental transduction behavior of the sensor, prior to normalization, resolves this conflict and reduces the dynamic range of each sensor output. An example of the improvement in separation (discriminability) of odors in multidimensional sensor space via linearization and normalization is shown in Figure 1.3.5. The outputs of seven surface acoustic wave sensors are recorded for 17 analytes across a range of concentrations between 20 and 40 P/P_{sat}, where P and P_{sat} are the partial pressure and saturated partial pressure of the analyte of interest, respectively. The projection of the sensor array response to these 17 analytes in principal component space (PC1 and PC2) is shown in Figure 1.3.5 a without any signal preprocessing and in Figure 1.3.5 b after linearization and normalization. Linearization forces the processed sensor outputs to approximate linear behavior, regardless of concentration and normalization reduces the influence of background influences over their total response range. To linearize, each sensor response is taken to its 0.75 power and to normalize, each sensor is divided by the output of the quartz reference sensor at a given moment in time. The results show clearly the significant improvement in separation generated by the signal preprocessing, which effectively reduces the influence of concentration on the variance of the sensor array output across its dynamic range.

Reduced dynamic range, as in biological systems, enables more of a limited odorant space to be consumed by differences between odorants rather than differences in their concentrations. Separation of the concentration and discrimination problems improves performance of signal processing in solving both of these problems. Biology ignores concentration (beyond a certain threshold) and focuses computational power on odor discrimination. Artificial chemical sensing

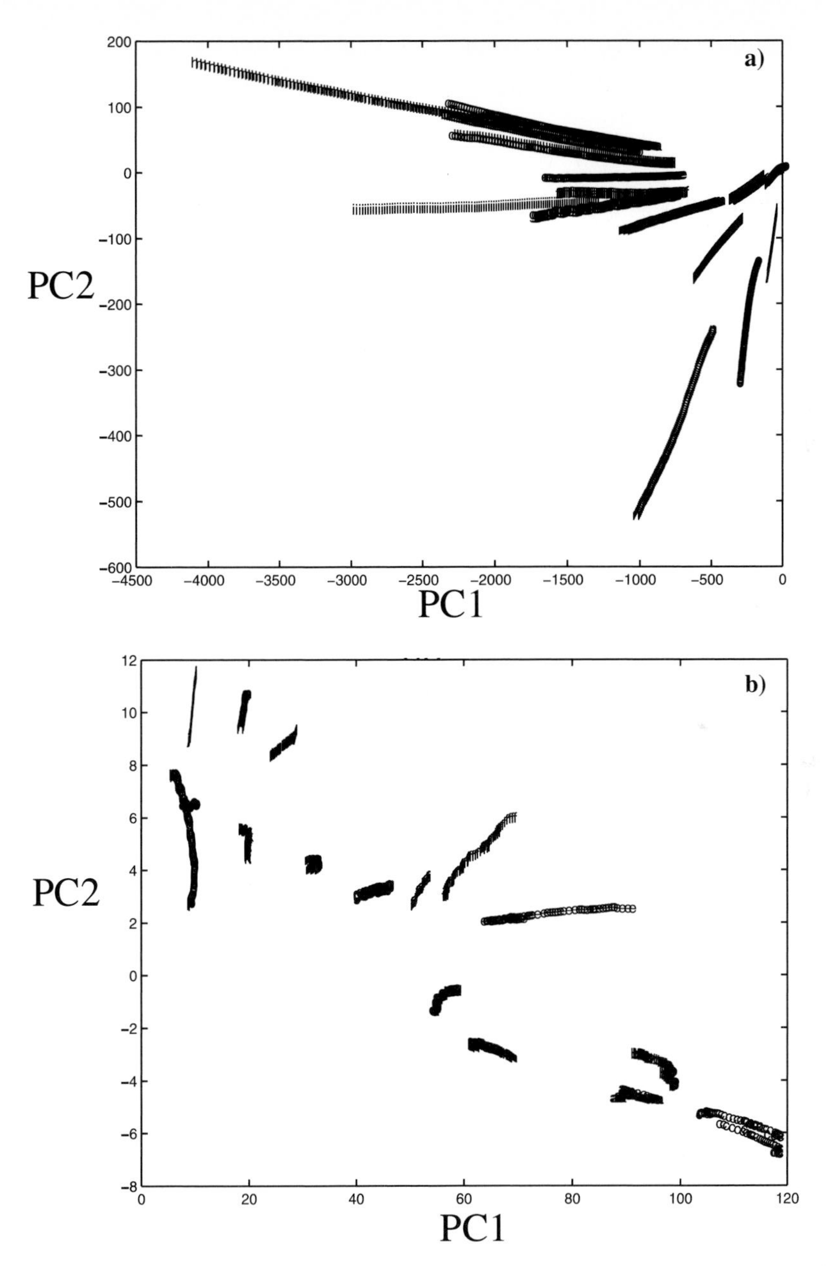

Figure 1.3.5. Modulation of concentration. The responses of seven surface acoustic wave sensors coated with materials of varying chemical sensitivity are exposed to 17 analytes across a range of concentrations. The (**a**) projection of this 7-dimensional response in principal component space shows significant overlap between analytes and the significant contribution of concentration to the overall variance of the first two principal components. After (**b**) linearization and normalization, concentration effects are reduced, resulting in significantly more separable clusters in principal component space.

systems, for the most part, must extract both concentration and discrimination information; better efficiency at both tasks is achieved by separation of these tasks in signal processing.

1.3.3.3 Summary

Neuromorphic as well as biologically inspired architectures have tremendous potential to enhance the capability of chemically diverse chemical sensor arrays and systems ('electronic noses'). Biology works with rapidly aging, broadly selective sensors to allow a broad odor discrimination capability at the brain level. Many sensor technologies used for 'electronic noses' have similar functional characteristics to biological receptors. In order to achieve comparable performance to the biological system using such sensor technologies, artificial chemical sensing systems based on biology can benefit from the following basic characteristics for achieving broad chemical diversity and recognition accuracy:

Sensor array composition and optimization
- Large heterogeneous array sizes: containing different kinds of receptors.
- Redundancy through homogeneous sub-arrays: multiple copies of one type of sensor.

Neuromorphic signal processing architectures
- Stimulus generation that compensates for baseline variation: to reduce the transmission of superfluous information to downstream signal processing centers.
- Concentration-independent odor discrimination: modulation of concentration to reduce dynamic range at the sensor transduction stage.
- Aggregation of similar but not identical signals prior to pattern identification: comparable to the function of the glomeruli.
- Feedback between sensors and groups of sensors: to reduce noise at the odor map ('image') stage.
- Communication among sensing nodes to enhance system robustness: comparable to the function of the tufted cells in interconnecting the two olfactory bulbs.

Signal processing
- Unsupervised correlation of odors to previously seen odors in content addressable memory, comparable to function of olfactory cortex.
- Temporal dynamics: incorporated into unsupervised organization scheme.

As understanding of the mammalian olfactory system improves in the biological research community, additional insight into the design of the 'electronic nose' using nose-like qualities will be achieved. It is essential to exploit these strengths of biological systems with prudence in artificial chemical sensing systems, often in combination with more traditional signal processing techniques.

1.3.4 Linear Signal Processing Techniques

Linear signal processing techniques are often used in situations where a signal processing problem is well understood in terms of underlying principles or when an initial understanding of a classification problem is needed. When clearly differentiable classes or acceptable error levels can be demonstrated from linear signal processing, no need exists to move to more complex nonparametric and nonlinear signal processing techniques as described in subsequent sections. In many chemical sensing problems, however, linear signal processing by itself is insufficient and is disregarded or used in addition to a nonlinear technique. The three most common techniques for processing chemical sensor signals in a linear manner are principal component and cluster analysis, discriminant function analysis, and linear regression techniques. Principal component and cluster analysis allows the projection of multidimensional data on to two- or three-dimensional space where relationships among groups of data may be more easily visualized and analyzed. Discriminant function analysis seeks, in a linear manner, the clearest way to differentiate classes or categories of interest and regression seeks to establish a linear relationship between input chemical parameters and output sensor responses in a manner that minimizes the error in this linear relationship. Each of these three techniques is described further in this section with some accompanying examples of prior research that effectively use these techniques alone or in conjunction with nonlinear, nonparametric techniques to solve chemical classification problems.

1.3.4.1 Principal Component Transformations: Gateway to Signal Processing

Principal component transformations have been heavily applied to the analysis of arrays of chemical sensor data, but in themselves do not provide a classification of analytes in multidimensional chemical sensor space. Instead, principal component analysis (PCA) provides a tool for understanding the signal processing problem sufficiently to (a) evaluate the impact of signal preprocessing; (b) advocate the need for additional signal preprocessing; and (c) identify a suitable parametric or nonparametric technique for pattern recognition. PCA re-orients axes in multidimensional sensor space, so that the direction of most variance in the sensor data forms the first principal component axis, the next larger variance is expressed in the second principal component dimension and so on. All new directions are perpendicular to one another just as in the original input data. Often, the first two or three principal components are used to visualize the multidimensional data to determine the next step in pattern classification and recognition. Because of its very nature, multidimensional sensor data must be normalized prior to a principal component transformation or else data with larger signal magnitude tend to dominate the expression of the data in principal component space. Normalization permits a valid comparison of variance for each dimension of sensor data in determining the principal components and source (direction) of primary variance. Other factors that are known a priori to be less relevant to the classification problem (eg, chemical concentration in a chemical discrimination problem) must also be addressed in signal preprocessing to optimize the use of principal component transformations in determining an appropriate decision-making model.

In the analysis of chemical sensor array data and construction of signal processing architectures for discrimination among single analytes, mixtures, and interferents, PCA has been used as a gateway for the design and demonstration of a wide variety of signal processing techniques. In some cases, separation of classes (of analytes) in principal component space has been provided as sufficient proof for the discrimination capability of chemical sensor arrays. For example, Gardner et al. presented sufficient separation of a toxic cyanobacterium and closely related non-toxic cyanobacteria in principal component space that further processing is not necessarily required at the research level [31]. Likewise, Guadarrama et al. used visual projection of the outputs of an eight polymer sensor array on to principal component space to argue the ability to differentiate among four different types of 'healthy' olive oil and four different types of 'off odor' olive oil. In this effort, however, clusters of olive oils in principal component space are not circular or highly separated, suggesting the need for an alternative signal processing algorithm to automatically (without human visual interpretation) classify the oil [32]. Gibson et al. used principal component transformations to optimize the parameters of a conductive polymer film array used to train and test an artificial neural network for distinguishing various types of bacteria and yeast [33]. In some cases, the use of the first two or three principal components is not the wisest choice to address a classification problem. Guadarrama et al. used the second and third principal components to distinguish among red wines because the predominant influence on the polymers used to sense these wines is the ethanol in each wine which does not provide sufficient variation across wines to facilitate chemical discrimination. Volatiles present in much more minute quantities are the key to separating the wines, hence the use of the principal components 2 and 3 rather than 1, which contains the variance caused by ethanol fluctuations [34]. In these and a wide variety of other research efforts, PCA has been used as a precursor to other parametric and nonparametric signal processing approaches to solving chemical classification problems.

PCA remains an important technique for projecting data into two- or three-dimensional space in a manner that facilitates increased understanding of the complexity of classification problems. PCA, and related cluster analysis, in less complex problems, can provide a means for identifying and separating chemicals, mixtures, and interferents in principal component space. Cluster analysis in PCA space involves locating centroids of clusters and calculating class membership by evaluating the distance (eg, Euclidean, Mahalanobis) between 'new' sensor array points and existing class centroids. In more complex problems, PCA allows the immediate localization of which classes in a particular problem are most difficult to separate. While PCA is an inherently linear technique, it provides visualization that encourages better choices of linear vs. nonlinear, parametric vs. nonparametric signal processing techniques for chemical discrimination and concentration detection.

1.3.4.2 Discriminant Function Analysis

Related to principal component cluster analysis, discriminant function analysis is another common technique for parametric analysis of multidimensional sensor data. This technique can be used to analyze two groups of sensor data (Fisher linear discriminant analysis) or multiple groups of sensor data in sensor space, in order to fit the data to a discriminant function that maximizes the differences among groups of data. In the simplest situation where only two groups are analyzed, the two groups are fitted to a single dependent variable, the identifier, in such a way that the distance (separation) between the two groups is maximized:

$$Identifier = a + b_1x_1 + b_2x_2 + \ldots + b_nx_n \tag{1}$$

where *Identifier* is a numerical label or score associated with the two groups of sensor data, the variables $x_1, x_2, \ldots, x_n$ are the sensor input signals, a is a constant and $b_1, b_2, \ldots, b_n$ are the regression coefficients. Maximizing the separation between the two groups can be done using least-squares techniques, maximum likelihood estimators, or a similar technique. Multiple groups are then analyzed in a quasi-pairwise fashion, where the first discriminant function is automatically chosen to provide the most discrimination among groups, the second discriminant function provides the second highest discrimination among groups, and so on. Discriminant function analysis provides orthogonal functions that are ordered in their group separation capability in an analogous manner to the way in which principal component analysis provides orthogonal functions that are ordered in their contribution to overall signal variance. Discriminant function analysis, by its very nature, tends to become less effective as the number and diversity of groups to be distinguished increases. In situations where the total number of analytes to be discriminated consist of multiple subsets, each containing closely related analytes, discriminant function analysis can likely fail to distinguish among closely related analytes in a subset, while still optimizing overall discrimination capability of the total set of analytes. Discrimination capability between closely related analytes within a subset will often be sacrificed for distinguishing across subsets of analytes or discriminating among analytes only within a certain subset. For example, Magan et al. successfully use discriminant factor analysis to distinguish between two types of yeasts (closely related analytes: *Candida pseudotropicalis* and *Kluyveromyces lactis*) and spoiled and unspoiled milk (subsets of analytes) but have difficulty distinguishing among bacteria (*Staphylococcus aureus*, *Bacillus cerus*, *Pseudomonas fluorescens*, *Pseudomonas aureofaciens*) that are also involved in the spoilage of milk samples [35]. Separate discriminant function analyses directed at smaller sets of analytes are successful in distinguishing between spoiled and unspoiled milk samples, between the two types of yeasts, and between two levels of spoilage.

Despite its limitations in discriminating among large numbers of analytes and in interpreting mixtures, discriminant function analysis has been successfully applied to such problems as the discrimination of oak barrel toasting levels [36] and the differentiation of carbon dioxide from forane R134A [37] using arrays of metal oxide sensors, as well as evaluating the quality of oranges with shear mode quartz resonator arrays [38]. In many cases, discriminant function analysis, like principal component and cluster analysis, has been applied as a precursor to determining the suitability of a more complex, nonlinear pattern recognition technique such as a neural network. Nonlinear signal processing techniques can be capable of stretching differences between closely related analytes and compressing differences between less related analytes to achieve more balance in successfully interpreting groups of diverse analytes.

1.3.4.3 Linear and Multilinear Regression

Linear regression assumes a linear relationship between input variables (chemical concentration) and output variables (chemical sensor response). A linear regression is a mapping of one input variable to one output variable in a manner that minimizes the least-squares error between the fitted line and the data points. Multilinear regression extends linear regression to multiple input variables:

$$y = w_0 + w_1 x_1 + \ldots + w_N x_N \tag{2}$$

where x_1, $x_2, \ldots, x_N$ are the input variables, y is the output variable, and w_1, $w_2, \ldots, w_N$ are the fitting parameters. It is not likely that the above equation will be fit exactly, so the actual relationship between the output variable y and the input variables x involves some error:

$$y = w_0 + w_1 x_1 + \ldots + w_N x_N + \text{error} \tag{3}$$

This error is minimized over the data set using a least-squares minimization procedure [39].

Poor results from a linear regression model tend to suggest that the relationship between input and output vectors is not linear or that peripheral influences on the input vector produce prohibitive variance around the line to which data are fitted. In both cases, high error levels in linear regression suggest the need for additional modifications to the decision-making model. These modifications can involve either (a) better preprocessing techniques to reduce variance in the data or (b) the replacement of linear regression with a nonlinear or nonparametric signal processing technique that is better suited to handling inherent nonlinearities in the input–output relationships. As an example of the impact of additional preprocessing, Boeker et al. demonstrated the use of multilinear regression on the detection of ammonia in agricultural emissions using six quartz microbalance (QMB) sensors integrated on to the same substrate. Initial multilinear regression on a sensor array operated in conventional, constant-temperature operating mode leads to very poor predictions of ammonia content. However, when the preprocessing of sensor signals is modified to include temperature modulation of the QMB films, the mean squared error of the multi-linear regression drops by an order of magnitude, allowing the prediction of ammonia content to within 16% [40]. Further reduction of the error rate could be accomplished by the use of additional preprocessing or the use of a nonlinear signal processing technique.

In chemical sensing systems, insufficient understanding of underlying relationships between input and output vectors often leads to the use of a linear regression model to make an initial determination as to the suitability of a linear over a nonlinear signal processing technique.

1.3.5 Nonlinear Signal Processing Architectures

Most nonlinear signal processing techniques applied to chemical analysis problems are nonparametric and well suited to interpreting interactions between arrays of chemical sensors and complex sensing environments where underlying relationships between input and output data are poorly understood. Research has shown that often nonparametric pattern recognition techniques can produce statistically equivalent classification results, given enough training time, memory, classification time, algorithm complexity, and flexibility to adapt to new data [41]. One notable exception to this general nature of nonparametric techniques in chemical sensing systems is the influence of sequence (or time) on the meaning of sensor response and characteristics. Many pattern recognition techniques capture only a moment in time, during training and classification, and consideration of time or sequence of sensor operation requires the choice of a pattern recognition algorithm that incorporates the past into computing the present output. In this section, we discuss the most popular nonparametric signal processing approaches that disregard time and sequence in the learning and classification of data as well as alternative approaches that consider the meaning and usefulness of time and sequence in solving chemical sensing problems.

The most popular choice for nonlinear signal processing of arrays of chemical sensor responses, parameters, and extracted features is the feed forward artificial neural network, where sequence and time do not influence training and classification. The artificial neural network is a nonparametric pattern recognition technique that uses interconnections among processing elements, called neurons, to enable the network to predict the output vector from an input vector after a training cycle. Success of the network is highly dependent on the appropriate level of diversity and size of the training data set as well as proper training time to reduce the error in the network's understanding of the training set. The most common of the artificial neural networks is the multilayer perception (MLP). The typical MLP contains N input neurons corresponding to N chemical sensors in an array, $2N$ hidden neurons, and M output neurons, where M is the number of analytes or chemicals to be identified by the network. The number of hidden neurons can vary widely, but $2N$ is a common starting point for the MLP architecture. Each of the hidden neurons produces an output that is a function of the weighted sum of all the input neurons. This function performed by the hidden neurons to produce their output is called an activation function and can be a hyperbolic tangent, a sigmoid, or similar function. The weights used to produce the weighted sum that is transferred into each hidden neuron's activation function are determined in an iterative training cycle. The training cycle commonly uses the back-propagation technique, to minimize the error between the desired output pattern and the actual output pattern produced by the network for each set of N inputs in the training set. The output neurons are summed, weighted, and activated in a similar manner to produce the final output of the network. The back-propagation training technique is the most popular choice for the MLP since it is guaranteed to converge in a three-layer network (input, hidden, and output layers) to a desired error level, given enough training time. While it is guaranteed to converge, the MLP trained by back-propagation can consume a prohibitive amount of training time, especially for large training sets or networks containing large numbers of neurons. This slow training time is often avoided or overcome via the use of the related radial basis function neural network.

The radial basis function neural network contains only two rather than three neuron layers and requires moderate training time when compared with the long training times of the MLP. In a radial basis function (RBF) neural network, the training patterns are first grouped into basis functions (typically radial Gaussian functions) using common techniques such as the K means clustering technique. K means clustering enables all of the training patterns to be clustered into a predetermined number of clusters that are chosen to group the most closely related input patterns together. These closely related patterns in each cluster are quantified and described via parameters associated with the radial basis function (such as the mean and standard deviation of a radial Gaussian function). Training of the two-layer neural network then considers the input training patterns in terms of these clusters and seeks to minimize the mean square error between the desired output pattern and the Gaussian cluster input. Training can be done using the familiar back-propagation technique or much faster least-squares techniques. By using the least-squares approach to training, the RBF neural network can produce comparable results to the MLP architecture, with much faster training times. The disadvantage of the RBF is that once it is trained, classification times, where an output is predicted based on an unfamiliar input vector, can be significantly longer than the MLP classification times [39].

When the time or sequence of sensor behavior is relevant and useful to solving the signal processing problem of interest, the Elman neural network is a useful choice. The Elman neural network incorporates the previous behavior of each sensor (transient response) into the training and classification capability of the network through the use of recurrent connections. Unlike the MLP and RBF, an Elman network architecture contains connections that feed activated and weighted sums of the hidden layer and output neurons back to the inputs during both training and classification. These recurrent connections ensure that the previous behavior of each sensor is incorporated into the computation of the output of the network in the present state of the input sensors. While useful in its treatment of time in understanding chemical classification problems, the Elman neural network can consume long training times and, during classification, is not guaranteed to 'pick the right answer' the first time; in other words, during a transient response, the network may initially choose an incorrect chemical before settling on the correct chemical classification. This potential increase in false alarm rates must be taken into account when choosing the Elman neural network, because even though a classification prediction may be computed sooner than other networks, the prediction has a higher probability of being incorrect during the initial stages of sensor response.

The MLP, RBF, and Elman neural networks all use supervised learning approaches, where the desired output for each input vector in the training set is known. In chemical discrimination problems, it is easy to assume that the M desired outputs of the pattern recognition engine correspond to the M analytes of interest in the sensing environment. This simple assumption, however, does not take into account the complexity of the sensing environment where concentration, presence of interferents, temperature, humidity, drift, and other factors may cause input patterns for a particular analyte to vary drastically, demanding more than one output category for a single analyte. One way to address the complexity of the sensing environment and the many competing influences on chemical sensors in an array is to use an unsupervised neural network where the only basic assumptions made regarding the classification of the outputs based on the training data set is the number of categories (or network outputs) into which the input data will assemble during training. The most common unsupervised learning technique applied to chemical sensing systems is the self-organizing map (SOM) trained using Kohonen techniques. Once the data self classify into a predetermined number of output categories, these outputs may then be injected into a supervised neural network to produce the final response of interest regarding which of M analytes is actually present in the sensing environment.

The MLP, by far, has dominated the literature for pattern recognition in chemical sensor arrays. The MLP has been applied to a wide variety of chemical sensor technologies and chemical classification tasks, including:

- Differentiation of aromatics and alcohols using optical fibers coated with chemically and photosensitive polymer dyes [42, 43].
- Discrimination of beer flavor and quality using 24 conducting polymer films on microelectrodes providing a 93% accuracy rate as compared with 87% for statistical fingerprinting techniques [44].
- Classification of ammonia, acetone, 2-propanol and their mixtures using nine tin oxide sensors, a humidity sensor, and a temperature sensor [45].
- Determination of chemical type and concentration (methane, propane, butane) using nine doped metal oxide films [2].
- Combining liquid- and vapor-phase measurements using potentiometric electrodes and quartz microbalance sensors, respectively, to make fine distinctions in red wine composition and quality. This effort uses a Levenberg–Marquardt algorithm (which uses least-squares techniques to optimize the weights) rather than the standard back-propagation training algorithm [46].
- Discrimination among bacteria that commonly cause diseases associated with ear, nose and throat using six MOS sensors in a Fox 2000 electronic nose at 100% (*S. aureus*) and 92% (*E. coli*) accuracy rates [31].
- Determination of growth phase of bacteria that commonly cause diseases with ear, nose, and throat using the same sensors as above at 80.7% average accuracy rate [31].
- Detection of ketosis in cow's breath using the same sensors as above at an 80–90% accuracy rate [31].
- Detection and discrimination of volatiles from various microorganisms (yeast and bacteria) using a 16-sensor array of conductive polymer chemiresistors [33].
- Differentiation of NO_2 and CO in mixtures using a 12-element tin oxide sensor array [47].

The RBF function has been a popular alternative to the MLP neural network and has been used for such classification tasks as detecting the freshness of fish using metal oxide sensors [48], discriminating acids from alcohols and ethers (in conjunction with a fuzzy *c* means modification to the neural network) [49], and the detection of wheat taint (mold) at 92% accuracy rates with a 32-sensor array [50]. The Elman neural network has been demonstrated on an array of metal oxide sensors to distinguish xylene, propane, and gasoline using the transient and steady-state response of the component sensors [51]. Finally, the SOM has great benefit when the organization of the input data is not well known; for example, input vectors corresponding to the same chemical analyte may be separable into further classes because of the impact of humidity, temperature, or concentration. Di Natale et al. have demonstrated the SOM as a preprocessor to a more conventional MLP network for distinguishing hydrogen, carbon dioxide, carbon monoxide, methane, and their mixtures using six metal oxide Taguchi-style sensors. The SOM, when used to organize classes and generate inputs to the MLP network, is more accurate than the MLP used alone [52]. Similar results have been demonstrated by this same group using the SOM for discriminating aromatic and spicy orders in a noisy environment [53].

Despite the dominance of the artificial neural network in addressing chemical classification problems, other techniques of interest used to process arrays of chemical sensors include fuzzy modifications to these artificial neural networks, fuzzy modelling techniques, hidden Markov models, and genetic evolutionary algorithms. Fuzzy modifications to neural networks have been used to determine more effective initial weights than the random weight selection used in most artificial neural networks. Gardner et al. have used this type of fuzzy modification [54] on arrays of tin oxide sensors to demonstrate up to 10% improvement in discriminating among different types of coffee and among different types of tainted water; specifically, these fuzzy methods enable coffees to be discriminated within 93% accuracy compared with 86% for MLP architectures trained using random weight initialization. The Fuzzy ARTMAP has also been used in several chemical classification efforts. The MAP layer provides supervised clustering to the Fuzzy ART concept by associating each of the clusters created in the organization of the Fuzzy ART with a known class (of chemicals). In this method, Fuzzy refers to the use of analog input vectors, ART refers to the self-organizing assignment of training vectors to a suitable number of clusters, and the MAP refers to the assignment of a useful chemical class to each cluster. In this way, significant variations in training vectors that correspond to a particular chemical (class) can be assigned to separate clusters, preventing over-generalization in the training process; over-generalization leads to oversized clusters and false alarms. Gardner et al. demonstrated the usefulness of the ARTMAP over standard back-propagation trained neural networks in the application of an array of six commercial chemical sensors (Fox 2000) to the detection of cyanobacteria in water. In this effort, the Fuzzy ARTMAP distinguishes between toxic and non-toxic versions of the bacteria at a 100% accuracy rate compared with the 97.1% accuracy rate generated by the standard neural network architecture. The Fuzzy ARTMAP has the additional advantage that training times are an order of magnitude less than the MLP approach [55].

Genetic algorithms have been used as precursors to other pattern classification schemes to optimize arrays of sensors for specific classification problems. Corcoran et al. use such genetic algorithms to optimize an array of chemical sensors to differentiate Chardonnay from Muscadet wine. Short, low-order, above average fitness array compositions in genetic evolutionary approaches propagate best to the next generation of optimization; cross-overs and mutation are also used in each generation to produce unexpected jumps in array composition from generation to generation that may lead to better solutions to the classification problem at hand [56]. Hidden Markov models, commonly used in speech recognition, have also been used to recognize sequences of chemical events that are important to chemical classification problems and are discussed in the context of a composite polymer chemiresistor array in Section 1.3.5.3. Custom nonlinear signal-processing algorithms have also achieved some success in various chemical classification tasks. Sandia National Laboratories has developed a unique pattern classification algorithm, called VERI, which relies on the human visual system to classify clusters in multidimensional sensor space. The program projects multidimensional sensor space into a three-dimensional format that is then trained and evaluated using psycho-physical information regarding the ability of the human visual system to differentiate these projections of multidimensional space on to comprehendable, three-dimensional formats. VERI has been successfully demonstrated on arrays of SAW sensors for the detection of chemical warfare agents and their simulants [57].

1.3.5.1 Example: Nonlinear Signal Processing Comparing MLP and RBF Neural Networks

This example demonstrates the use of an MLP and an RBF neural network to distinguish seven odorants: acetone, ammonia, beer, formaldehyde, 2-propanol, vodka, and red wine. An array of 30 tin oxide sensors is used, consisting of three different sensor dopant types, each operating at 10 different temperatures. Sensor responses are measured at specific elapsed times after the onset of odor presentation. The data are processed and recognition is attempted using (1) a multilayer perceptron (MLP) with 30 inputs, 40 hidden layer neurons, and seven outputs, and (2) a radial basis function (RBF) neural network with 30 inputs, 56 hidden neurons, and seven outputs. Details of the network architectures and the training and testing methodologies are available in [58]. A number of parameters are varied, including the type of prefiltering, level of rank-order filtering, and the length of time elapsed from onset of odor presentation. PCA visualization of this classification problem is shown in Figure 1.3.6.

Overall, the results show little distinction between the performance of the two types of neural networks. The MLP has a recognition rate of 95% versus 94% for the RBF under optimized conditions. The RBF has improved performance early in the sensor transient response compared with the MLP, but neither offers recognition rates above 90% until about 8 min of odor presentation. The RBF, as described earlier, however, requires significantly less training time than its MLP counterpart while providing comparable classification accuracy on unfamiliar data.

1.3.5.2 Example: Nonlinear Signal Processing Using Recurrent Neural Networks

Response characteristics of an array of metal oxide sensors contain information relevant in time and in space. In most implementations, sensor response is allowed to stabilize before a recognition decision is made. In order to reduce the

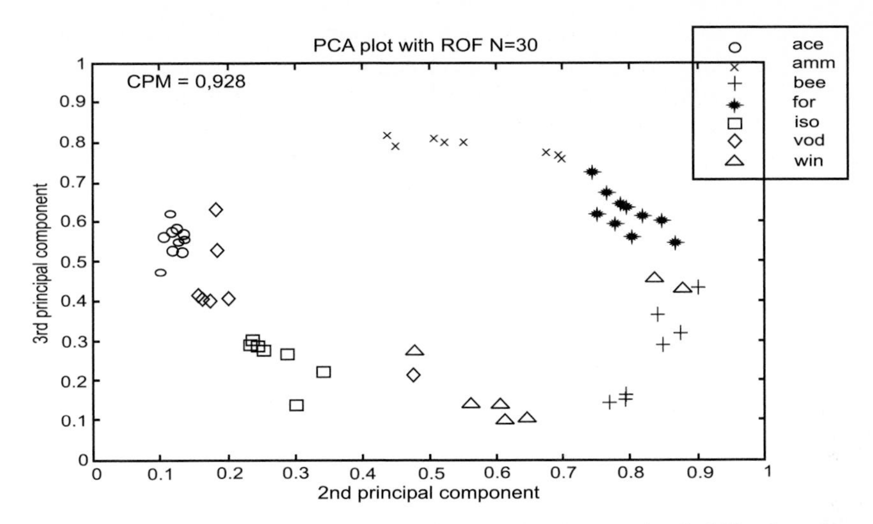

Figure 1.3.6. PCA projection of data used for MLP/RBF example. Stabilized sensor outputs are used for analysis. The data are preprocessed by prefiltering and rank-order filtering the sensor responses across the 30 elements of the array. The second and third principal components represent best the most relevant distinctions among sensor responses that must be evaluated by the neural network during training and subsequent classification.

recognition time for an array of chemical sensors, it is of interest to determine whether features in the transient response can be extracted to make a comparably accurate decision sooner and a more accurate decision in the same amount of time required to wait for the sensors to reach steady state. Pattern recognition techniques that are able to account for time and sequence in learning and generalizing patterns are few in number. A neural network-based approach to capturing time and sequence is a recurrent neural network architecture called the Elman network [59]. The Elman neural network uses connections in the network architecture that feed back information from the previous state in time to the inputs, to be considered in the calculation of the present state.

An example of using the Elman network for effective and improved recognition of chemicals is described in detail in [51]. An array of 15 tin dioxide sensors having various selective responses (determined by catalyst type and temperature of operation) is exposed to three vapors: gasoline, paint thinner, and xylene (G, P, X). Three independent presentations of each vapor are made to the sensor array. The responses of each sensor are thresholded to yield a binary result. That is, if the sensor output is above threshold, the output is 1, otherwise the output is zero. This yields a set of image maps such as those shown in Figure 1.3.7 in which the value 1 is shown as white and the value 0 is shown as black. One presentation (the second column of transient response maps) is used to train the Elman network, and it is tested on all three presentations. During testing, the input data is presented as a sequential stream, so the network only has access to the data up to the current time step.

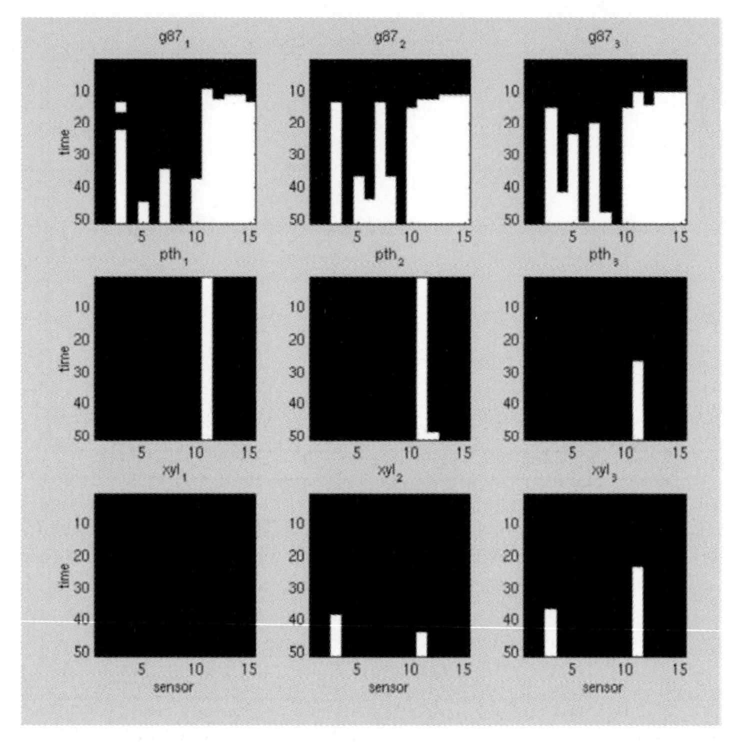

Figure 1.3.7. Temporal maps. The binary-thresholded representation of the response of 15 tin dioxide sensors to nine odor presentations. There are three presentations each of three samples: row 1, 87 octane gasoline ($g87_1$, $g87_2$, $g87_3$); row 2, paint thinner (pth_1, pth_2, pth_3); row 3, xylene (xyl_1, xyl_2, xyl_3). In each subplot, the horizontal axis is the sensor number (1–15). The vertical axis is time, running from $t=0$ at the top to $t=51$ sample intervals at the bottom. The sensor output voltages (raw data) were sampled approximately every 0.5 s. The raw data (0 to 5 V) were thresholded at 2 V, so a white pixel indicates that a given sensor output exceeds 2 V.

The results of using an Elman neural network to train on these temporal sequences of data clearly show the inherent dynamic nature of the Elman network, and suggest its value in processing temporal data. Figure 1.3.8 shows the response of the Elman network when tested on its training data in the original sequence of presentation. The response is correct at all time steps.

However, when the trained data sets are presented in reverse order, the neural network response is particularly interesting. As shown in Figure 1.3.9, the network output eventually reaches the correct result for each vapor, but only after an initial period of confusion and readjustment. Other experiments investigate the response to various sequential presentations of the available data, with generally the same result, and an average recognition accuracy of 94%. The less 'familiar' the network is with the input data, based on its training, the longer is the time required to make the correct identification. However, in every case the correct

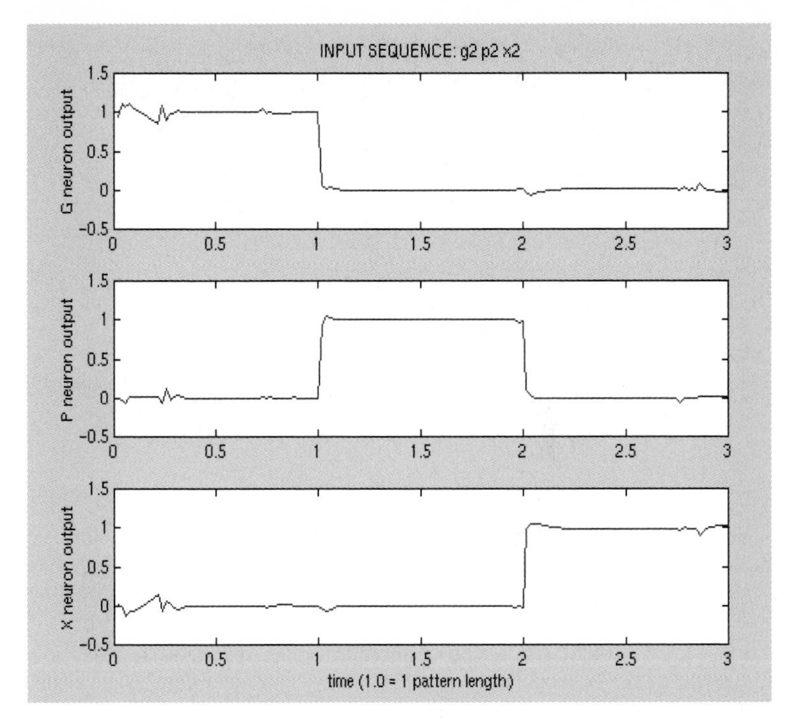

Figure 1.3.8. Elman results: the result of presenting the training data (second column of image maps shown in Figure 1.3.7) sequentially to the trained Elman network. The response is essentially perfect if an output value of 0.5 is established as the binary decision threshold for each neuron. The horizontal axis is time measured in units of pattern length, where one pattern length is 51 samples (ie, the length of time one odor is presented to the net). For example, the input sequence (g2 p2 x2) is 3 pattern lengths long, and is a shorthand way of indicating that the odors are presented in the order $g87_2$, pth_2, xyl_2. Each subplot shows the output of the indicated output neuron.

identification can be made, based on early transient response of the sensors, in a time interval approximately an order of magnitude shorter than required for the sensor outputs to stabilize. The recurrent neural network, then, has the potential to provide accurate responses sooner, using a richer and possibly more robust spectrum of input data than its 'snapshot' counterparts which evaluate only a single moment in the sensor response at a time. The disadvantage of these networks, of course, is the potential increase in false alarm rate caused by premature, inaccurate responses.

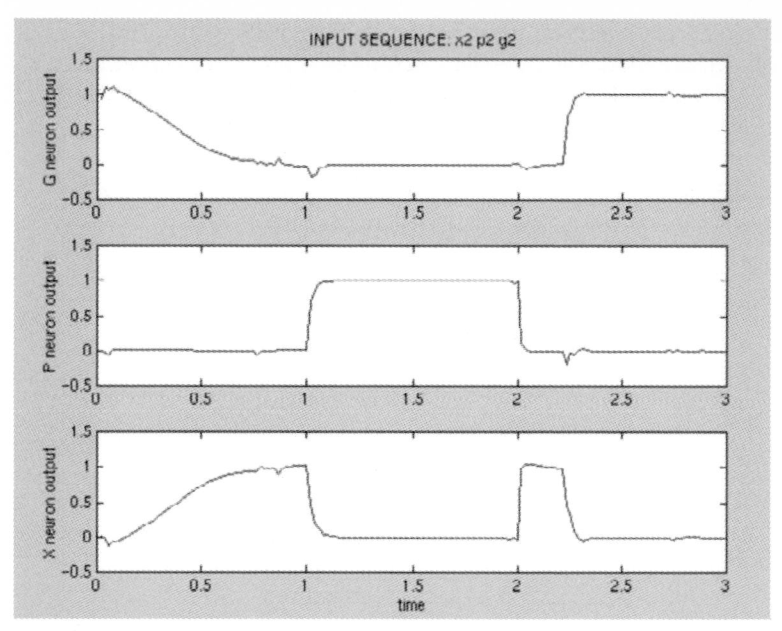

Figure 1.3.9. Elman results: the result of presenting the vapors in reverse order to the trained Elman network. Nomenclature is the same as used in Figure 1.3.8. The network initially outputs the trained response G, as it does not have enough input data to determine otherwise. As more data become available for the first vapor, the G and X outputs quickly begin to shift toward the correct result X. During the second pattern length interval, P is properly recognized very quickly. During the third interval, gasoline is recognized, although the recognition time is nontrivial. Generalization to unfamiliar data takes additional time.

1.3.5.3 Example: Nonlinear Signal Processing Using Hidden Markov Models

The hidden Markov model is a common choice in automated speech recognition because of its ability to identify sequences in the input streams of multiple features simultaneously. The hidden Markov model contains internal states connected by transitions and probabilities of transitions determined through training, as in other nonparametric pattern recognition techniques. The hidden states of the model through which an input stream sequences are not directly observable, hence the name hidden. Two major architectures for the hidden Markov model are currently in use. The first, a left–right model, allows the input data to move only forward through the model and does not facilitate returning to a state once the input vector has passed through that state. This type of model is well suited to problems where the input stream is not periodic and does not cycle through the same behavior more than once to reach its steady-state condition. The other common type allows data to pass back to states from which it has already transitioned and is well suited to problems that may contain some level of periodicity in the input feature vectors. For chemical sensors, where the transient response is not at all periodic, the left–right model is best suited. Further details regarding the hidden Markov model, its training procedure, and architecture can be found in [60].

Here, an example of a left–right model with four states trained on 10 types of composite polymer chemiresistors is presented. The objective of the model is to establish the health and validity of each sensor as it responds to an analyte. This problem requires 10 models, one for each type of sensor, and the data set is divided equally into training and testing sets. All data sets are tested on each model, and the output of each model represents the likelihood that the model produced the input vector under evaluation. Representative results from this sensor validity evaluation are shown in Figure 1.3.10. In all but one case, sensors of type 1 pass through model 1 with a high likelihood, indicating that they are representative of that model. Likewise, sensors that are not of type 1 do poorly in model 1; their transient response is not consistent with those of sensor 1 and, therefore, the model is less confident that it produced responses from sensors not of type 1. One sensor of type 1 is particularly noisy and does not produce a high likelihood in model 1, indicating that its validity for use in a system is inade-

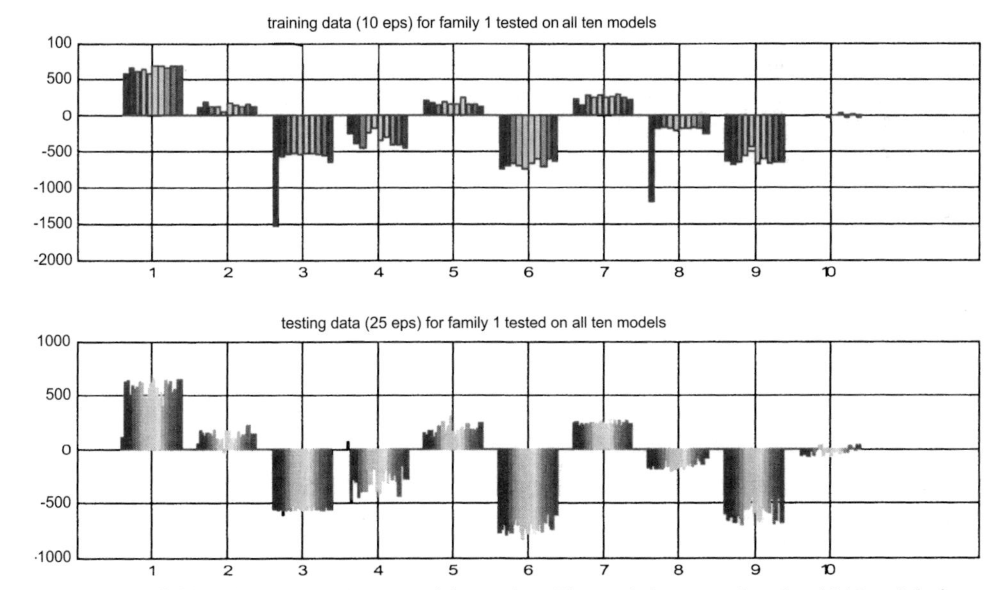

Figure 1.3.10. Hidden Markov model results. The training set for the hidden Markov model consists of a subset of responses for sensor of type 1. After traininig, the model for sensor 1 is used to test all training data sets (upper graph) and all testing data sets (lower graph) for all ten sensor types. The *y*-axis is the likelihood that the model for sensor 1 produced the response under evaluation. With one exception, the model recognizes the transient response of sensor 1 accurately. Other like sensors (sensors 2, 5, and 7) receive moderately high scores, indicating their fundamental physical similarity to sensor type 1 (same types of polymers, but different percentage composition in the chemiresistor films).

quate. The hidden Markov model can be used to analyze sensors for health and validity in order to flag the need for calibration and replacement, as shown in this example; alternatively, the hidden Markov model can be used for more typical chemical pattern recognition and classification tasks, where artificial neural networks currently dominate.

1.3.6 Opportunities for Future Research

Signal processing of chemical sensing systems has been dominated by the use of highly generalized signal processing architectures applied to chemical classification and discrimination. The use of highly generalized algorithms, especially the artificial neural network, is a result of (a) the complexity of typical chemical sensing environment and a poor understanding of how variations in these environments influence sensor behavior, (b) the competitive nature of chemical interactions with sensor surfaces and incomplete quantitative understanding of chemistry of all aspects of these interactions, and (c) the complicated, difficult to visualize, multidimensional nature of chemical sensor array responses. The generic nature of the artificial neural network and similar signal processing algorithms has allowed a partial solution of a wide variety of chemical classification problems using a broad variety and number of chemical sensor technologies and topologies. The disadvantage of these highly generalized algorithms, however, is that they often do not reach targeted accuracies of recognition because uncertainties in the complex chemical classification problems quickly consume the multidimensional sensor response space as well as the computational capability of these systems. More customized signal-processing algorithms, although much more difficult to develop and not nearly as transferable across sensor technology platforms, can offer the following distinct advantages:

- ability to use additional embedded information in the sensor response (data distribution, noise characteristics, transient response, etc.) to improve the robustness and accuracy of the pattern recognition process;
- capacity for identifying sensors which require re-calibration or replacement;
- ability to optimize arrays for particular chemical classification problems;
- faster training or classification speeds.

Additional customization of nonparametric pattern recognition for chemical sensor arrays is a necessary next step for improving system performance and decision-making capability. Customization requires more a priori knowledge of the system. Humidity and temperature, because of their strong influence on the performance of many chemical sensors, should at least be monitored or controlled in the sensing process, and used as an input to the pattern recognition process.

Preprocessing hardware can be used to optimize the resolution of information transferred to pattern recognition processes. Compensation for baseline variations, noise filtering, outlier detection and removal, redundant sensor operation, and operating parameter control are just a few techniques for improving the quality of signal used in subsequent stages of signal processing.

Multi-stage signal processing architectures are necessary to address the complexities of chemical sensing, classification, and especially recognition. The first stage of a good chemical sensing system is known control of sampling and sensor operating parameters including air flow, temperature, and local concentration gradients. Sensor transduction occurs next in a multi-stage architecture and should be done, consciously, at as low a noise level as possible taking into account electrode geometries, control parameters and other factors to minimize the amount of noise added to the process by transduction itself. After transduction, preprocessing electronics can be used to normalize, autoscale, average, aggregate, remove outliers, and filter the resulting electronic signals in preparation for pattern recognition. Pattern recognition architectures can also be multi-stage to optimize the use of the computational space in understanding and interpreting the complex multidimensional sensor space. Initial stages of processing should perform group classification, followed by more finely tuned classification processes within a group or family of chemicals. At each stage of processing, first principles as well as empirical models should be used to tailor the signal-processing to the chemical classification process at hand. The design of these customized signal processing architectures and algorithms can be automated to relieve research and development time, but the move toward more customization is necessary, given the computational limitations of pattern recognition algorithms and the profound complexity of multidimensional chemical sensor problems.

1.3.7 References

[1] Gardner, J.W., Pearce, T.C., Friel, S., Bartlett, P.N., Blair, N., *Sens. Actuators B*, **18/19** (1994) 240–243.

[2] Lee, D.-S., Jung, H.-Y., Lim, J.-W., Lee, M., Ban, S.-W., Huh, J.-S., Lee, D.-D., *Sens. Actuators B*, **71** (2000) 90–98.

[3] Apsel, A., Stanford, T., Hasler, P., in: *Int. Symp. on Circuits and Systems;* Monterrey, CA, May 1998, Vol. 3, pp. 107–110.

[4] McKennoch, S., Wilson, D., presented at IEEE Sensors; Orlando, FL, 2002.

[5] McKennoch, S., Wilson, D., *Proceedings of the SPIE International Symposium on Environmental and Industrial Sensing*; Boston, MA, 2001.

[6] Ortega, A., Marco, S., Perara, A., Sundic, T., Pardo, A., Samitier, J., *Sens. Actuators B*, **78** (2001) 32–39.

[7] Wilson, D.M., Warren, M., Booksh, K., Obando, L., presented at Eurosensors 2002; Prague, Czech Republic, September 2002.

[8] Tan, B.P., Wilson, D.M., *IEEE Trans. Circuits and Syst. II: Analog Digital Signal Process.*, **48** (2001) 198–205.

[9] Wilson, D.M., Dunman, K., Roppel, T., Kalim, R., *Sens. Actuators B*, **62** (2000) 199–210.
[10] Wilson, D.M., Roppel, T., in: *SPIE Proceedings Photonics East*; Boston, MA, September 20–22, 1999.
[11] Keller, P.E., in: *SPIE Conf. on Applications and Science of Computational Intelligence II*: Orlando, FL, April 1999, pp. 144–152.
[12] Pearce, T.C., in: *Neurocomputing: the 8th Annual Computational Neuroscience Meeting*; Amsterdam, July 18–22, 1999, pp. 941–952.
[13] Caltech Electronic Nose Project, *http://www.micro.caltech.edu/micro/research/electronic_nose.html.*
[14] Wilson, D.M., Roppel, T.A., Kalim, R., *Sens. Actuators B*, **64** (2000) 107–117.
[15] Di Natale, C., Macagnano, A., Paolesse, R., Tarizzo, E., D'Amico, A., Davide, F., Boschi, T.O., Faccio, M., Ferri, G., Sinesio, F., Bucarelli, F.M., Moneta, E., Quaglia, G.B., in: *Transducers '97;* Chicago, IL, June 1997, pp. 1335–1338.
[16] Doleman, B.J., Lewis, N.S., *Sens. Actuators B*, **72** (2001) 41–50.
[17] Keller, P.E., in: *Int. Symp. on Intelligent Control/Intelligent Systems and Semiotics*; Cambridge, MA, September 15–17, 1999, pp. 447–451.
[18] Dodd, J., Castellucci, V.F., in: *Principles of Neural Science*, Kandel, E.R., Schwarz, J.H., Jessels, T.M. (eds.); New York: Elsevier, 1991, 400–418.
[19] Kauer, J.S., Neff, S.R., Hamilton, K.A., A.R., in: *Olfaction: a Model System for Computational Neuroscience*, Davis, J.L., Eichenbaum, H. (eds.); Cambridge, MA: MIT Press, 1991, Ch. 2.
[20] Keverne, E.B., in: *The Senses*, Barlow, H.B., Mollon, J.D. (eds.); Cambridge: Cambridge University Press, 1988, Ch. 18.
[21] Getchell, T.V., Getchell, M.L., in: *Neurobiology of Taste and Smell*, Finger, T.E., Silver W.L. (eds.); New York: Wiley, 1987, Ch. 5.
[22] Scott, J.W., Harrison, T.A., in: *Neurobiology of Taste and Smell*, Finger, T.E., Silver, W.L. (eds.); New York: Wiley, 1987, Ch. 7.
[23] Price, J.L., in: *Neurobiology of Taste and Smell*, Finger, T.E., Silver, W.L. (eds.); New York: Wiley, 1987, Ch. 8.
[24] Kauer, J.S., in: *Neurobiology of Taste and Smell*, Finger, T.E., Silver, W.L. (eds.); New York: Wiley, 1987, Ch. 9.
[25] Roppel, T.A., Wilson, D.M., *Pattern Recog. Lett.*, **21** (2000) 213–219.
[26] Pearson, R.J., Alexander, J.R., in: *Intelligent Engineering Systems Through Artificial Neural Networks Proc*; St. Louis, MO, November 7–10, 1999, pp. 51–57.
[27] Grade Tavares, V.M., Principe, J.C., in: *Proc. IEEE International Conference on Electronics, Circuits and Systems*; Lisbon, September 7–10, 1998, pp. 131–134.
[28] Shoemaker, P.A., Hutchens, C.G., Patil, S.B., *Analog Integrated Circuits Signal Process*, **2** (1992) 297–311.
[29] Wilson, D.M., Roppel, T., in: *SPIE Proceedings Photonics East*; Boston, MA, September 20–22, 1999.
[30] Wilson, D.M., Garrod, S.D., *IEEE Sens. J.*, in press.
[31] Gardner, J.W., Shin, H.W., Hines, E.L., *Sens. Actuators B*, **70** (2000) 19–24.
[32] Guadarrama, A., Rodriguez-Mendez, M.L., de Saja J.A., de Saja J.L., Olias, J.M., *Sens. Actuators B*, **69** (2000) 276–282.
[33] Gibson, T.D., Prosser, O., Hulbert, J.N., Marshall, R.W., Corcoran, P., Lowery, P., Ruck-Keene, E.A., Heron, S., *Sens. Actuators B*, **44** (1997) 413–422.
[34] Guadarrama, A., Fernandez, J.A., Iniguez, M., Souto, J., de Saja, J.A., *Sens. Actuators B*, **77** (2001) 401–408.
[35] Magan, N., Pavlou, A., Chrysanthakis, I., *Sens. Actuators B*, **72** (2001) 28–34.

[36] Chatonnet, P., Dubourdieu, D., *J. Agric Food Chem.*, **47** (1999) 4319–4322.
[37] Sarry, F., Lumbreras, M., *Sens. Actuators B*, **57** (1999) 142–146.
[38] Di Natale, C., Macagnano, A., Martinelli, E., Paolesse, R., Proietti, E., D'Amico, A., *Sens. Actuators B*, **78** (2001) 26–31.
[39] Kennedy, R. L., Lee, Y., Van Roy, B., Reed, C. D., Lippmann, R. P., *Solving Data Mining Problems Through Pattern Recognition*; Unica Technologies, 1997, Ch. 10.
[40] Boeker, P., Horner, G., Rosler, S., *Sens. Actuators B*, **70** (2000) 37–42.
[41] Lee, Y., Lippmann, R., in: *Advances in Neural Information Processing Systems II*; Morgan Kaufmann, 1990.
[42] Dickinson, T. A., Chadha, S., Walt, D. R., *Proc. SPIE*, **2676** (1996) 308–310.
[43] White, J., Kauer, J. S., Dickinson, T. A., Walt, D. R., *Anal. Chem.*, **68** (1996) 2191–2202.
[44] Gardner, J. W., Pearce, T. C., Friel, S., Bartlett, P. N., Blair, N., *Sens. Actuators B*, **18/19** (1994) 240–243.
[45] Keller, P. E., *Proc. SPIE*, **3722** (1999) 144–152.
[46] Di Natale, C., Paolesse, R., Macagnano, A., Mantini, A., D'Amico, A., Ubigli, M., Legin, A., Lvova, L., Rudnitskaya, A., Vlasov, Y., *Sens. Actuators B*, **69** (2000) 342–347.
[47] Martin, M. A., Santos, J. P., Agapito, J. A., *Sens. Actuators B*, **77** (2001) 468–471.
[48] Hammond, J., Misna, T., Smith, D., Fruhberger, B., *Proc. SPIE*, **3856** (1999) 88–96.
[49] Ping, W., Jun, X., *Sens. Actuators B*, **37** (1996) 169–174.
[50] Evans, P., Persaud, K. C., McNeish, A. S., Sneath, R. W., Hobson, N., Magan, N., *Sens. Actuators B*, **69** (2000) 348–358.
[51] Roppel, T., Dunman, K., Padgett, M., Rixey, C. A., Wilson, D., Lindblad, T., in: *Proceedings of the Industrial Electronics Conference IECON '97*; New Orleans, November 9–14, 1997, pp. 218–221.
[52] Di Natale, C., Davide, F. A. M., D'Amico, A., Gopel, W., Weimar, U., *Sens. Actuators B*, **18/19** (1994) 654–657.
[53] Davide, F. A.M., Di Natale, C., D'Amico, A., *Sens. Actuators B*, **18–19** (1994) 244–258.
[54] Signh, S., Hines, E. L., Gardner, J. W., *Sens. Actuators B*, **30** (1996) 185–190.
[55] Gardner, J. W., Shin, H. W., Hines, E. L., Dow, C. S., *Sens. Actuators B*, **69** (2000) 336–341.
[56] Corcoran, P., Anglessa, J., Elshaw, M., *Sens. Actuators A*, **76** (1999) 57–66.
[57] Osbourn, G. C., Martinez, R. F., *Pattern Recognit.*, **28** (1995) 1793–1806.
[58] Dunman, K. L., *MS Thesis*, Auburn University, 1999.
[59] Elman, J. L., *Cognit. Sci.*, **14** (1990) 179–211.
[60] Rabiner, L. R., *Proc. IEEE*, **77** (1989) 257–285.

List of Symbols and Abbreviations

Symbol	Designation
a	constant
b	regression coefficient
N	degree of redundancy
P	partial pressure of analyte
P_{sat}	saturated partial pressure of analyte
w	fitting parameter
x	input variable
x	sensor input signal
y	output variable

Abbreviation	Explanation
ChemFET	chemical field effect transistor
CMOS	complementary metal oxide semiconductor
MLP	multilayer perceptron
PCA	principal component analysis
QMB	quartz microbalance
RBF	radial basis function
SAW	surface acoustic wave
SOM	self-organizing map

1.4 CMOS Single Chip Gas Detection Systems – Part I

C. HAGLEITNER, A. HIERLEMANN, O. BRAND and H. BALTES,
Physical Electronics Lab, ETH, Zürich, Switzerland

Abstract

The current trend to control indoor air-quality and to monitor environmental pollution has created a strong demand for miniaturized and inexpensive gas sensors for volatile organic compounds (VOCs). Gas sensor arrays based on industrial CMOS-processes combined with post-CMOS micromachining (CMOS MEMS) are a promising approach to low-cost sensor devices. In this article, the state of research of CMOS-based gas sensor systems is reviewed, and a platform technology is described, which provides the possibility of monolithically integrating several different transducers on a single chip. A design environment, batch-fabrication processes, and fast testing procedures were developed to realize an example single-chip gas detection system. The chip includes the transducers, their biasing circuitry, reference elements, a digital interface, and a temperature sensor. The three polymer-based transducers and their interface electronics will be detailed in the second part of this article [1].

Keywords: VOCs, chemical sensors, monolithic gas sensor, integrated sensor, temperature sensor, interface circuitry, CMOS MEMS, polymer

Contents

1.4.1 Introduction

The evolution of information technology (IT) over the last years has created a growing demand for all kinds of sensors and actuators that enable automated interaction of machines with their environment or human beings. This trend has been fueled by the miniaturization of formerly bulky computers, sensors or actuators to pocket-size devices and an associated dramatic production-cost reduction through the use of batch fabrication. Owing to this technology push, the field of microelectromechanical systems (MEMS) has recently attracted significant interest. The worldwide market for MEMS products for 2000 was estimated to be a total of 14.2 billion dollars and is expected to double by 2004 despite the current slow-down in the semiconductor industry.

While pressure sensors and accelerometers are well established on the market, the development of gas sensors based on MEMS technology is still at an early stage. Table 1.4.1 shows a comparison of the different evolution stages of several types of micromachined sensors [1]. It also shows that the average time between the initial discovery of a device effect and its full commercialization is up to 25 years. This is about three times more than currently needed in the semiconductor integrated circuit (IC) industry.

In addition to MEMS-related miniaturization efforts, two major trends govern current gas sensor research: (a) the search for highly selective (bio)chemical layer materials and (b) the use of arrays of different partially selective sensors with subsequent pattern recognition and multi-component analysis [2–5].

This chapter discusses the use of industrial complementary metal oxide semiconductor (CMOS) technology with the aim of reducing costs, size and time to market of array-based gas detection systems. After the Introduction, Section 1.4.2 gives an overview on IC technology-based gas sensors. Section 1.4.3 describes the design, fabrication and test of a single-chip gas detection system comprising three different micromachined transducers (mass-sensitive, capacitive and calorimetric). All sensors rely on commercially available polymeric layers to detect ppm levels of airborne volatile organic compounds (VOCs) such as alcohols or benzenes. The technology platform presented here can be easily extended to other transducers depending on the specific requirements of a given application (eg, metal oxide sensors for the detection of CO and NO_x). An on-chip digital

Table 1.4.1. Product evolution of some micromachined sensor types [1]

Product	Discovery	Evolution	Cost reduction	Commercialization
Pressure sensors	1954–1960	1960–1975	1975–1990	1990
Accelerometers	1974–1985	1985–1990	1990–1998	1998
Bio/chemical sensors	1980–1994	1994–2000	2000–2004	2004
Gas sensors	1986–1994	1994–1998	1998–2005	2005

bus interface combined with a novel packaging approach allows the assembly of modular arrays of identical sensor chips coated with different sensitive layers. The array approach further enhances the discrimination performance of the system for applications where a single chip is not sufficient.

Section 1.4.4 details the circuitry building blocks, which are generally needed to realize a monolithic gas detection system: bus interface, digital controller, bias current generators, voltage references and a temperature sensor. Circuitry details of the three transducers are discussed in [6]. Section 1.4.5 shows two application examples of the multisensor chip, and Section 1.4.6 contains the outlook.

1.4.2 CMOS-MEMS for Gas Sensing Applications

Chemical sensors usually consist of a sensitive layer or coating and a transducer [7–13]. Upon interaction with a chemical species, the physicochemical properties of the coating such as its mass, volume, optical properties or resistance reversibly change. These changes are detected by the respective transducer and translated into an electrical signal such as frequency or current, which is then read out and subjected to further data treatment and processing.

A variety of transducers have been devised. We will follow a classification scheme suggested by Janata [8, 9] to give a brief overview of CMOS-based gas sensors:

1. Chemomechanical or mass-sensitive sensors
2. Thermal sensors
3. Optical sensors
4. Electrochemical sensors

We will briefly address each of those four sensor categories and describe the most important CMOS-based devices.

1.4.2.1 Chemomechanical or Mass-sensitive Sensors

Chemomechanical or mass-sensitive sensors are in the simplest case gravimetric sensors responding to the mass of species accumulated in a sensing layer [14, 15]. Some of the sensor devices additionally respond to other mechanical properties such as changes of polymer elastic moduli or viscosity [14, 15], which will not be discussed here. Any species that can be immobilized on the sensor can, in principle, be sensed. Mass changes can be monitored by either deflecting a micromechanical structure due to stress changes or mass loading, or by assessing the frequency characteristics of a resonating structure or a traveling acoustic wave upon mass loading. Both, deflection and resonance frequency vary in proportion to stress changes or mass loading on the device; for details, see [14] and [15].

1.4.2.1.1 Resonant Cantilever

The most common CMOS technology-based mass-sensitive sensor is a micromachined cantilever, which is a layered structure composed of, eg, silicon, silicon oxide/nitride, and, finally, metallizations. The cantilever base is firmly attached to a silicon support (chip). The free-standing cantilever end is coated with a sensitive layer (see Figure 1.4.5).

There are two fundamentally different operational methods: (a) *static mode*: measurement of the cantilever deflection upon stress changes or mass loading by means of, eg, a laser via laser light reflection on the cantilever [16–19]; (b) *dynamic mode*: excitation of the cantilever in its fundamental resonance mode and measurement of the frequency change upon mass loading [20–23] in analogy with other mass-sensitive transducers. The excitation of the cantilever is performed by applying piezoelectric materials (eg, ZnO) [24] or by making use of the different temperature coefficients or mechanical stress coefficients of the cantilever layer materials (bimorph effect) [21–23, 25]. The different materials give rise to a cantilever deflection upon heating. Cantilever deflection or resonance frequency changes can be detected by embedding piezoresistors in the cantilever base [19–23]. The absolute mass resolution of the cantilevers is in the range of a few picograms [16–23].

Typical cantilever applications include the detection of VOCs or humidity in the gas phase by using polymeric layers [16–23].

1.4.2.2 Thermal Sensors

Calorimetric or thermal sensors can be used to determine the presence or concentration of a chemical compound by measurement of an enthalpy change produced by the chemical to be detected [7, 8, 26]. Any chemical reaction or absorption/desorption process releases or absorbs from its surroundings a certain quantity of heat. This thermal effect shows a transient behavior: continuous liberation/abstraction of heat occurs only as long as the reaction proceeds. However, there will be no heat release and hence no measurable signal at thermodynamic equilibrium ($\Delta G=0$), in contrast to mass-sensitive, optical or electrochemical sensors. The various types of calorimetric sensors differ in the way that the evolved heat is transduced [7, 8, 26].

1.4.2.2.1 Catalytic Thermal Sensors (Pellistors)

The catalytic thermal sensor measures the evolved heat during the controlled combustion of flammable gaseous compounds in ambient air on the surface of a hot catalyst by means of a resistance thermometer in proximity with the catalyst. The heated catalyst permits oxidation of the gas at reduced temperatures and at concentrations below the lower explosive limit (LEL). The term 'pellistor' originally refers to a device consisting of a small platinum coil embedded in a ceramic bead, which is impregnated with a noble metal catalyst [27].

CMOS-based sensor structures include surface-micromachined, free-standing, Pt-coated polysilicon micro-filaments separated from the substrate by an air gap [28, 29] or micromachined membranes [30–34]. Heat losses to the silicon frame are minimized in both designs. By passing an electric current through the micro-filament or a meander heater on the membrane, the active area is heated to a temperature sufficient for the Pt surface to catalytically oxidize the combustible mixture; the heat of oxidation is then measured as a resistance variation in the Pt. The combustion of, eg, methane generates 800 kJ/mol of heat.

Typical applications include monitoring and detection of flammable gas hazards such as methane [32, 35], hydrogen [28–30, 32] or carbon monoxide [30, 33] in industrial, commercial and domestic environments at concentrations below the lower explosive limit (LEL).

1.4.2.2.2 Thermoelectric or Seebeck-effect-based Sensors

This type of sensor relies on the thermoelectric or Seebeck effect [26, 36]: When two different semiconductors or metals are connected at a hot junction and a temperature difference is maintained between this hot junction and a colder point, then an open-circuit voltage is developed between the different leads at the cold point. This thermovoltage is proportional to the difference of the Fermi levels of the two materials at the two temperatures and thus proportional to the temperature difference itself [26, 36]. This effect is used in thermal sensors by placing the hot junction on a thermally isolated structure such as a membrane, bridge, and the cold junction on the bulk chip with the thermally highly conducting silicon underneath [23, 36–39]. To achieve a higher thermoelectric voltage, several thermocouples are connected in series to form a thermopile. The membrane structure supporting the hot junctions is covered with a sensitive or chemically active layer liberating or abstracting heat upon interaction with an analyte. The resulting temperature gradient between hot and cold junctions then generates the measured thermovoltage.

CMOS-based thermoelectric sensors include polysilicon/aluminum thermocouples with a Seebeck coefficient of, eg, 111 μV/K [23, 39, 40]. Typical applications include the detection of different kinds of VOCs in the gas phase by using polymeric layers [23, 39–41].

1.4.2.3 Optical Sensors

Optical techniques offer a great deal of selectivity already inherent in the various transduction mechanisms in comparison with other chemical sensing methods. Characteristic properties of the electromagnetic waves such as amplitude, frequency, phase, and/or state of polarization can be used to advantage [42–46]. Geometric effects (scattering) can provide additional information. Furthermore, optical sensors, like any other chemical sensor, can capitalize on all the selectivity effects originating from the use of a sensitive layer.

The generation of light in CMOS or silicon devices is very difficult, since there is no first-order transition from the valence band to the conduction band without the involvement of a phonon (lattice vibrations) [47]. Consequently, integrated optical sensors and systems nowadays mostly are made of III–V semiconductors such as gallium arsenide (GaAs) or indium phosphide (InP), which allow for light generation due to first-order radiative electron-hole recombinations with high quantum efficiency [45]. The detection of light is possible with either silicon-based devices (photodiodes) or other semiconducting materials (GaAs, InP).

1.4.2.3.1 Bioluminescent Bioreporter Integrated Circuits (BBIC)

This technique employs bioluminescent bacteria placed on an application-specific optical integrated circuit [48–50]. The bacteria have been engineered to luminesce when a target compound such as toluene is metabolized. Chemical energy is thus directly converted into light energy in most cases without additional heat generation (cold luminescence).

The microluminometer uses the p-diffusion (source and drain diffusions of p-channel MOSFETs, PMOS) in the n-well as the photodiode. The shallow p-diffusion has a strong response to the 490 nm bioluminescent signal. The entire sensor including all signal processing and communication functions can be realized on a single chip (Figure 1.4.1). The integrated circuitry includes units to detect the optical signal and convert the pixel currents to a frequency-signal (Figure 1.4.1) [48–50].

Depending on the integration time of the device, trace amounts of toluene and naphthalene were detected in the gas phase using engineered cell colonies of *Pseudomonas putida* [48–50].

1.4.2.3.2 Microspectrometers

Photodiode or phototransistor structures such as CMOS charge-coupled device (CCD) detector imagers have been used to develop microspectrometers for biochemical analysis [51, 52] or smart optical sensor systems to measure light intensity and color [53, 54]. Fabry-Perot-based single-chip microspectrometers (16 addressable Fabry-Perot étalons) that can be used, eg, for chemical analysis are detailed in [54] and [55].

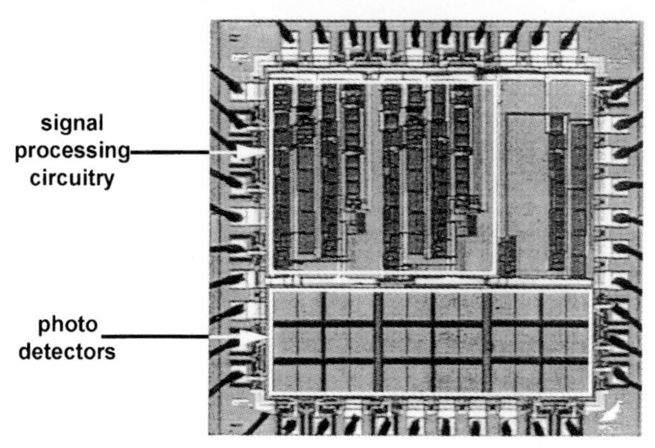

Figure 1.4.1. Micrograph of a bioluminescent bioreporter integrated circuit (BBIC). The circuitry blocks are in the upper part and the photodetectors in the lower part of the chip. Reprinted from [50] with permission.

Typical microspectrometer applications include gas sensors to monitor carbon monoxide (characteristic absorption wavelength: 4.7 μm), carbon dioxide (4.2 μm) and hydrocarbons (3.3 μm) [54, 55]. The radiation source in most cases is a light bulb or light-emitting diode (LED), which cannot be integrated on a CMOS substrate.

1.4.2.4 Electrochemical Sensors

Electrochemical sensors constitute the largest and oldest group of chemical sensors and make use of electrochemical or charge-transfer reactions, ie, charge transfer from an electrode to a solid or liquid sample phase or vice versa. Chemical changes take place at the electrodes or in the probed sample volume and the resulting charge or current is measured. Electrode reactions and charge transport in the sample are both subject to changes by chemical processes upon analyte exposure [7–13, 56].

An electrochemical sensor is always composed of at least two electrodes with two electrical connections: one through the probed sample and the other via transducer and measuring equipment. The charge transport in the sample can be ionic, electronic or mixed, whereas that in the transducer branch is always electronic.

Electrochemical sensors can be classified according to electronic components [7–13]: there are *chemoresistors*, *chemocapacitors* and *chemotransistors*. The following brief overview of these devices will be restricted to CMOS-based systems.

1.4.2.4.1 Chemoresistors

Chemoresistors rely on changes in the electric conductivity of a film or bulk material upon interaction with an analyte. Chemoresistors are usually arranged in a *metal electrode 1/sensitive layer/metal electrode 2* configuration [7–13, 56]. The measurement is done either via a Wheatstone bridge arrangement or by recording the current at an applied voltage in a DC (direct current) mode or in a low-amplitude, low-frequency mode to avoid electrode polarization. The contact resistance should be much lower than the sample resistance and be minimized, so that the bulk contribution dominates the measured overall conductance.

There are two major classes of chemoresistors: (1) high-temperature chemoresistors (200–600 °C) with semiconductor metal oxide coatings and (2) low-temperature chemoresistors (room temperature) with polymeric and organic sensitive coatings. The focus here will be on the technologically more challenging high-temperature sensors; the second type will be mentioned with the applications.

The sensitive materials used with high-temperature chemoresistors include wide-bandgap semiconducting oxides such as tin oxide, gallium oxide, indium oxide or zinc oxide, all of which can only be operated as sensing materials at high temperature (>200 °C). In general, gaseous electron donors (hydrogen) or acceptors (nitrogen oxide) adsorb on the metal oxides and form surface states which, at high temperature, can exchange electrons with the semiconductor. An acceptor molecule will extract electrons from the semiconductor metal oxide and thus decrease its conductivity. The opposite holds true for an electron-donating surface state. The reaction between gases and oxide surface depends on the sensor temperature, the gas involved and the sensor material [8, 57–60].

Semiconductor metal oxide sensors usually are not very selective, but respond to almost any analyte. One method to modify the selectivity pattern includes surface doping of the metal oxide with catalytic metals such as platinum, palladium, gold and iridium [8, 9, 57–60].

The device requirements for a high-temperature chemoresistor include a thermally well-isolated stage such as a membrane, which allows the sensing materials to be kept at high temperature without heating the bulk chip, an integrated heater and a temperature sensor [61]. Figure 1.4.2 shows a back-side-etched micro-hotplate on a CMOS substrate [61]. Tin dioxide is then deposited on top of the metal electrodes. Typical micro-hotplate applications include the detection of inorganic gases such as hydrogen [62, 63], oxygen [62, 63], nitrogen oxide [64], carbon monoxide [61–64] and VOCs [62, 63, 65], using predominantly tin oxide as the sensitive layer.

Several classes of organic materials are used for application with chemoresistors at room temperature (electrode spacing typically 5–100 μm, applied voltage 1–5 V). Conducting polymers such as polypyrroles, polyaniline and polythiophene are used to monitor a variety of polar VOCs such as ethanol or methanol [66–69]. Conducting carbon black can be dispersed in nonconducting polymers so that if the polymer absorbs vapor molecules and swells, the particles are, on average, further apart and the conductivity of the film is reduced (conductivity by particle-to-particle charge percolation) [70]. Applications also include organic solvents such as hydrocarbons, chlorinated compounds and alcohols [70, 71].

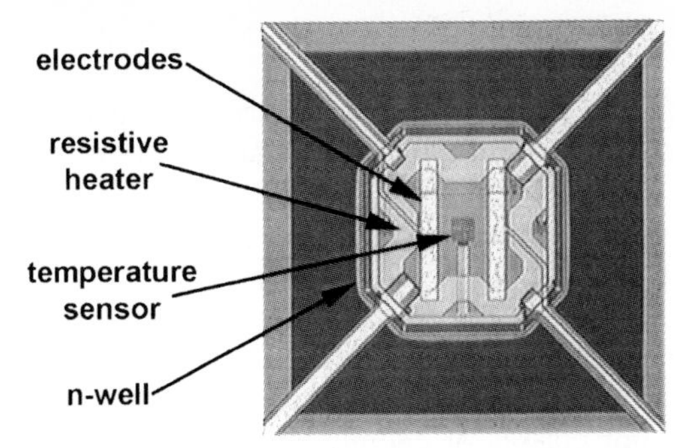

Figure 1.4.2. Micrograph of a micro-hotplate with an n-well underneath the heated area featuring a temperature sensor, a resistive ring heater and gold electrodes.

1.4.2.4.2 Chemocapacitors

Chemocapacitors (dielectrometers) rely on changes in the dielectric properties of a sensing material upon analyte exposure. Two effects change the capacitance of, eg, a polymeric sensitive layer upon absorption of an analyte: (i) swelling and (ii) change of the dielectric constant due to incorporation of the analyte molecules into the polymer matrix [23, 72, 73]. Interdigitated electrode structures are predominantly used for capacitance measurements [74, 75]. The devices usually are operated at a frequencies from a few kHz up to a few MHz.

Since the nominal capacitance of microstructured capacitors (eg, electrode width and spacing: 1.6 μm, footprint: 800×800 μm^2) is of the order of 1 pF and the expected capacitance changes are in the attofarad range, an integrated solution with on-chip circuitry is required [72].

Typical chemocapacitor applications include humidity sensing with polyimide films [72, 74, 75], since water has a high dielectric constant of 78.5 (liquid state) at 298 K, leading to large capacitance changes. More recent applications also include the detection of VOCs in the gas phase using polymeric layers [23, 72, 73].

1.4.2.4.3 Chemotransistors

Field-effect-based transistors, which are the most common electronic components on modern IC logic chips, rely on modulation of the charge carrier density in the semiconductor surface space-charge region through an electric field perpendicular to the device surface: The source-drain current is controlled by an isolated gate electrode.

The MOSFET (metal oxide semiconductor field-effect transistor) as used for chemical gas sensing (Figure 1.4.3) has a p-type silicon substrate (bulk) with two n-type diffusion regions (source and drain). The structure is covered with a silicon dioxide insulating layer on top of which a metal gate electrode is deposited [76, 77].

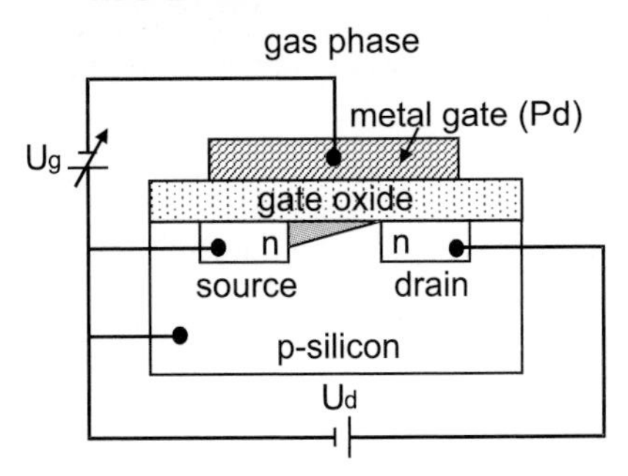

Figure 1.4.3. Schematic representation of a MOSFET structure. V_g denotes the gate voltage and V_d the drain voltage.

When a positive voltage (with respect to the silicon) is applied to the gate electrode, electrons, which are the minority carriers in the substrate, are attracted to the surface of the semiconductor. Consequently, a conducting channel (n-channel) is created between the source and the drain, near the silicon dioxide interface. The conductivity of this channel can be modulated by adjusting the strength of electrical field between the gate electrode and the silicon, perpendicular to the substrate surface.

Palladium (Pd) gate field effect transistor (FET) structures were demonstrated to function as a hydrogen sensor by Lundström et al. [76]. Hydrogen molecules readily absorb on the gate metal (platinum, iridium, palladium) and dissociate into hydrogen atoms. These H-atoms can diffuse rapidly through the Pd and absorb at the metal/silicon oxide interface partly on the metal and partly on the oxide side of the interface [77, 78]. Owing to the absorbed species and the resulting polarization phenomena at the interface, the threshold voltage (U_d) is shifted. The voltage shift is proportional to the concentration or coverage of hydrogen at the oxide/metal interface. Sensitivity and selectivity patterns of gas-sensitive FET devices hence depend on the type and thickness of the catalytic metal used, the chemical reactions at the metal surface and the device operation temperature.

MOSFETs have been predominantly used to detect hydrogen [79, 80]. Applications also include the detection of ammonia [77, 78], amines and any kind of molecule that gives rise to polarization in a thin metal film (hydrogen sulfide, ethene, etc.) or causes charges/dipoles on the insulator surface [77, 78].

1.4.2.5 Monolithic versus Hybrid Designs

Key features of sensors in consumer-product applications include low costs, low power consumption, small size, on-chip calibration possibility and device robustness. This typically requires a large amount of control and signal processing functions to be monolithically integrated with the sensor elements. In some cases, auxiliary sensors such as temperature sensors are needed to deal with cross-sensitivities of the sensing device. Since the first papers on micromachined sensors appeared in the 1970s, the question of whether it is advantageous to use monolithic systems combining CMOS circuitry and sensors on the same chip (CMOS-MEMS) or whether it is better to use hybrid designs with optimized sensor processes and external electronics, has been intensely discussed. Table 1.4.2 gives an overview of the technological options for hybrid and monolithic designs. For hybrid designs, most micromachining techniques are available through inexpensive multi-project-wafer (MPW) prototyping services. However, only a limited selection of MEMS techniques are available for monolithic designs. For fabless design houses, which rely on MPW services for prototyping, this selection is reduced to a small number of pre-CMOS and post-CMOS micromachining options.

Micromachined chemical sensors are not yet established on the market. For a comparison of the commercial success of hybrid versus monolithic designs, we hence analyze the situation for well-established acceleration and pressure sensors. The commercially available products provide no clear answer as to which approach offers more competitive advantages: While the products of Analog Devices [81], Infineon [82] and Motorola [83] are based on monolithic integration, the sensor systems offered by SensoNor [84], Intersema [85], Delphi [86] and NovaSensors [87] are based on hybrid designs. Some companies such as Bosch [88] offer monolithic and hybrid products. A trend towards monolithic solutions can be identified in the case of larger production volumes and more severe cost restrictions.

For chemical sensor arrays, a monolithic solution offers several advantages despite of the high initial costs and the limitations in the choice of the micromachining techniques:

- The full performance of capacitive, resonant and calorimetric sensors is only exploited in monolithic implementations [89] because, eg, the influence of parasitic capacitances and crosstalk-effects can be reduced.
- The number of electrical connections significantly contributes to the overall system costs. The monolithic implementation of a single-chip gas sensor (see section on monolithic CMOS gas sensor systems later) with three micromachined sensors requires only seven connections (three for supply voltages, one for a clock signal, one for reset and two for the serial interface). Up to 16 chips can be connected without adding any additional communication lines due to the use of a digital bus-interface. A hybrid approach would require at least 30 pads for the three sensors and a total of 480 connections, if 16 sensors of each type were to be combined.
- Hybrid implementations require complex packages to reduce sensor interference, to minimize electric cross-talk and to optimize the critical connections. This further complicates the already difficult task of chemical sensor packaging.
- The use of a standardized digital bus interface makes the system more scalable, because an additional sensor chip can be added without changing the system architecture or the packaging scheme.

Table 1.4.2. Exemplary overview of micromachining techniques for hybrid and monolithic designs: the techniques are compared with respect to their availability for fabless design centers and their use in the field of chemical sensors

Type	Micromachining technique	Foundry/University	MPW available	Academic/ commercial	Gas sensors
Hybrid design	Bulk micromachining from backside (KOH, TMAH, Deep-RIE)	PEL [90], Tronics, SensoNor [84], Standard MEMS [91]	Yes	A/C	Yes
	Bulk micromachining from frontside (EDP, TMAH)	PEL, NIST [92], Tronics Standard MEMS	Yes	A	Yes
	One-level or multilevel polysilicon micromachining + sacrificial layer etching	Sandia [93], Bosch [88]	Yes	A/C	Yes
	Piezoelectric layer deposition	Delft	No	A/C	Yes
Pure CMOS	Intrinsic etching steps of CMOS process	PEL, Infineon, Austria microsystems	Yes	A/C	Yes
Pre-CMOS	Polysilicon surface micromachining	Sandia	Yes	A	No
	Deep RIE with oxide refill	PEL	No	A	No
Intermediate CMOS	Polysilicon microstructures + sacrificial layer etching	Analog Devices [81], Texas Instruments [94], Infineon [82]	No	C	No
Post-CMOS	Anisotropic etching from the backside	PEL, Univ. of Warwick, Austria microsystems, Sensirion [95], Stanford [96]	No	A/C	Yes
	Anisotropic etching from the frontside	NIST	No	A	Yes
	Micromachining of CMOS metal layers	Carnegie Mellon [97]	No	A	No

- The typical development time until full commercialization of hybrid micromachined sensors is more than 20 years from idea to off-the-shelf products. A large part of this time is consumed by establishing a dedicated and reliable fabrication technology. The production equipment and cleanrooms require large investments. CMOS-MEMS technology relies on established industrial CMOS technology to fabricate the read-out circuitry and the basic sensor structures. The fabrication equipment and technology development is reduced to a few post-processing steps that are characteristic for the sensor. This reduces development time and enables the business model of fabless MEMS companies.
- The response time of a gas sensor is determined by the volume of the measurement chamber and the flow rate. Using the monolithic approach and a suitable packaging technique (eg, flip-chip packaging), the volume of the measurement chamber can be kept very small.
- Auxiliary sensors such as temperature or flow sensors, which are often needed with chemical sensors, can be co-integrated. Calibration and self-test features also can be realized on-chip.

The main disadvantage of a monolithic CMOS-MEMS solution is the restriction to CMOS-compatible materials and micromachining processes. For post-CMOS micromachining, processing steps requiring, eg, high temperatures (>400 °C) cannot be used, because the aluminum metallization is destroyed and the transistor parameters are altered owing to changes in the thermal budget of the CMOS process.

1.4.3 Single-chip Gas Detection System

In this section, a versatile multisensor chip is described, which was developed for the detection of VOCs. Industrial CMOS technology combined with post-CMOS micromachining provides a modular technology platform, which allows the assembly of a tailored solution for a given application by combining several different transducers on a single chip (three transducers in this example). Arrays of identical sensor chips coated with different sensitive layers are packaged on a common ceramic substrate. This further enhances the detection performance of the system for applications where a single chip is not sufficient.

1.4.3.1 The System Architecture

Figure 1.4.4 shows a block diagram of the multisensor chip. It comprises three transducers that respond to fundamentally different analyte properties. All sensors rely on commercially available polymeric layers to detect airborne VOCs. The first transducer is a micromachined cantilever (Section 1.4.2.1.1) oscillating

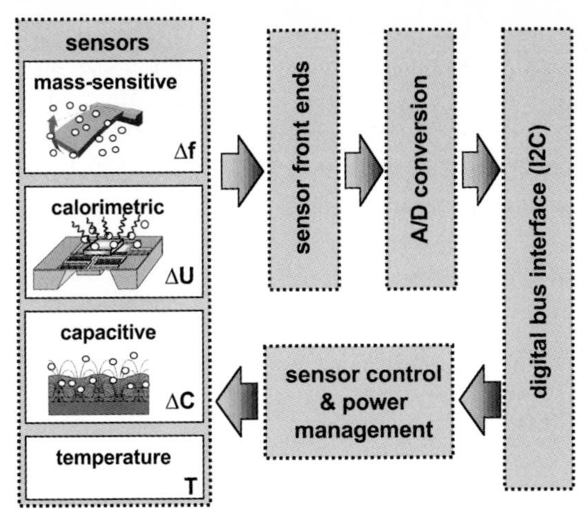

Figure 1.4.4. Block diagram of the single-chip gas detection system.

at its resonance frequency (see Figure 1.4.5). The n-well of the CMOS process forms the bulk of the cantilever, which is covered by the dielectric layer stack (oxide layers, which insulate the conducting layers of the CMOS process and silicon nitride passivation) and the polymer layer. The absorption of analyte in the chemically sensitive polymer causes shifts in resonance frequency as a consequence of changes in the oscillating mass. The cantilever acts as the frequency-determining element in an oscillator circuit and the resulting frequency changes can be easily read out by an on-chip counter.

The second transducer, a planar capacitor (Section 1.4.2.4.2) with polymer-coated interdigitated electrodes, is shown in Figure 1.4.6. This transducer monitors changes in the dielectric constant of the polymer upon absorption of the analyte into the polymer matrix. The capacitance of a passivated reference is subtracted in order to reduce the influence of temperature and flow-induced effects. The sensor-response is read out as a differential signal between the coated sensing capacitor and a reference capacitor, which is converted into a digital signal by a second-order $\Sigma\Delta$ modulator.

The third transducer is a thermoelectric calorimeter (Section 1.4.2.3.2), which detects enthalpy changes upon absorption (heat of condensation) or desorption

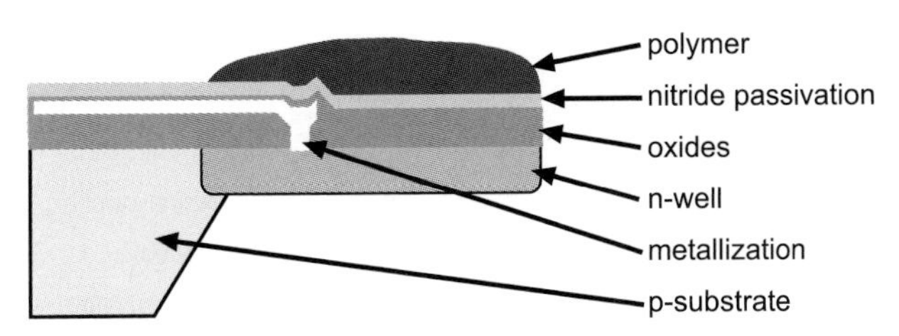

Figure 1.4.5. Cross-section of the cantilever resonator.

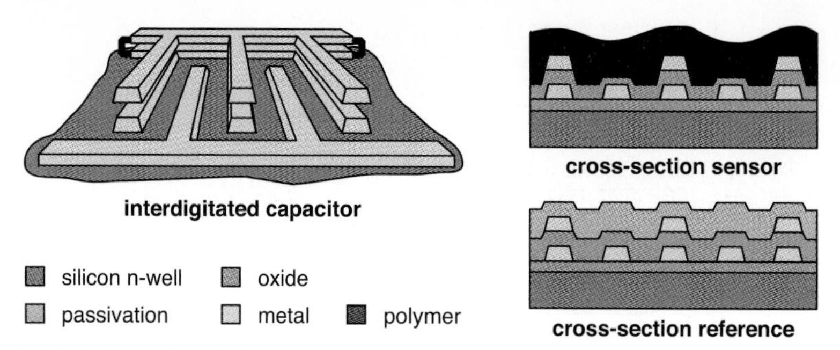

Figure 1.4.6. Schematic of the sensing and reference capacitor made from the two metal layers of the CMOS process. The sensitivity is increased by maximizing the polymer volume in the region with strong electric field.

(heat of vaporization) of analyte molecules in a polymer film located on a thermally insulated membrane (see Figure 1.4.7). The enthalpy changes cause transient temperature variations in the polymer film and hence a temperature gradient between the membrane and the silicon substrate. The calorimeter made from polysilicon/aluminum thermocouples is used to generate a voltage that is proportional to the temperature difference between the isolated membrane and the silicon substrate.

The chip additionally includes a temperature sensor since bulk physisorption of volatiles in polymers is strongly temperature dependent; a temperature increase of 10 °C decreases the fraction of absorbed analyte molecules by approximately 50% and thus leads to a drastic reduction in the sensor signal. Therefore, the operating temperature must be known exactly to permit quantitative measurements. The temperature sensor relies on the linear temperature dependence of the base-emitter voltage of the vertical bipolar transistor available in the CMOS process.

The analog sensor front-ends provide the sensors with the necessary bias voltages and currents and acquire the small sensor signals with minimal additional noise. Where necessary, the signals are amplified to render the noise performance of subsequent stages less critical. The sensor signals are then converted into digital values and decimated to the final data rate. The serial bus interface delivers the values to the signal processing unit upon request.

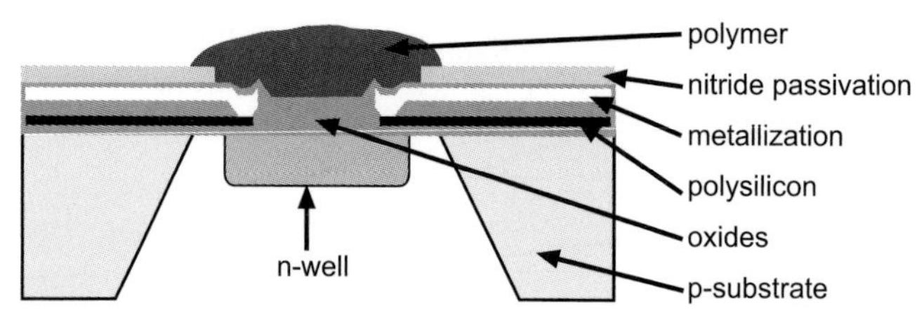

Figure 1.4.7. Cross-section of the calorimetric transducer.

The controller receives and interprets the commands from the signal processing unit. Each sensor has a power-down mode and can be individually addressed. The power consumption of the chip is only 40 μW when all sensors are switched off. The controller also manages the self-test functions of the sensors and handles the different programmable parameters such as the amplification factor of the chopper amplifier for the calorimetric sensor. The controller finally manages the access of the single sensors to the serial interface when simultaneous data transmission is necessary.

1.4.3.2 Chip/System Interface

A critical issue is to partition the system into on-chip and off-chip components [98]. The best solution depends on the size, costs and target production volume of a given sensor system. Therefore, the conclusions that are derived in this section for the case of a gas detection system based on an array of identical multisensor chips cannot be generalized. The lower limit for integration of on-chip functions is determined by the following boundary conditions:

- The packaging of chemical sensor-arrays requires the number of electrical connections and their complexity to be reduced to a minimum, because standard multilayer printed circuit boards (PCBs) cannot be used (see Section 1.4.3.4.3).
- The transfer of analog signals over large distances in a noisy environment increases the power consumption and reduces the performance of the system.
- The multisensor chip has to perform self-test and calibration functions. This requires either a digital interface or a large number of connections.
- The system includes a total of 18 sensors. The bandwidth of the single signals is below 1 kHz but on-chip filtering is needed to remove the out-of-band noise components and to reduce the amount of data that needs to be transmitted.

On-chip integration of pattern recognition and classification capabilities is not efficient, because these would be replicated on every single chip, while only one device finally performs the recognition. The pattern recognition and other computation-intensive tasks such as communicating with a personal computer and visualizing results on a liquid crystal display (LCD) are better done with a microcontroller that can be bought off the shelf.

In conclusion, a chip with a serial interface, digital transmission of all sensor values and no external analog reference voltages or currents is the solution of choice for this application.

1.4.3.3 Design Flow

At present, no single commercial software package is available that can handle all aspects of CMOS-MEMS design. For the lower levels in the design hierarchy, specialized software packages can be used to perform the particular tasks (assuming that the interfaces to import and export data are properly specified). The difficult step is the top-level design, where the layouts of transducers, analog circuitry and digital circuitry have to be joined, verified and simulated [99–101].

1.4.3.3.1 Layout

In addition to the commercial tools, various non-commercial tools for circuit design and MEMS design are available [102, 103]. Furthermore, there are widely accepted standard formats to exchange data between the different software packages (GDS format, CIF format) and to import the different fragments of the design.

1.4.3.3.2 Simulation

The simulation of a complete smart sensor system (sensor elements, analog circuitry, digital circuitry) using FEM-simulation tools is very complex and computationally intensive. Therefore, behavioral models of the transducers are required. The generation of these models from the layout or from the result of a 3-D finite element method (FEM) simulation is not straightforward. Senturia [99] discusses the advantages and disadvantages for macromodels based on lumped-circuit elements and hardware description languages (HDLs). For some structures (mostly comb-structures for, eg, accelerometers), the generation of macromodels is supported by academic [104, 105] and commercial tools [106, 107]. A suitable simulator must be able to combine macromodels of the sensor, transistor-level netlists and various levels of HDL representations of analog and digital circuitry. Furthermore, analog and digital libraries of the CMOS process must be available. Most foundries only provide parameters for the tools supplied by the large electronic design automation (EDA) companies (Cadence, Mentor, Synopsys). This limits the choice to the mixed-signal simulators delivered with those packages (eg, SPECTRE-Verilog, SPICE-Verilog, SABER).

1.4.3.3.3 Verification and Post-layout Simulation

Most CMOS foundries deliver rule-files to perform design rule checks (DRCs) and extraction of the layout for the layout-versus-schematic (LVS) check. As for the simulation parameters, verification rules for MEMS design software are not provided. As each software package uses different formats for the rules, conversion is a difficult task. For micromachined sensors formed by intrinsic CMOS processing steps in combination with post-CMOS processing, there are no rule-files available. The most convenient way, therefore, is to extend the rule-files provided for the circuitry elements with rules for the micromachined devices.

1.4.3.3.4 Design Flow of the Multisensor Chip

The design flow used for the single-chip gas sensor microsystem is depicted in Figure 1.4.8. Various simulation tools were used to simulate and optimize the transducers [108–110]. Cadence was employed to draw and simulate the analog schematics, while Synopsys and Modelsim were used to synthesize and simulate the digital part. The automatic layout of the digital part was performed with Cadence.

The top-level design was done in Cadence. Simple lumped-circuit models for the transducers were developed to include the transducers into the top-level simulation. The transducers and the contact-network for the electrochemical etch stop have design-rules, which deviate from those of the circuitry. Therefore, an extension to the standard DRC was developed that defines the design rules for the transducers and accepts certain violations of the CMOS design rules in their vicinity. Owing to the design-rule violations caused by the transducers, the standard extraction of the layout and the subsequent LVS is not possible. The problem was solved by adapting the extraction rules in order to recognize and extract the electrical features of the transducers (eg, the poly-resistors of the piezoresistive Wheatstone bridge on the resonant gas sensor). This allows the verification of the top-level design by comparing the final layout with the simulated top-level schematic and avoids wiring errors at this level.

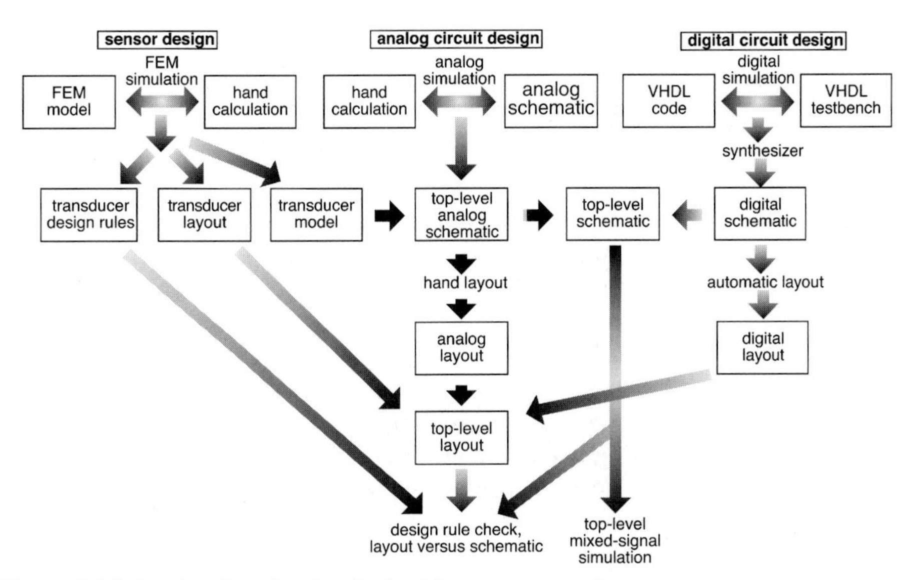

Figure 1.4.8. Design flow for the single-chip gas sensor microsystem.

1.4.3.4 Fabrication Process

1.4.3.4.1 CMOS Fabrication

The circuitry and the basic sensor elements (thermocouples, heating resistors, piezoresistive Wheatstone bridge, etc.) are fabricated in an unaltered industrial 0.8 μm CMOS-technology provided by austria*micro*systems, Austria. Processing steps of the CMOS process are used to reduce the amount of post-processing, eg, the pad-etch is used to remove the passivation on top of the capacitive sensor [108].

The core of the CMOS process (implantation steps, thermal budget, etc.) is left unchanged. An add-on to the basic CMOS process was developed to enable wafer-level anisotropic etching from the back of the wafer with an etch stop at the n-well of the CMOS process [111, 112]:

- A modified starting material is used to improve the quality of the etch grooves (wafers with low oxygen content in the bulk, with or without epi-layer [112]).
- The lithography of the metal layers was modified to achieve a wafer-level contact network needed for the electrochemical etch stop.

The fully processed wafers are thinned to a thickness of 380 μm, and a silicon-nitride layer that serves as a mask for the subsequent KOH-etching is deposited on the backside. Figure 1.4.9 shows a micrograph of the fabricated chip and indicates the key areas.

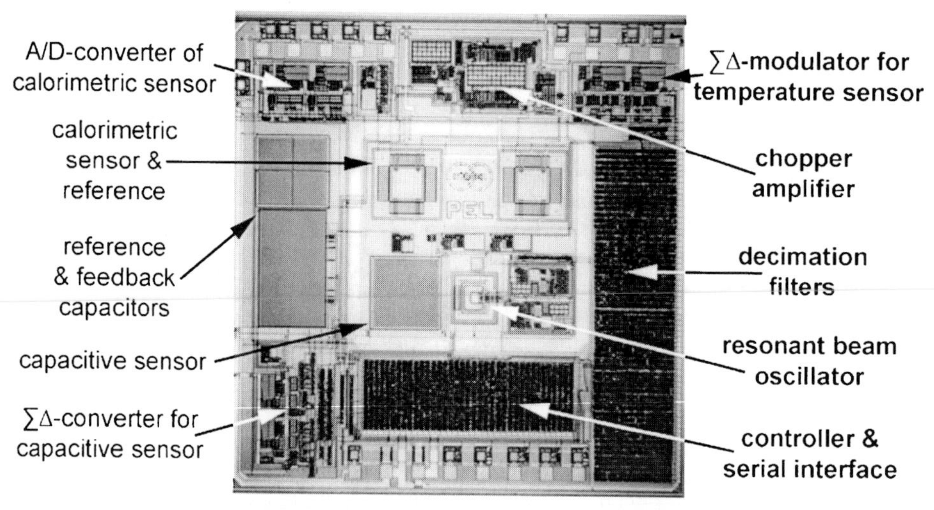

Figure 1.4.9. Micrograph of the single-chip gas sensor microsystem layouted for flip-chip packaging.

1.4.3.4.2 Post-CMOS Micromachining

After completion of the industrial CMOS process sequence, the n-well membrane for the mass-sensitive device and the thermally insulated island structure for the calorimetric sensor are released simultaneously (see Figure 1.4.10 b). This is done by anisotropic silicon etching with KOH from the backside of the wafer with an electrochemical etch-stop technique that stops at the n-well of the CMOS process. Then, the silicon cantilever is released by two reactive ion etching (RIE) steps as shown in Figure 1.4.10 c. The wafers are subsequently diced

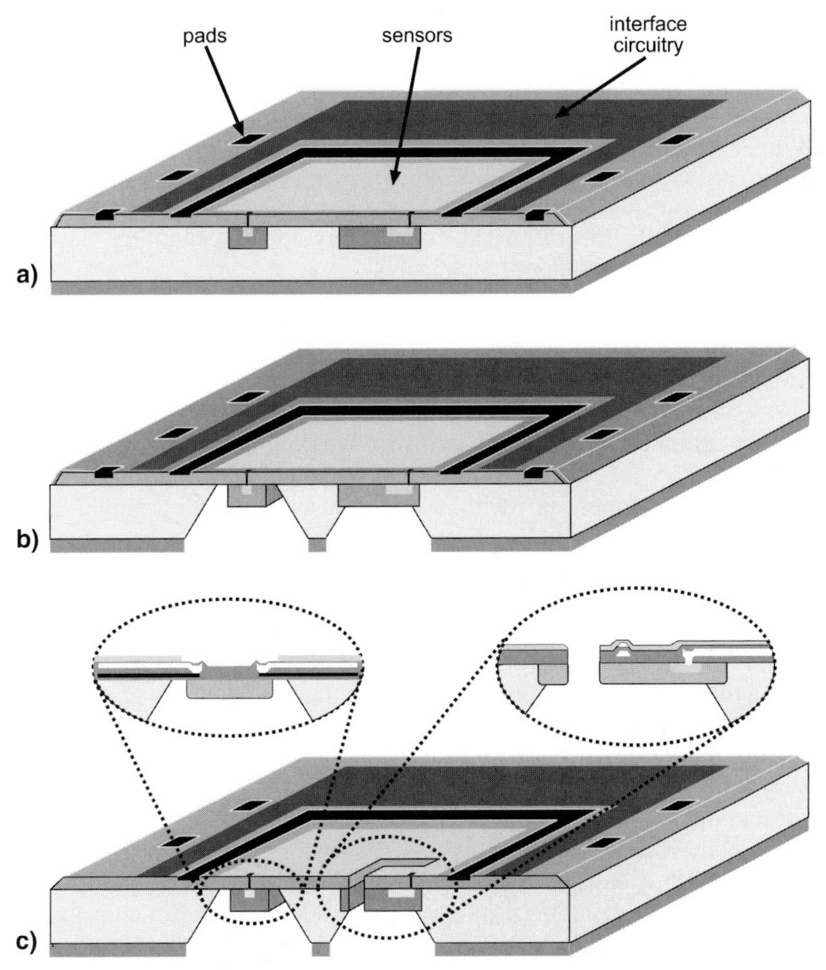

Figure 1.4.10. Post-CMOS processing sequence of the single-chip gas detection system. (**a**) Thinned CMOS wafer with silicon nitride layer on the backside. (**b**) KOH etching from the backside with electrochemical etch stop at the n-well of the CMOS process. (**c**) Cantilever is released using RIE.

using a protective foil over the microstructures. After exposure to UV radiation, the foil has no adhesion to the silicon microstructures and can be removed without damage. Three masks are needed for the silicon micromachining, one for the KOH etching and two for releasing the cantilevers.

1.4.3.4.3 First-level Packaging and Polymeric Coating

The packaging of microsensors is often underestimated despite the fact that the package determines to a large extent the reliability of a microsystem and is a major cost component of commercial systems [113]. Just as CMOS-MEMS technology takes advantage of reliable industrial CMOS processing, the packaging of microsystems benefits from the knowledge of IC packaging. For most sensor applications, specific and customized solutions have to be developed in order to provide access of the measurand to the sensor surface. For gas sensors, the two most significant problems include [114]:

- Window to the outside world: the gas flow must have direct access to the sensing structures, while the electrical connections and the circuitry have to be shielded from the environment (chemicals).
- Inert packaging: the packaging material selection must be restricted to components that do not outgas possible interfering analytes and are inert to the (sometimes aggressive) analytes.

Simple 28-pin ceramic DIL packages and wire bonding have been used in laboratory experiments. For the hand-held unit (see Section 1.4.5.2), a first-level packaging method that combines six identical multisensor chips on a single substrate had to be developed. Two different packaging solutions have been realized: a 'chip-on-board' solution using an epoxy protection (glob top) and a flip-chip packaging solution. The advantage of the packaging with glob top protection is its simplicity. The flip-chip packaging reduces the volume of the gas flow system, hermetically shields the circuitry from the gas-flow and provides improved reliability.

The packaging also has some implications on the top-level layout of the multisensor chip. For glob top protection, all electrical connections of the chip and the sensitive parts of the circuitry should be located on one side of the chip, while the sensors occupy the other side as shown in Figure 1.4.11. This is necessary because the epoxy cannot be applied with micrometer precision and still flows and expands somewhat during the process of drying. The flip-chip solution requires a layout with a balanced number of connections on each side of the chip (see Figure 1.4.9). The bonding pads have to be adapted for flip-chip packaging because the minimum area required for a reliable flip-chip connection is $150 \times 150\ \mu m^2$ compared to only $80 \times 80\ \mu m^2$ for a regular bonding pad. The sensors have to be placed in the center of the die and a metal frame is designed around the transducers in order to shield the sensitive parts of the circuitry from the gas flow.

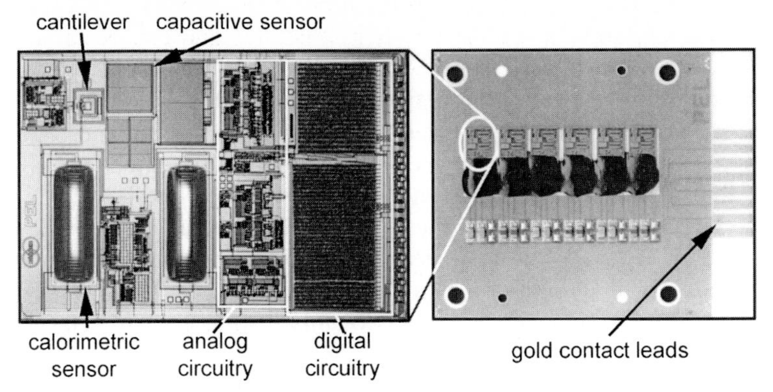

Figure 1.4.11. Ceramic substrate with six multisensor chips and glob top protection is shown on the right-hand side. On the left, a close-up view of a single chip is shown.

Chip-on-Board Packaging

To realize this packaging solution, six identical chips ($5\times7\ mm^2$) are die-attached on a common ceramic substrate (see Figure 1.4.11). Electrical interconnects are made by wire bonding. The wire bonds and the circuitry are protected with glob top. Finally, the sensing structures of each chip are spray-coated, each with a different polymer using a shadow mask.

Flip-chip Packaging

Figure 1.4.12 shows schematically the flip-chip packaging solution. Laser cutting is used to open a window for the sensors in the ceramic substrate. Then, the electric connections are screen-printed on the ceramic. The soft solder paste for the flip-chip packaging is applied to the ceramic using stencil printing. Before dicing the wafers, the metallic frame surrounding the sensors and the pads are covered with nickel/gold bumps. Then, a glass wafer is glued to the backside of the wafer to prevent gas flow through the opening of the cantilever. After dicing, six chips are flip-chip mounted on the ceramic substrate and a reflow is performed at 230 °C. Finally, an epoxy-based underfill is applied and cured at 160 °C. The sensors are then coated with different polymers using a drop-coating method.

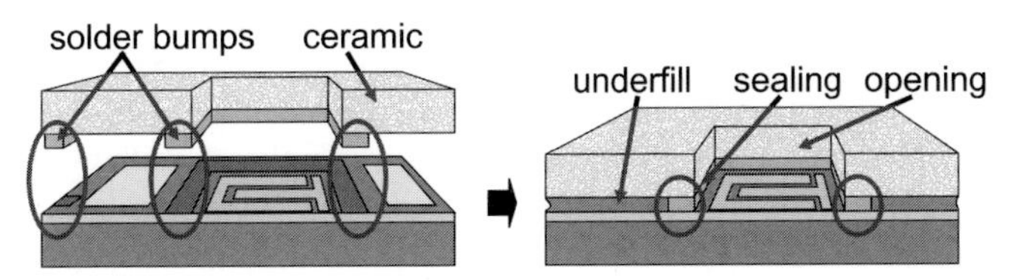

Figure 1.4.12. Schematic drawing of the flip-chip packaging approach for the single-chip gas sensor.

1.4.3.5 Testing

There are two different types of tests for a mixed signal IC: functionality and production tests. The functionality test verifies whether the performance of every single building block of the chip is in accordance with the specifications and simulations. For the single-chip gas detection system, test chips were fabricated that have only some building blocks and an increased number of pads to provide access to internal nodes. The final chip was extensively tested by sequentially setting all parameter values and analyzing the output of uncoated and coated chips. Finally, gas tests were performed and evaluated. Functionality testing is a time-consuming task which, however, has to be done only once and only for a few chips.

For the production test it is assumed that the functionality of the design has already been verified. The goal is to verify that a given single chip has no production defects such as shortcuts, open connections or excessive contact resistance. The ITRS'99 roadmap for IC testing predicted that the testing costs will exceed the production costs of a microprocessor chip by 2012. Even though this picture has been revised for the roadmap of 2001 [115], the basic statement is still valid: While the fabrication costs per transistor continue to scale according to Moore's law, the testing costs per transistor are almost constant.

For digital designs, the test structures are automatically included during the design process (design for testability) by modern HDL compilers. The techniques and algorithms are well defined (eg, scan path, boundary scan). For mixed signal designs, the situation is less comfortable, because there are no standardized test procedures, and the test features have to be included manually [116, 117]. For some designs, direct access to the nodes of the analog circuitry is not available. Furthermore, the test of slow analog circuitry, eg, filters with low cutoff frequencies, is time-consuming. The test of CMOS-MEMS designs is even more challenging, because it also includes the testing of the transducers.

In the case of the application-specific sensor system (AS^3, see Section 1.4.5.2), six multisensor chips are flip-chip packaged on to a ceramic board. All chips are connected to the same serial interface, and hence any defective chip renders the complete board unusable. Therefore, the chips must be tested before packaging to avoid an unacceptable decrease in the yield. The goal is to test the complete chip in approximately 1 s.

Dedicated commercial and automated test equipment for volume production is costly and is not required for a 'proof of concept'. Therefore, a less expensive testing solution with similar functional units has been developed. The schematic of the testing setup is depicted in Figure 1.4.13. Electrical connections to the chip are established by using an automated wafer prober and a custom-designed probe card. The digital communication is done by a pattern generator in combination with a logic analyzer. This is supported by a small PCB for level shifting and supply.

The test is performed at wafer level after all micromachining steps have been completed, but before dicing. Defects that occur during dicing or coating of the sensors (eg, cracks in membranes) have to be identified by subsequent optical inspection.

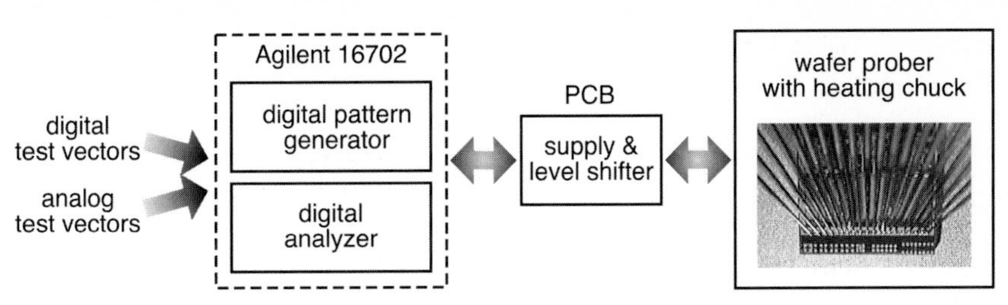

Figure 1.4.13. Setup for the test of the single-chip gas detection system.

1.4.3.5.1 Test of the Digital Controller and Filter

The digital part of the chip is tested using the well-established scan-chain architecture [118]. Scan chain generation is supported by commercial digital synthesis tools. These tools also automatically generate the test vectors needed to guarantee a fault coverage of 99% for the digital part. The fault coverage can be further increased by performing I_{DDQ} testing (monitoring of the current consumption) in parallel. Using a test clock at 1 MHz, the 351 patterns require a testing time of 190 ms.

1.4.3.5.2 Test of the Sensors and Analog Circuitry

The flip-chip packaging approach reduces the number of pads to the absolute minimum. Therefore, there are no test pads providing access to the analog signals and the sensors as well as the analog circuitry have to be tested through the digital part. The straightforward approach is to program the respective parameters via the serial interface, start the test sequence and then read out the results. A different approach can be applied to reduce the overall testing time:

1. The chip is switched to test mode and the initial (digital) state of the controller is prepared by clocking it into the scan-chain. A small collection of software tools has been developed to automatically translate a given simulation status or controller status into a corresponding test vector.
2. The chip is switched to normal operation and operated for as long as needed to allow for settling of the analog signals and their conversion to digital values by the A/D converters.
3. The chip is switched to the test mode. The results contained in the registers at the outputs of the A/D converters are collected via the scan chain. At the same time, step 1 is repeated and the next 'test vector' is clocked in.

This method allows for a reduced testing time, because the three sensors can be tested in parallel. This is possible, because each sensor (with the exception of the temperature sensor) has its own dedicated A/D converter. The calculated testing times for the transducers are less than 1 s. After some additional modifications of the current design, it is possible to reduce the total time to 1 s.

1.4.4 Circuitry Building Blocks

The three transducers on the multisensor chip and the dedicated circuitry that was developed to read their small output signals will be described in [6]. In addition to the transducers, there are several building blocks that are needed in almost every smart sensor system based on CMOS-MEMS technology. A digital controller and a serial interface are necessary to reduce the number of bonding pads, to control the units on the chip and to enable either simple readout by a microcontroller or direct displaying of the results. Reference voltages are required for the A/D converters and are used to bias the gas sensors. Finally, a temperature sensor has to be included in order to deal with the strong temperature dependence of the gas absorption process in the sensitive layer.

1.4.4.1 Controller and Serial Interface

Figure 1.4.14 shows a schematic drawing of the controller and the I^2C interface, which is included on the multisensor chip. Both have been designed and simulated using VHDL. The synthesis and layout were performed according to the design flow in Section 1.4.3.3.

1.4.4.1.1 Serial Interface

The I^2C bus [119] offers some advantages over other serial interfaces such as the CAN bus (mostly used in the automotive industry), the SPI bus by Motorola or the standardized IEEE 1451 bus interface:

- Simple, serial protocol: the I^2C protocol clearly defines the most important features (addressing, word length, master/slave communication, start/stop condition) without adding expensive overhead that requires large amounts of on-chip memory.
- Small number of connections: only two wires for clock and data are needed.
- Bus-enabled: up to 127 devices can be connected on the same bus. A simple arbitration algorithm solves the problem of data collision.

The multisensor chips in the final gas detection system are identical except for their sensitive coatings. The address of each chip is defined via four pads. The binary pattern defining the address is hardwired on the ceramic board. Therefore, the position where the chip is mounted determines its address.

The I^2C bus offers two possible modes of operation:

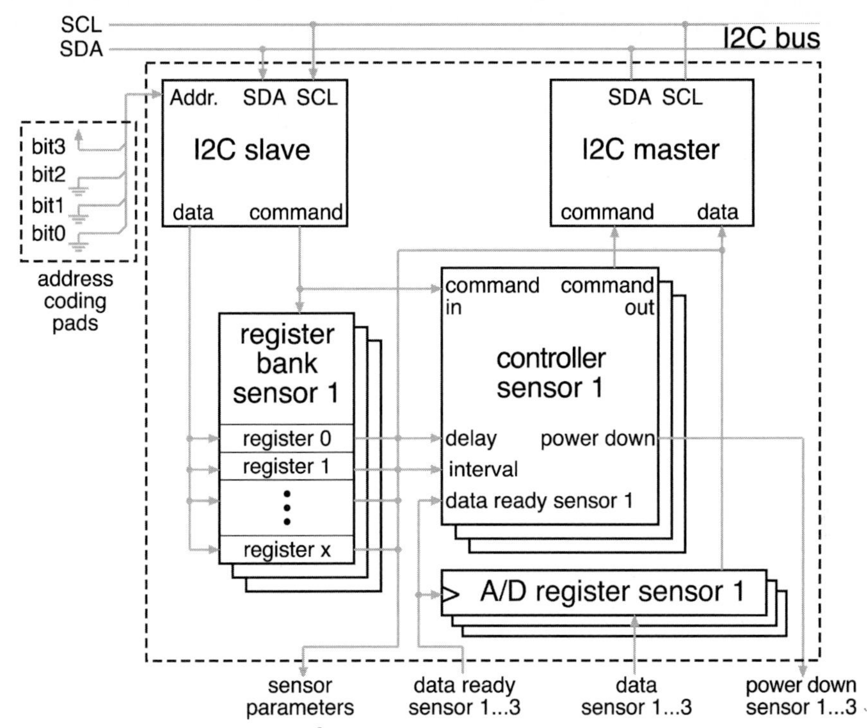

Figure 1.4.14. Schematic of the I^{2}C bus interface and the digital controller. The controller, register bank and the register for the digital output values are replicated for each sensor.

- Single-master: an external unit polls the multisensor chips and collects the data from the chips.
- Multi-master: every single chip can initiate a data transfer.

The multi-master mode was chosen to reduce the data traffic on the bus. The data recording unit only initializes the multisensor chips and sets their timing parameters. The multisensor chips then send their data in regular intervals without being polled by the data recording unit. Besides reducing the traffic on the bus, this also leaves more flexibility for future extensions, eg, threshold limiting values where one sensor triggers the system, when its signal exceeds a certain threshold.

1.4.4.1.2 Digital Controller

The digital controller (see Figure 1.4.14) interprets the commands coming from the data-recording unit, stores the sensor parameters and manages the access of the single sensors to the I^{2}C bus. Each sensor on the chip can be individually addressed via the serial interface. A minimal set of commands was implemented for the controller: power on/off, read data, set/read parameters.

1.4.4.2 On-chip Voltage and Current References

For an array of monolithic gas sensor chips, it is neither convenient nor cost-effective to supply each chip with a number of accurate external reference voltages or currents. Therefore, all reference voltages and currents have to be generated on-chip. The only stable and reliable reference available in standard CMOS processes is the silicon bandgap voltage. Taking the difference between the base-emitter voltages (V_{BE}s) of two identical bipolar transistors biased at different currents leads to a voltage, which is proportional to the absolute temperature (PTAT) and is often used for temperature sensors. On the other hand the weighted sum of a V_{BE} and a PTAT voltage generates a temperature-independent reference voltage, which is approximately equal to the silicon bandgap voltage.

The first designs of integrated bandgap references and temperature sensors were published in the late 1960s and early 1970s [120, 121]. A few years later, the first CMOS implementations were presented. Some CMOS designs employ the temperature dependence of the gate-source voltage of MOS transistors [122]. It is possible to generate PTAT voltages using MOS transistors operating in the weak inversion region [123], but it is very difficult to generate an accurate and reliable reference voltage. While the bipolar versions are based on the universal silicon bandgap voltage, the MOS references depend on the implants defining the threshold voltage which renders them process-dependent and hardly reproducible.

Therefore, most CMOS designs are based on the parasitic bipolar transistors available in CMOS processes. The standard designs [124] have limited accuracy owing to mismatch and offset problems. Chopping techniques can be employed to reduce the errors due to amplifier offset [125]. If no continuous-time reference voltage is needed, switched-capacitor techniques and dynamic element matching can be employed to further reduce errors due to mismatches [125–127].

The references and the temperature sensor presented here use the vertical pnp-transistor available in p-substrate CMOS processes. The first part of this section contains an analysis of the basic equations. Then, current and voltage references are presented. Chapter 1.4.4.3 describes an accurate temperature sensor based on switched-capacitor techniques.

1.4.4.2.1 Temperature Dependence of V_{BE}

Basic Equation Based on Extrapolated Bandgap Voltage

The base-emitter voltage V_{BE} of a bipolar transistor as a function of the collector current I_C is accurately described by

$$V_{BE}[T] = \frac{kT}{q} \ln\left(\frac{I_C[T]}{I_S[T]}\right) \tag{1.4.1}$$

where T denotes the absolute temperature, q the electron charge and k the Boltzmann constant. The current $I_S[T]$ is given by

$$I_S[T] = \frac{kT}{N_B} A n_i^2[T] \bar{\mu}[T] \tag{1.4.2}$$

where N_B is the Gummel number, A the base-emitter junction area, $n_i[T]$ the intrinsic carrier concentration and $\bar{\mu}[T]$ the effective mobility of minority carriers in the base. The temperature dependence of $n_i[T]$ is described by

$$n_i^2[T] = KT^3 e^{-\frac{qV_G[T]}{kT}} \tag{1.4.3}$$

where K is constant with respect to temperature and $V_G[T]$ is the silicon bandgap voltage.

The effective mobility $\bar{\mu}[T]$ is given by

$$\bar{\mu}[T] = LT^{-n} \tag{1.4.4}$$

where L and n are constants. Unfortunately, it is not possible to calculate all the constants from process data with sufficient accuracy. Therefore, the parameters A, K, L and N_B are eliminated from the equation by using the base-emitter voltage $V_{BE}[T_R]$ measured at a reference current I_R and a reference temperature T_R ($T_R = 300$ K). The temperature-dependent terms are exclusively retained and they are all relative to T_R. This leads to the equation

$$V_{BE}[T] = V_G[T] - \frac{T}{T_R}(V_G[T_R] - V_{BE}[T_R]) + \frac{T}{T_R} V_{tR} \ln\left(\frac{I_C[T]}{I_C[T_R]}\right) - (4-n)\frac{T}{T_R} V_{tR} \ln\left(\frac{T}{T_R}\right) \tag{1.4.5}$$

where $V_{tR} = \dfrac{kT_R}{q} = 25.85$ mV.

Equation (1.4.4) is only an approximation of the temperature dependence of the mobility, and the temperature dependence of the bandgap voltage is nonlinear [128]. In the majority of publications [125–127, 129] the bandgap voltage $V_G[T]$ is approximated with a linearly extrapolated bandgap voltage V_{G0R}, and a further fit-parameter η is introduced to replace the term (4–n) in Equation (1.4.5).

$$V_{BE}[T] = V_{G0R} - \frac{T}{T_R}(V_{G0R} - V_{BE}[T_R]) + \frac{T}{T_R} V_{tR} \ln\left(\frac{I_C[T]}{I_C[T_R]}\right) - \eta \frac{T}{T_R} V_{tR} \ln\left(\frac{T}{T_R}\right) \tag{1.4.6}$$

Vertical PNP Transistors in n-Well CMOS Technology

Two parasitic bipolar transistors are available in standard n-well CMOS technology. The lateral pnp-transistor shown in Figure 1.4.15 offers the advantage that

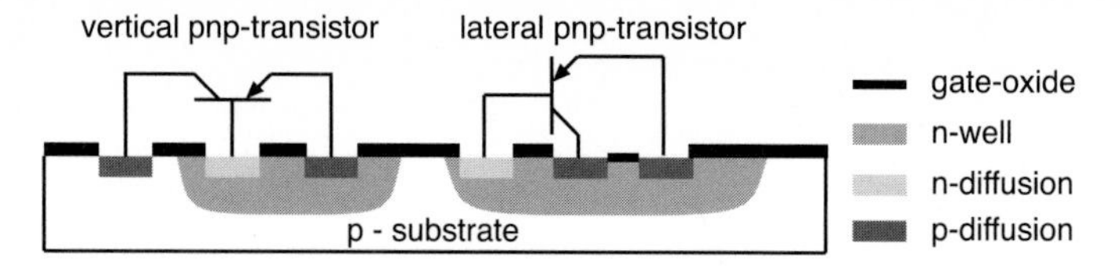

Figure 1.4.15. Bipolar transistors available in a standard n-well CMOS process: vertical substrate transistor (left) and lateral pnp-transistor with n-well as base (right).

the circuits known from bipolar designs can be re-used. Unfortunately, the lateral pnp-transistor has large process spread and exhibits nonidealities due to the use of a gate to separate emitter and collector. It is hardly used for bandgap references or temperature sensors.

The vertical pnp-transistor (see Figure 1.4.15) has better performance and less process spread but exhibits one major disadvantage: the collector, which is formed by the p-substrate, is tied to the negative supply voltage, and hence the collector current is not accessible. Furthermore, the current gain is lower than for common bipolar transistors. Most of the CMOS reference and temperature sensors rely on this transistor, but compared with bipolar designs some additional parameters have to be considered: the base current and resistance, the Early effect, and high-level and low-level injection.

Since only the emitter current of the vertical pnp-transistor can be controlled, the base current has to be subtracted before Equation (1.4.6) can be applied. Assuming a constant emitter current $I_E[T]$ equal to $I_C[T_R]$, the respective term in Equation (1.4.5) can be rewritten as

$$\ln\left(\frac{I_C[T]}{I_C[T_R]}\right) = \ln\left(\frac{I_E[T]\left(\frac{\beta[T]}{\beta[T]+1}\right)}{I_C[T_R]}\right) = \ln\frac{\beta[T]}{\beta[T]+1} \quad (1.4.7)$$

The current amplification factor $\beta[T]$ is small for the vertical pnp-transistors available in CMOS processes and is strongly dependent on temperature and current. Figure 1.4.16 shows measurements of $\beta[T]$ and the contribution to the base-emitter voltage by the term in Equation (1.4.7) for $I_E = 10\ \mu A$.

The base of the vertical pnp-transistor is formed by the n-well of the CMOS process. This leads to a large base resistance that has to be considered owing to its temperature dependence.

In some of the designs the base of the bipolar transistor is not tied to the substrate but to the common-mode voltage of the circuit. In this case, the Early effect leads to another correction in the collector current term of Equation (1.4.5):

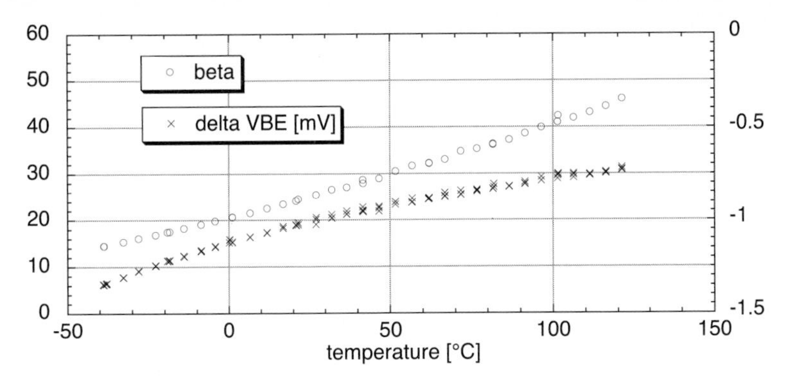

Figure 1.4.16. Temperature dependence of the current-gain factor β and the resulting difference in the base-emitter voltage ΔV_{BE}, when the transistor is biased with constant emitter current instead of constant collector current.

$$\ln\left(\frac{I_C[T, V_{CE}]}{I_C[T_R]}\right) = \ln\left(\frac{I_C[T]\left(1 - \dfrac{V_{CE}}{V_E}\right)}{I_C[T_R]}\right) = \ln\left(\frac{I_C[T]}{I_C[T_R]}\right) + \ln\left(1 - \frac{V_{CE}}{V_E}\right) \quad (1.4.8)$$

where V_E denotes the Early voltage. A description of the temperature dependence of the Early effect could not be found. As a result, this effect is treated as a constant contribution to the linear term of Equation (1.4.6).

The effects of high-level and low-level injection on the $V_{BE}[I_C]$ characteristics have been studied [130]. It was shown that these effects can be neglected, if an appropriate range is chosen for the collector currents.

An Equation for $V_{BE}[I_E, T]$ of Vertical Bipolar Transistors

Since the collector of a CMOS vertical bipolar transistor is not accessible, an equation for the temperature dependence of the base-emitter voltage as a function of the emitter current is needed to design temperature sensors and bandgap references in CMOS processes. The biasing currents are all in the range of 1–50 μA. In this range, the empirical Equation (1.4.6) can be used to fit $V_{BE}[I_E, T]$:

$$V_{BE}[T] = V_{GOR} - \frac{T}{T_R}\left(V_{GOR} - V_{BE}[T_R] - V_{tR}\ln\left(\frac{I_E[T]}{I_E[T_R]}\right)\right) - \eta\frac{T}{T_R}V_{tR}\ln\left(\frac{T}{T_R}\right) \quad (1.4.9)$$

The temperature dependence of the base current and a constant value for the voltage drop over the base resistance are then included in the fit parameters V_{GOR} and η. The remaining errors introduced by the bias-current dependence of β and the base resistance can then be neglected. Figure 1.4.17 shows the fit error for a

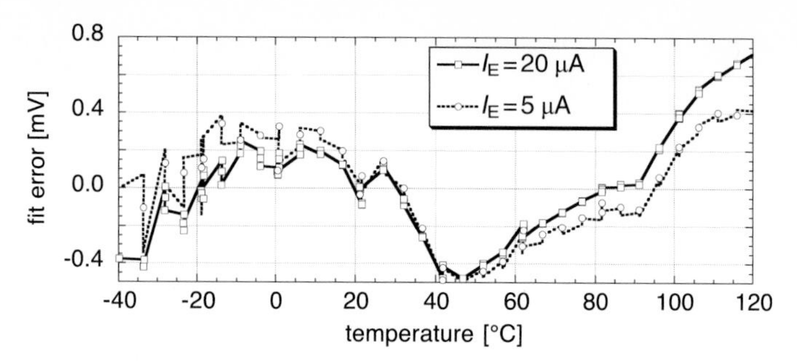

Figure 1.4.17. Fit error for $V_{BE}[I_E, T]$ approximated using Equation (1.4.9).

transistor biased at 5 and 20 μA. The error is smaller than 1.2 mV over the whole temperature range. In the remainder of this section, Equation (1.4.9) has been used for $V_{BE}[I_E, T]$. More details on the analysis of the temperature characteristics of bipolar transistors can be found in [130, 131].

1.4.4.2.2 Reference Voltages

For many applications, a standard CMOS bandgap reference as described in [124] can be adapted. Variations of a few millivolts over the specified temperature range from –40 to 120 °C can be tolerated owing to the strong temperature dependence of the absorption process, which renders calibration mandatory. In the case of the multisensor chip, the output of the capacitive sensor is independent of the reference voltage, and the absolute value of the reference voltage for the A/D converter is not crucial for the calorimetric sensor. Switching techniques can be used for sensors that require better accuracy of the reference voltage (eg, the temperature sensor described in Section 1.4.4.3).

A schematic of the CMOS bandgap reference is shown in Figure 1.4.18. Transistor T1 consists of N unit transistors T2. The reference voltage V_{ref} is generated relative to the common-mode voltage V_{CM}.

Only the linear terms in Equation (1.4.9) are compensated. The nonlinear curvature term is linearized around T_R, and the current through the transistors T1 and T2 is assumed to be PTAT. V_{BE1} is then described by

$$V_{BE}[T] = V_{G0R} - \frac{T}{T_R}(V_{G0R} - V_{BE}T_R + V_{tR}(\eta - 1)) \tag{1.4.10}$$

As the currents through T1 and T2 are identical, ΔV_{BE} is given by

$$\Delta V_{BE} = \frac{kT}{q}\ln(N) \tag{1.4.11}$$

In order to compensate for the linear term in Equation (1.4.10), the resistor ratio must be

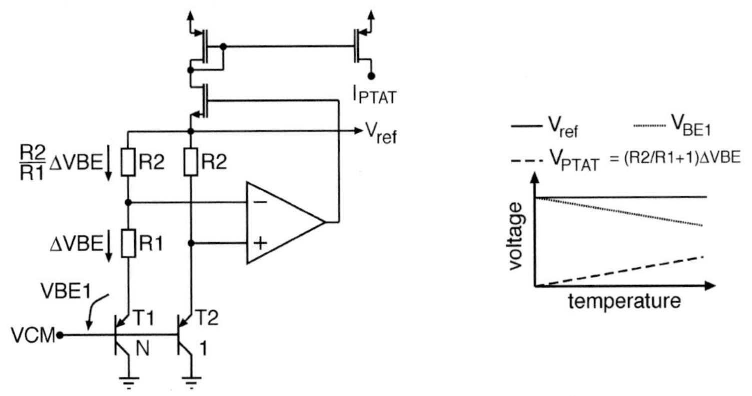

Figure 1.4.18. Standard CMOS bandgap reference [135]. The linear temperature dependence of V_{BE} is corrected by adding a scaled PTAT voltage.

$$\frac{R2}{R1} = \frac{V_{G0R} - V_{BE}[T_R] + V_{tR}(\eta - 1)}{V_{tR}\ln(N)} \tag{1.4.12}$$

A value of 7.23 was calculated for $N = 19$. In addition to the nonlinearities, the major sources of error in this approach include the offset voltage of the operational amplifier, the nonideal PTAT current and the mismatch of the two bipolar transistors. Figure 1.4.19 shows the simulated results for ideal components and an offset voltage of 1 mV for the operational amplifier.

As the variations in the absolute value are removed by calibration, only the temperature variations in the target range of the system from –10 to 40 °C have to be taken into account. They are less than 0.5 mV and can be tolerated for capacitive and calorimetric sensors.

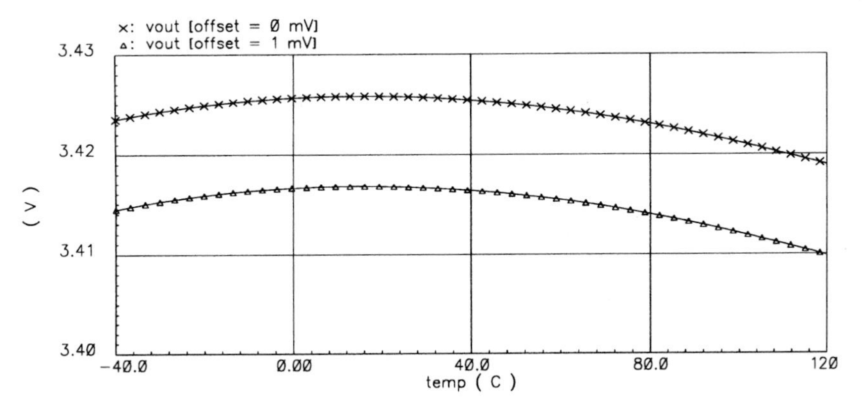

Figure 1.4.19. Simulated results for the standard bandgap reference with ideal components and under the assumption of 1 mV input offset for the operational amplifier.

1.4.4.2.3 Current Sources

The bias currents for the different analog blocks must be generated on the chip. While only limited accuracy of the currents is required, their temperature behavior should be well known. Two types of currents are needed:

- Proportional to absolute temperature (PTAT).
- Constant over temperature.

PTAT Current Source

The PTAT currents are generated by a circuit with the same topology that is used for the bandgap references (see Figure 1.4.18). As the voltage ΔV_{BE} is PTAT, the current through transistor T1 is also approximately PTAT. This is sufficient for most biasing applications. If the accuracy requirements are high (eg, for the temperature sensor), the temperature coefficient of the polysilicon resistor has to be considered:

$$I_{PTAT} = \frac{T}{T_R} \frac{V_{tR} \ln(N)}{R_0(1 + TCR(T - T_R))} \tag{1.4.13}$$

As the temperature coefficient TCR of the polysilicon amounts to 1000 ppm/°C, the deviation from an ideal PTAT current over the temperature range from –50 to +120 °C is ±10%. Known techniques such as combining resistors with different temperature coefficients [135] cannot be applied, because a material with negative temperature coefficient and sufficient resistivity is not available in a standard CMOS process.

PTAT currents are used to bias the bipolar transistors of the temperature sensor. A collector current which is proportional to absolute temperature leads to a prefactor $(\eta - 1)$ in the curvature term of Equation (1.4.9). If Equation (1.4.13) is used, an additional term is obtained:

$$V_{tR} \frac{T}{T_R} \ln(1 + TCR(T - T_R)) \tag{1.4.14}$$

Omitting this term in Equation (1.4.9) leads to an error of 5 mV between –50 and +120 °C. It can be accounted for by a correction factor in either the linear term or the curvature term in Equation (1.4.9). The fit error over the full temperature range is <200 µV, if the curvature term is adjusted compared with 1 mV for adjustment of the linear term. For biasing with a PTAT current source, Equation (1.4.9) can be rewritten:

$$V_{BE}[T] = V_{G0R} - \frac{T}{T_R}(V_{G0R} - V_{BE}[T_R]) - (\eta - 0.69)\frac{T}{T_R} V_{tR} \ln\left(\frac{T}{T_R}\right) \tag{1.4.15}$$

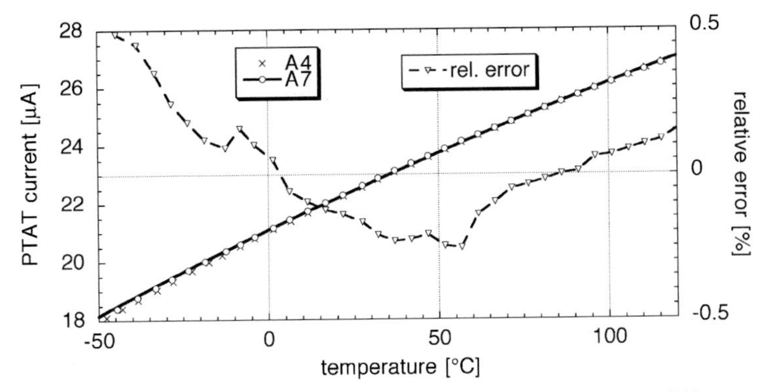

Figure 1.4.20. Measurement of the PTAT current obtained from two different chips and relative error to a fit with Equation (1.4.13).

Small changes of the circuit in Figure 1.4.18 were made to simplify the PTAT current source and reduce the area consumption:

- The VCM node is tied to ground, because no signal relative to VCM is needed.
- The ratio of the two resistor values *R2* and *R1* does not play a significant role for the PTAT current source. A small ratio reduces the area consumed by the resistors. Therefore, a ratio of unity was chosen.

Figure 1.4.20 shows a measurement of a PTAT current source with a nominal value of 25 µA at 300 K. As the tolerances of poly1-resistors are on the order of ±20%, accurate matching to the designed value cannot be expected. The measured value of 22.6 µA is within the specified range. Comparing Equation (1.4.13) with the measured current (the measured current at 300 K was used as a fit parameter for Equation (1.4.13)) leads to a maximum relative error of +0.5/–0.2%.

Constant-current Source

In addition to the PTAT currents, temperature-independent currents are also needed to bias the temperature sensor and some other analog blocks. A temperature-independent current results from adding a PTAT current to a current with negative temperature coefficient, which can be obtained by biasing a poly-resistor with constant voltage as shown in Figure 1.4.21.

The current through transistor T4 is then described by

$$I_{\text{const}} = \Delta V_{\text{BE}} \left(\frac{2R_3 + R_1}{R_2 R_3} \right) + \frac{V_{\text{BE2}}}{R_3} \tag{1.4.16}$$

A linear fit is used to describe the temperature dependence of the base-emitter voltage and the poly-resistors:

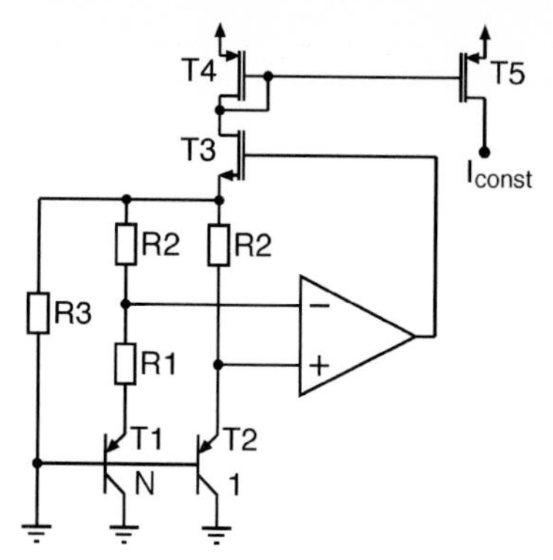

Figure 1.4.21. Schematic of the constant-current generator used to bias the different analog building blocks and the $\Sigma\Delta$ temperature sensor.

$$V_{BE2} = V_{BE}[0](1 - TC_{BG}T) \tag{1.4.17}$$

$$R_x = R_x[0](1 + TC_R T) \tag{1.4.18}$$

Equations (1.4.11), (1.4.17) and (1.4.18) are inserted in Equation (1.4.16):

$$I_{const} = \frac{V_{BE}[0]}{R_3} \frac{1 + \left(\frac{2R_3 + R_1}{R_2} \frac{k\ln(N)}{q} \frac{1}{V_{BE}[0]} - TC_{BG}\right)T}{1 + TC_R T} \tag{1.4.19}$$

R_3 is used to define the absolute value of the desired current. R_1, R_2 and N are used to cancel the linear terms in Equation (1.4.19). The large number of available parameters makes it possible to find a solution where the resistors have integer ratios. This improves the matching, because unit resistors can be used for the layout. The values of the components are summarized in Table 1.4.3.

Figure 1.4.22 compares the currents calculated using Equation (1.4.6) for V_{BE} and a linear fit for the resistors to the measured results of two chips from different wafers. Since the spread of the absolute resistor values is large owing to process tolerances, all currents have been normalized to unity. The maximum difference between simulation and measurement is 0.5%. The main reason for these errors include deviations of the resistance from the first-order behavior assumed in the calculations. Second-order terms have to be taken into account in order to achieve better results over an extended temperature range of 160 °C.

Table 1.4.3. Component values for constant current source

I_{const}	*R1*	*R2*	*R3*	*N*
30 μA	5 kΩ	5 kΩ	55 kΩ	8

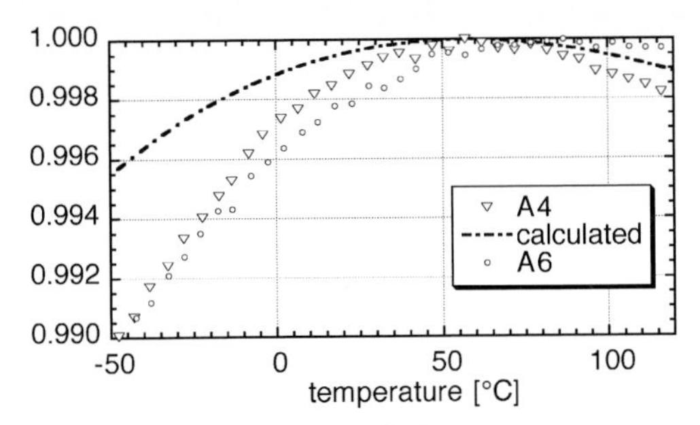

Figure 1.4.22. Calculation and measurement of the constant current source (current normalized to unity, nominal value 30 μA). The maximum difference between calculated values and measurements is 0.5%.

1.4.4.3 Temperature Sensor

Integrated bandgap references and temperature sensors are closely related, because they are all based on the approximately linear temperature dependence of the base-emitter voltage (V_{BE}) of diode-connected bipolar transistors. A 10-bit temperature sensor for a temperature range from –40 to 120 °C can only be realized if both, the PTAT voltage and the reference voltage, meet the linearity requirements for this resolution. This cannot be achieved with the continuous-time designs presented above owing to mismatches and process spread. Better accuracy is achieved in implementations that use a single bipolar transistor combined with current-switching techniques and sum the contributions to the PTAT and reference voltages in order to eliminate errors due to mismatches. The first designs were published by Williams [123] and Leung and Mok [136]. Improved designs of temperature sensors that are based on this approach have been published by Tuthill [127] and Bakker [125].

1.4.4.3.1 ΣΔ Temperature Sensor

The designs by Tuthill and Bakker use a configuration as shown in Figure 1.4.23. One block generates the PTAT voltage, and a second block is needed to provide the A/D converter with the reference voltage.

The aim of the design presented here was to generate all the signals from a single bipolar transistor, to reduce the number of building blocks needed and to increase the accuracy of the reference voltage and its matching with the measured voltage. This can be achieved when all elements are incorporated into a switched-capacitor $\Sigma\Delta$ modulator. The design is derived from a conventional first-order $\Sigma\Delta$ modulator as shown in Figure 1.4.24a: in each clock cycle, the reference voltage V_{ref} is added to or subtracted from the input voltage V_{in}. The sign of the reference voltage depends on the comparator output at the end of the previous clock cycle. The resulting difference is then summed in the integrator.

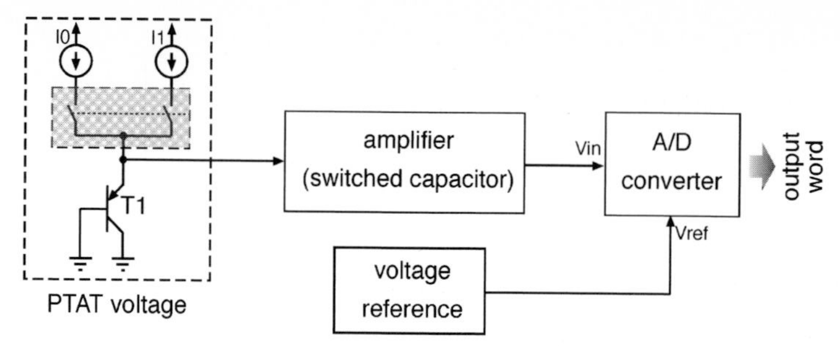

Figure 1.4.23. Block diagram of a temperature sensor based on switching techniques.

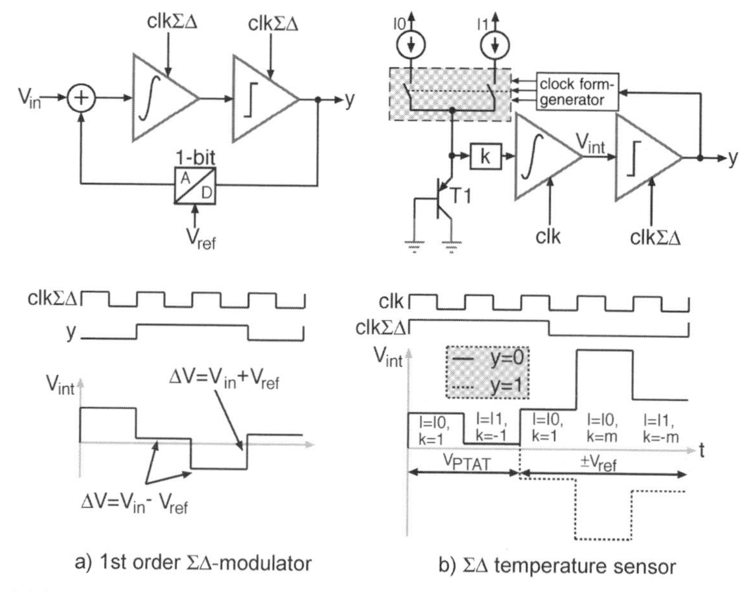

Figure 1.4.24. Basic structure of the $\Sigma\Delta$ temperature sensor.

The block diagram in Figure 1.4.24b illustrates the principle of the $\Sigma\Delta$ temperature sensor. V_{in} and V_{ref} of a temperature sensor are both sums of weighted base-emitter voltages. As the input stage of a $\Sigma\Delta$ modulator is an integrator, these sums can be generated inside the modulator, because a switched-capacitor integrator sums its input voltages at each falling edge of the clock cycle. If the clock cycle of the modulator is subdivided into several phases, the single contributions (V_{BEx}) to V_{PTAT} and V_{ref} can be accumulated inside the integrator. If a switched-capacitor $\Sigma\Delta$ modulator is used, the different scaling factors of the single voltage contributions (see, eg, Equation (1.4.14)) can be implemented by arrays of unit capacitors (factor k in Figure 1.4.24b). The clock cycle $clk\Sigma\Delta$ of the first-order $\Sigma\Delta$ modulator is subdivided into five phases. In each phase, one weighted base-emitter voltage is added to the voltage V_{int} of the integrator. The resulting voltage change during one clock cycle is $\Delta V_{int} = V_{PTAT} \pm V_{ref}$. This is identical with a first-order $\Sigma\Delta$ modulator.

There are some shortcomings of this design that have to be solved for a practical implementation:

- The bandgap reference voltage in the example shown above uses linear temperature compensation, as was introduced in Section 1.4.4.2.2. This is not accurate enough for the temperature sensor.
- The voltages $mV_{BE}[I_0]$ and $-mV_{BE}[I_1]$ are added in phases 4 and 5. If $m = 7.23$ as derived in Equation (1.4.24), the resulting voltage is far beyond the output swing of the integrator.
- Only a small fraction of the modulator's dynamic range is used, because the maximum PTAT voltage is ca 80 mV, while V_{ref} is larger than 1 V.

The following paragraphs are dedicated to some changes and extensions of the basic design, which address the aforementioned problems.

1.4.4.3.2 PTAT- and Base-emitter-voltage Generator

PTAT Voltages

If the biasing currents are small enough to avoid high-level injection, and if appropriate current ratios (integer ratios are preferable) are chosen, the specifications for the PTAT voltage can be met either by a one-point calibration or by taking into account the effective emission coefficient [130]. Figure 1.4.25 shows a measurement for bias currents of 1 and 10 μA. With an effective emission coefficient of 1.015, the relative error is smaller than 10^{-3}. This is sufficient for an accuracy of 10 bits.

When only one single bipolar transistor is used in the input stage of the temperature sensor, two large base-emitter voltages (approx. 1 V) must be sequentially applied to generate the small PTAT voltages (approx. 50 mV). This can lead to errors in the integrator due to incomplete charge transfer. Therefore, a design with two transistors as shown in Figure 1.4.26 was chosen. The errors in the PTAT voltage

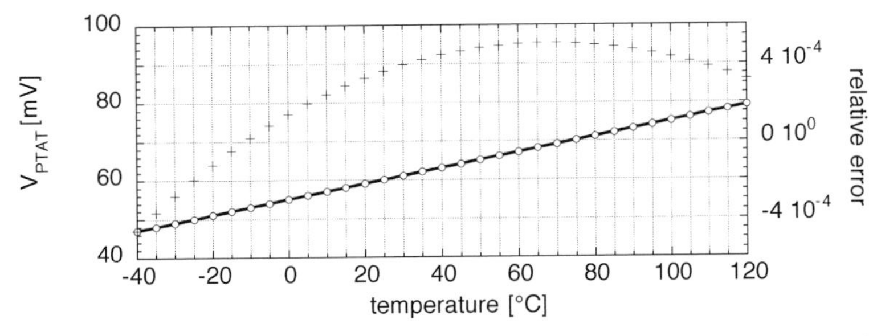

Figure 1.4.25. Measured PTAT voltage of a transistor biased at 1 and 10 μA and relative error compared with the ideal value.

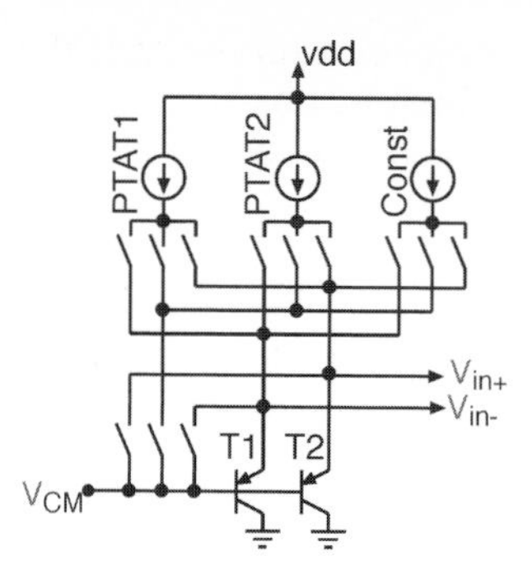

Figure 1.4.26. PTAT and base-emitter voltage generator for $\Sigma\Delta$ temperature sensor. Every current can be supplied to each bipolar transistor.

due to mismatches in V_{BER} of T1 and T2 are eliminated by swapping the currents of the transistors in the subsequent clock phase. The resulting voltage after two clock phases is

$$\begin{aligned}&(V_{\mathrm{BE1}}[I_1] - V_{\mathrm{BE2}}[I_2)] - (V_{\mathrm{BE1}}[I_2] - V_{\mathrm{BE2}}[I_1])\\&= ((V_{\mathrm{BE1}}[I_1] - V_{\mathrm{BE1}}[I_2]) + (V_{\mathrm{BE2}}[I_1] - V_{\mathrm{BE2}}[I_2])) = V_{\mathrm{PTAT1}} + V_{\mathrm{PTAT2}}\end{aligned} \tag{1.4.20}$$

where V_{BE1} is the base-emitter voltage of transistor T1 and V_{BE2} is that of transistor T2. In this way, the accuracy of a single-transistor design is maintained, while the voltages are doubled, and the problems due to charge transfer efficiency [131, 137] are reduced.

Curvature-corrected Reference Voltage

The acceptable error of the reference voltage is 0.85 mV (for an accuracy of 10 bits over the specified temperature range). The curvature term in Equation (1.4.6) has a maximum value of 50 mV. A linear approximation of this term as it is used in Section 1.4.4.2.2 still leads to an error of 5 mV. Different techniques for higher-order compensation of the curvature term in bandgap references have been presented [125, 126]. They all use an approximation to reduce the curvature term. An exact compensation was first proposed by Gilbert [135]. His technique is based on the bias-current dependence of the curvature term. Gilbert starts from the original configuration (see Section 1.4.4.2.2) and subtracts a third base-emitter voltage of a transistor biased with a constant current.

If this technique is combined with the switched-capacitor techniques described above, the resulting reference voltage is given by

$$V_{\mathrm{ref}} = (l+2)V_{\mathrm{BE}}[I_{\mathrm{PTAT}}] + 2mV_{\mathrm{PTAT}} - 2V_{\mathrm{BE}}[I_{\mathrm{const}}] \tag{1.4.21}$$

The factors of two are necessary to reduce the errors due to mismatches by swapping the currents of the two branches, as has already been shown for the

PTAT voltage in Equation (1.4.20). Assuming that the temperature dependence of the bias currents can be described by

$$I_E[T] = I_E[T_R]\left(\frac{T}{T_R}\right)^k \tag{1.4.22}$$

The curvature for different currents can then be calculated

$$\frac{T}{T_R}V_{tR}\ln\left(\frac{I_E[T]}{I_E[T_R]}\right) - \eta\frac{T}{T_R}V_{tR}\ln\left(\frac{T}{T_R}\right) = -(\eta - k)\frac{T}{T_R}V_{tR}\ln\left(\frac{T}{T_R}\right) \tag{1.4.23}$$

The factor k in Equation (1.4.23) has a value of 0.69 for the PTAT current generated by the circuit in Figure 1.4.18; k is zero for a constant current. The voltage V_{PTAT} is the base-emitter voltage difference of a transistor sequentially biased with two different PTAT currents:

$$V_{PTAT} = V_{tR}\ln\left(\frac{I_{PTAT\,high}}{I_{PTAT\,low}}\right) \tag{1.4.24}$$

The resulting reference voltage is then given by

$$\begin{aligned} V_{ref} = {} & lV_{G0R} - l\frac{T}{T_R}(V_{G0R} - V_{BE}[I_{ER}, T_R]) \\ & - \frac{T}{T_R}V_{tR}\left(-2m\ln\frac{I_{PTAT\,high}}{I_{PTAT\,low}} - (l+2)\ln\frac{I_{PTAT\,high}}{I_{ER}} + 2\ln\frac{I_{const}}{I_{ER}}\right) \\ & - (2\eta - (l+2)(\eta - 0.69))\frac{T}{T_R}V_{tR}\ln\left(\frac{T}{T_R}\right) \end{aligned} \tag{1.4.25}$$

The factor l to cancel the curvature term can be calculated from Equation (1.4.25):

$$\underbrace{2\eta\frac{T}{T_R}V_{tR}\ln\left(\frac{T}{T_R}\right)}_{\text{current constant}} \underbrace{-(l+2)(\eta - 0.69)\frac{T}{T_R}V_{tR}\ln\left(\frac{T}{T_R}\right)}_{\text{PTAT current}} = 0 \tag{1.4.26}$$

$$l = \frac{2\eta}{\eta - 0.69} - 2 \tag{1.4.27}$$

Finally, the factor m necessary to eliminate the linear terms in Equation (1.4.25) is determined:

Table 1.4.4. Important parameters of the $\Sigma\Delta$ temperature sensor

	V_{ref} (mV)	l	m	$I_{PTAT\ high}$ (μA)	$I_{PTAT\ low}$	I_{const} (μA)
Calculated	330	0.295	1.051	22.5	$I_{PTAT\ high}/8$	30
Approximated	331	3/10	1.000	22.5	$I_{PTAT\ high}/8$	30

$$m = \frac{l(V_{G0R} - V_{BE}[I_{ER}, T_R]) - V_{tR}\left((l+2)\ln\dfrac{I_{PTAT\,high}}{I_{ER}} - 2\ln\dfrac{I_{const}}{I_{ER}}\right)}{2V_{tR}\ln\dfrac{I_{PTAT\,high}}{I_{PTAT\,low}}} \quad (1.4.28)$$

Table 1.4.4 summarizes the calculated values for the variables in Equation (1.4.21). The weights of the base-emitter voltages (factor k in Figure 1.4.24) are realized by capacitor ratios in the input stage of the switched-capacitor integrator. Accurate ratios are only obtained for arrays of unit devices. Therefore, the best integer approximations of the calculated values were identified with simulations. The resulting weights are also given in Table 1.4.4.

The switching sequence to generate V_{ref} can now be derived from Equation (1.4.21). One prefactor ($l/2$, realized using unit capacitor ratios) and six clock phases are necessary to generate the temperature-independent reference voltage:

$$V_{ref} = \underbrace{2(V_{BE}[I_{PTAT}] - V_{BE}[I_{const}])}_{\text{Phase 1, 2}} + \underbrace{2V_{PTAT}}_{\text{Phase 3, 4}} \underbrace{-2\frac{lV_{BE}}{2}[I_{PTAT}]}_{\text{Phase 5, 6}} \quad (1.4.29)$$

1.4.4.3.3 Second-order $\Sigma\Delta$ Modulator

Figure 1.4.24 shows a schematic of the second-order switched-capacitor $\Sigma\Delta$ modulator, which was chosen to realize the $\Sigma\Delta$ temperature sensor. Besides the novel switching scheme that was described in Section 1.4.4.3.2, additional modifications to the conventional modulator were developed in order to improve the performance of the temperature sensor:

- The input range of the conventional modulator is $\pm V_{ref}$. For the temperature sensor, a range of $[0, +V_{ref}]$ is more appropriate, because only positive PTAT voltages have to be converted. This is done by dividing the V_{ref} by two and shifting the range by $V_{ref}/2$. The output word and the updates of the integrator voltages in each clock cycle are then given by

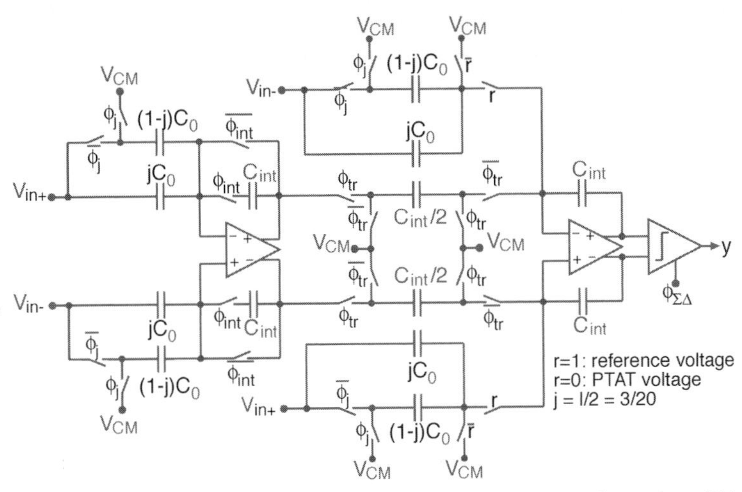

Figure 1.4.27. Second-order switched-capacitor $\Sigma\Delta$ modulator configured as $\Sigma\Delta$ temperature sensor.

$$\text{Out} = \frac{V_{\text{PTAT}} - \frac{V_{\text{ref}}}{2}}{\frac{V_{\text{ref}}}{2}} \Rightarrow \begin{array}{l} y = 1;\ \Delta V_{\text{in}} = V_{\text{PTAT}} - V_{\text{ref}} \\ y = 0;\ \Delta V_{\text{int}} = V_{\text{PTAT}} \end{array} \tag{1.4.30}$$

The variable y denotes the output of the modulator and ΔV_{int} is the change of the voltage on the first integrator after the next clock cycle.

- The PTAT voltage for an emitter-current ratio of 8 reaches a maximum value of 80 mV at a temperature of 120 °C. Owing to the clocking scheme in Section 1.4.4.3.2, which swaps the currents of the two bipolar transistors, this voltage is doubled. The reference voltage of 330 mV allows for another doubling of the PTAT voltage without exceeding the dynamic range of the modulator. This is done by repeating the switching sequence generating the PTAT voltage. Two additional phases are required. This leads to a total of 10 clock phases for one conversion cycle of the $\Sigma\Delta$ modulator: six phases for V_{ref} and four phases for V_{PTAT}.
- The sampling switches used in the input stage of the conventional $\Sigma\Delta$ modulator can be saved because the voltages are switched in the PTAT and base-emitter voltage generator (see Figure 1.4.27).

Sampling Frequency

The maximum sampling frequency $f_{\Sigma\Delta}$ is determined by the transient behavior of the PTAT- and reference-voltage generator. The smallest current that is used in the PTAT- and reference-voltage generator has a nominal value of 3.125 µA at 300 K. If the process tolerances and the minimum temperature of –40 °C are taken into account, this current is reduced to 1.9 μA. The $\Sigma\Delta$ modulator was designed for full settling (99.9%) within one clock phase in order to avoid any sensitivity to clock jitter. A maximum clock frequency $f_{\Sigma\Delta}$ of 100 kHz was obtained from corner simulations.

1.4.4.3.4 Decimation Filter

For most gas sensing applications, the temperature information is only required every few seconds. Therefore, a simple decimation counter can be employed to decimate the output bitstream. This reduces the overall chip area. In case a higher sampling ratio for the temperature is desirable (eg, for transient measurements), a different filter is needed. Many digital decimation filter architectures (eg, [138]) are not well suited for hardwired filters, because they make excessive use of multipliers. Multipliers consume a lot of area, if accuracies of 12 bits and better are required. Filters based on multiplier-free topologies [139] and comb-filters are preferred in order to achieve an area-efficient solution. Candy [140] showed that a simple sinc^l filter can be used for decimation down to four times the output-word rate f_0 without degrading the signal-to-noise ratio. The order l of the filter must equal the order of the $\Sigma\Delta$ modulator plus one. For the subsequent decimation to f_0, a finite-impulse response (FIR) filter with a higher order cutoff is required. The details of the design and implementation can be found in [6].

1.4.4.3.5 Experimental Results

Figure 1.4.28 shows a micrograph of a test chip bearing the temperature sensor. The separate test chip allows for measurements of the various analog and digital signals, which are not accessible on the multisensor chip.

Figure 1.4.29 shows a measurement of the output of the $\Sigma\Delta$ temperature sensor with curvature-corrected reference voltage. The error of the linear fit is less than ±0.2 °C in the measured temperature range of –40 to 120 °C. The nonlinearity is a combination of errors from the measurement setup, the current generators, the generated reference voltage and the inaccuracy of the temperature measurement (~0.1 °C). The matching of the error curve is excellent for all measured chips. Therefore, an empirical correction was extracted from one chip and applied to all other chips. A reduced nonlinearity with a worst-case standard deviation of 0.18 °C for the seven measured chips was obtained.

The performance of the temperature sensor is summarized in Table 1.4.5. The output of the sensor is proportional to absolute temperature. The offset is calculated by normalizing the data for 0 °C and extrapolating the measured data to 0 K. The discrepancy between 'designed' and measured sensitivity cannot be explained.

An accuracy of 0.1 °C over a temperature range of 50 °C was targeted for the single-chip gas detection system. Owing to the sensitivity mismatch, a maximum error of 0.4 °C results if the sensitivity is not calibrated. Therefore, a two-point

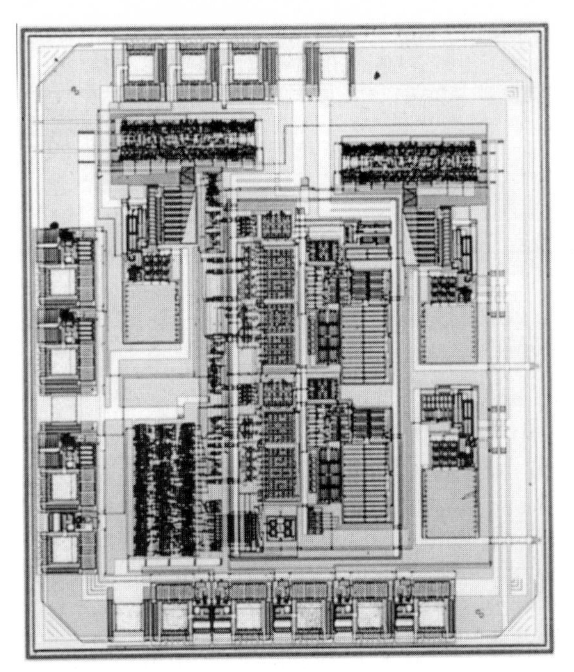

Figure 1.4.28. Micrograph of the temperature sensor test chip.

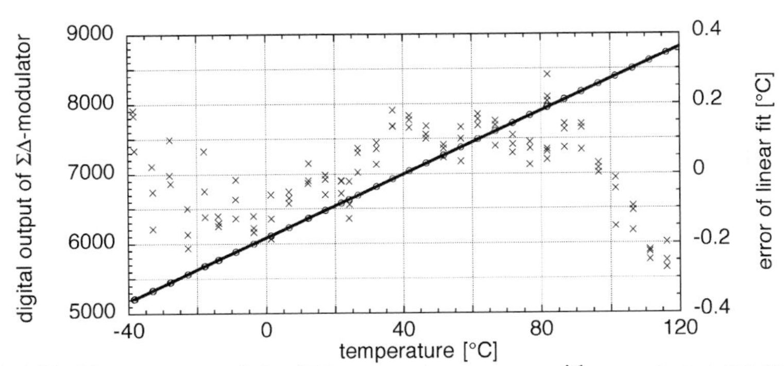

Figure 1.4.29. Measurement of the $\Sigma\Delta$ temperature sensor with curvature-corrected reference voltage. The error of the linear fit is below $\pm 0.2\,^\circ$C over the full temperature range from –40 to 120 °C.

calibration at, eg, 0 °C and room temperature is necessary to achieve the desired accuracy.

The measurements are in good agreement with the theoretical values. Depending on the requirements of a given gas-sensing application, the stress sensitivity has to be addressed in more detail: packaging- and temperature-induced stress affects the saturation current of a bipolar transistor owing to the piezojunction effect [141, 142]. The resulting degradation of the performance has to be simulated and experimentally verified.

Table 1.4.5. Performance of the $\Sigma\Delta$ temperature sensor with curvature-corrected reference voltage

–40 to 120 °C	Mean	Standard deviation	Unit
Sensitivity	1.02	0.008	–
Offset at 0 K	6.5	1.5	°C
Nonlinearity	0	0.11	°C

1.4.5 Application Examples

1.4.5.1 Laboratory Experiments

The first application includes laboratory measurements to investigate effects that are difficult to monitor with distributed macro-sensor arrays. The small volume of the measurement chamber enables, eg, fast switching of transients and truly simultaneous recording of data from three different transducers.

Figure 1.4.30 shows simultaneously recorded sensor signals of the three transducers upon exposure to 1200 and 3000 ppm of ethanol and 1000 and 3000 ppm of toluene (at 301 K). The sensors were alternately exposed to analyte-loaded gas and pure carrier gas. The polymeric coating consisted of poly(etherurethane), at a thickness of approx. 4 µm. Figure 1.4.30 a shows the measured signals (output of the sigma-delta converter) of the capacitor. Ethanol with a dielectric constant of 24.5, which is higher than the value of 2.9 for poly(etherurethane), causes a ca-

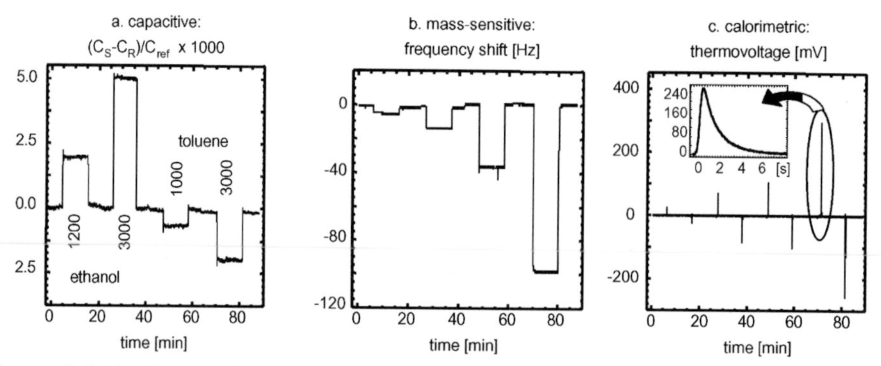

Figure 1.4.30. Sensor signals simultaneously recorded from all three polymer-coated transducers upon exposure to 1200 and 3000 ppm of ethanol and 1000 and 3000 ppm of toluene at 301 K: (**a**) $\Sigma\Delta$ converter output of the capacitor, (**b**) frequency shifts of the resonant cantilever and (**c**) thermovoltage transients of the calorimetric sensor. The close-up shows the development of the calorimetric transient within 6 s.

pacitance increase and hence positive frequency shifts. Toluene (with a dielectric constant of 2.4) causes a capacitance decrease and hence negative frequency shifts. The limits of detection are 3–5 ppm for ethanol and 5–8 ppm for toluene. Figure 1.4.30b displays the cantilever response. Ethanol shows rather low signals (compared with toluene) owing to its lower molecular mass and its lower enrichment (partitioning) in the polymeric phase. The extent of physisorption is, to first order, proportional to the boiling temperature of the analyte (ethanol, 351 K; toluene, 360 K): the lower the boiling temperature, the less enrichment there is in the polymer. Here, the limits of detection are 10–12 ppm of ethanol and 1–2 ppm of toluene, which is comparable to those for other mass-sensitive transducers.

The calorimetric results in Figure 1.4.30c represent a superposition of partitioning and heat-budget change due to analyte ab-/desorption. The absorbing analyte liberates heat (predominantly heat of condensation), causing a positive transient signal (positive peak), whereas the desorbing analyte abstracts vaporization heat from the environment, generating a negative transient signal (negative peak upon purging). The inset in Figure 1.4.30c shows the time-resolved response during the first 6 s of exposure to 3000 ppm of toluene. The heat of condensation/vaporization of ethanol is higher than that of toluene (42.3 compared with 38.0 kJ/mol at 298 K) owing to directional interaction in the liquid phase (even stronger for water), whereas the polymer partitioning of toluene is stronger owing to its higher boiling point. The limits of detection of the calorimetric method in this measurement are higher than those of the other transducers (40–50 ppm of toluene, 100–150 ppm of ethanol).

1.4.5.2 The Application-specific Sensor System (AS^3)

The ceramic board containing six multisensor chips has been assembled into a hand-held application-specific sensor system (AS^3) by adding a microcontroller board (signal processing unit) developed at the University of Bologna, a miniaturized gas-flow system developed at the University of Tübingen, and a battery supply. Figure 1.4.31 shows the functions of the signal processing unit. It

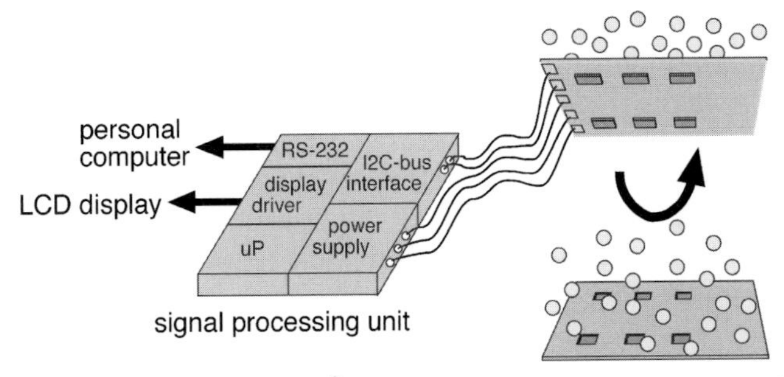

Figure 1.4.31. Block diagram of the AS^3 excluding the gas-flow unit, based on six multisensor chips.

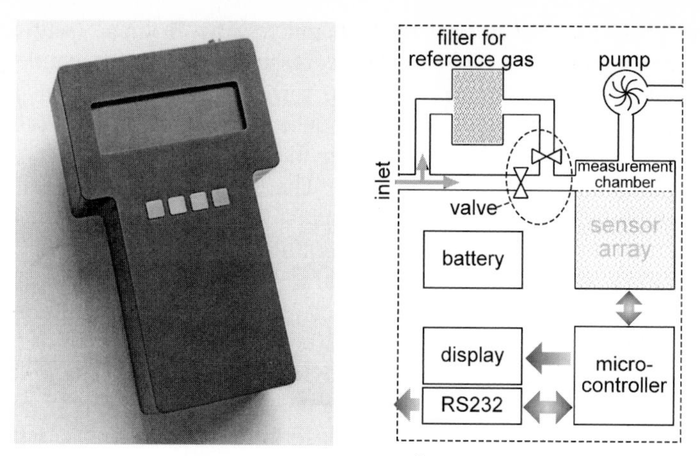

Figure 1.4.32. Photograph and schematic of the AS^3 hand-held sensor unit.

acquires the signals from the multisensor chips on the ceramic board and provides the chips with stable supply voltages of 2.2 and 5 V. The signals are evaluated by the microprocessor, and the results are then either transmitted to a personal computer via a standard RS232 serial interface or directly visualized on an LCD.

Figure 1.4.32 shows a photograph of the completely assembled AS^3. The AS^3 has been successfully used to classify solvents under various environmental conditions, eg, at trade fairs, public exhibitions and in a laboratory environment.

1.4.6 Outlook

The single-chip chemical microsensor system presented here is a first step towards application-specific gas-sensor systems based on CMOS-MEMS technology. The multisensor chip can be seen as a platform, which demonstrates the co-integration of various transduction principles and all necessary electronics on a single CMOS chip. Future work has to be focused on two main aspects:

- Extension of the system platform to other transduction principles such as metal oxide gas sensors and the development of novel signal-conditioning building blocks. The main challenge is not the transducer or the signal-conditioning algorithm itself but rendering it compatible with the CMOS process, post-CMOS micromachining steps, coating technology and packaging technology.
- Identification of key applications that provide the large quantities (in excess of 100000 units per year) needed to justify the high fixed costs of industrial CMOS technology and post-CMOS micromachining. The system still has to

be tailored for each application by optimizing the transducers and their circuitry, minimizing the cost-relevant chip area and selecting the most appropriate packaging approach.

1.4.7 Acknowledgments

The authors acknowledge Dirk Lange, Andreas Koll, Nicole Kerness, Flavio Heer, Michael Morent, Markus Graf, Wan Ho Song, and Donat Scheiwiller for their contributions and the Körber Foundation, Hamburg, Germany for financial support of the project.

1.4.8 References

[1] Grace, R.H., Salomon, P., *mst-news* **5/01**, Nov. (2001), 3–7.
[2] Massart, D.L., Vandeginste, B.G.M., Deming, S.N., Michotte, Y., Kaufman, L., *Chemometrics: a Textbook*, Data Handling in Science and Technology, Vol. 2; Amsterdam: Elsevier, 1988.
[3] Brereton, R.G. (ed.), *Multivariate Pattern Recognition in Chemometrics*, Data Handling in Science and Technology, Vol. 9; Amsterdam: Elsevier, 1992.
[4] Riemannian, A., Schweizer-Berberich, M., Weimar, U., Kraus, G., Pfau, A., Göpel, W., Pattern Recognition and Multi-component Analysis, in: *Sensors Update*, Vol. 2, Baltes, H., Göpel, W., Hesse, J. (eds.); Weinheim: Wiley-VCH, 1996, pp. 119–180.
[5] Snow, A.W., Barger, W.R., Klusty, M., Wohltjen, H., Jarvis, N.I., *Langmuir* **2** (1986) 513–519.
[6] Hagleitner, C., Lange, D., Hierlemann, A., Brand, O., Baltes, H., CMOS Single-chip Gas Detection Systems: Part II, in: *Sensors Update*, Vol. 12, Baltes, H., Hesse J., Korvink, J.G. (eds.);, Weinheim: Wiley-VCH, in press.
[7] Göpel, W., Hesse, J., Zemel, J.N. (eds.), *Sensors: A Comprehensive Survey, Vol. 2/3, Chemical and Biochemical Sensors*; Weinheim: VCH, 1991; Baltes, H., Göpel, W., Hesse, J. (eds.), *Sensors Update*, Vols. 1–9; Weinheim: Wiley-VCH.
[8] Janata, J., *Principles of Chemical Sensors*; New York: Plenum Press, 1989.
[9] Janata, J., Huber, R.J. (eds.), *Solid State Chemical Sensors*; San Diego: Academic Press, 1985.
[10] Eggins, B.R., *Chemical Sensors and Biosensors*; Chichester: Wiley, 2002.
[11] Kress-Rodgers, E. (ed.), *Handbook of Biosensors and Electronic Noses*; Boca Raton, FL: CRC Press, 1997.
[12] Madou, M.J., Morrison, S.R., *Chemical Sensing with Solid State Devices*; Boston: Academic Press, 1989.
[13] Gardner, J.W., *Microsensors*; Chichester: Wiley, 1994.
[14] Grate, J.W., Frye, G.C. in: *Sensors Update*, Vol. 2, Baltes, H., Göpel, W., Hesse, J. (eds.); Weinheim: VCH, 1996, pp. 37–83.

[15] Ballantine, D.S., White, R.M., Martin, S.J., Ricco, A.J., Frye, G.C., Zellers, E.T., Wohltjen, H., *Acoustic Wave Sensors: Theory, Design and Physico-Chemical Applications*; San Diego: Academic Press, 1997.
[16] Gimzewski, J.K., Gerber, C., Meyer, E., Schlittler, E.E., *Chem. Phys. Lett.* **217** (1994) 589–594.
[17] Fritz, J., Baller, M.K., Lang, H.P., Rothuizen, H., Vettiger, P., Meyer, E., Güntherodt, H.J., Gerber, C., Gimzewski, J.K., *Science*; **288** (2000) 316–318.
[18] Lang, H.P., Baller, M.K., Berger, R., Gerber, C., Gimzewski, J.K., Battiston, F., Fornaro, P., Ramseyer, J.P., Meyer, E., Güntherodt, H.J., *Anal. Chim. Acta* **393** (1999) 59–65.
[19] Boisen, A., Thaysen, J., Jesenius, H., Hansen, O., *Ultramicroscopy* **82** (2000) 11–16.
[20] Jesenius, H., Thaysen, J., Rasmussen, A.A., Veje, L.H., Hansen, O., Boisen, A., *Appl. Phys. Lett.* **76** (2000) 2615–2617.
[21] Lange, D., Hagleitner, C., Hierlemann, A., Brand, O., Baltes, H., *Anal. Chem.* **74** (2002) 3084–3095.
[22] Lange, D., Hagleitner, C., Brand, O., Baltes, H., in: *Transducers '99: Digest of Technical Papers*; Sendai, Japan, 1999, pp. 1020–1023.
[23] Hierlemann, A., Lange, D., Hagleitner, C., Kerness, N., Koll, A., Brand, O., Baltes, H., *Sens. Actuators B* **70** (2000) 2–11.
[24] Lee, S.S., White, R.M., *Sens. Actuators A* **52** (1996) 41–45.
[25] Hagleitner, C., Lange, D., Brand, O., Hierlemann, A., Baltes, H., in: *Proc. IEEE ISSCC 2001*; San Francisco, 2001, pp. 246–247.
[26] Meijer, G.C.M., van Herwaarden, A.W., *Thermal Sensors*; Bristol: Institute of Physics Publishing, 1994.
[27] Baker, A.R., Combustible Gas-detecting, Electrically Heatable Element, *Br. Patent* 892 530, 1962.
[28] Manginell, R.P., Smith, J.H., Ricco, A.J., Moreno, D.J., Hughes, R.C., Huber, R.J., Senturia, S.D., in: *Technical Digest of Solid State Sensor and Actuator Workshop*; Hilton Head Island, SC, 1996, pp. 23–27.
[29] Manginell, R.P., Smith, J.H., Ricco, A.J., in: *Proc. 4th Annual Symposium on Smart Structures and Materials SPIE*; 1997, pp. 273–284.
[30] Zanini, M., Visser, J.H., Rimai, L., Soltis, R.E., Kovalchuk, A., Hoffman, D.W., Logothetis, E.M., Bonne, U., Brewer, L., Bynum, O.W., Richard, M.A., *Sens. Actuators A* **48** (1995) 187–192.
[31] Aigner, R., Dietl, M., Katterloher, R., Klee, V., *Sens. Actuators B* **33** (1996) 151–155.
[32] Krebs, P., Grisel, A., *Sens. Actuators B* **13/14** (1993) 155–158.
[33] Gall, M., *Sens. Actuators B* **4** (1991) 533–538.
[34] Gall, M., *Sens. Actuators B* **15–16** (1993) 260–264.
[35] Accorsi, A., Delapierre, G., Vauchier, C., Charlot, D., *Sens. Actuators B* **4** (1991) 539–543.
[36] Van Herwaarden, A.W., Sarro, P.M., *Sens. Actuators* **10** (1986) 321–346.
[37] Sarro, P.M., van Herwaarden, A.W., van der Vlist, W., *Sens. Actuators A* **41/42** (1994) 666–671.
[38] Van Herwaarden, A.W., Sarro, P.M., Gardner, J.W., Bataillard, P., *Sens. Actuators A* **43** (1994) 24–30.
[39] Kerness, N., Koll, A., Schaufelbühl, A., Hagleitner, C., Hierlemann, A., Brand, O., Baltes, H., in: *Proc. IEEE MEMS 2000*; Myazaki, Japan, 2000, pp. 96–101.
[40] Koll, A., Schaufelbühl, A., Brand, O., Baltes, H., Menolfi, C., Huang, H., in: *Proc. IEEE MEMS '99*; Orlando, FL, 1999, pp. 547–551.

[41] Lerchner, J., Seidel, J., Wolf, G., Weber, E., *Sens. Actuators B* **32** (1996) 71–75.
[42] Wolfbeis, O., Boisde, G.E., Gauglitz, G., Optochemical Sensors, in: *Sensors: A Comprehensive Survey*, Vol. 2, Göpel, W., Hesse, J., Zemel, J.N. (eds.); Weinheim: VCH, 1991, pp. 573–646.
[43] Gauglitz, G., Opto-Chemical and Opto-Immuno Sensors, in: *Sensors Update*, Vol. 1, Baltes, H., Göpel, W., Hesse, J. (eds.); Weinheim: VCH, 1996, pp. 1–49.
[44] Brecht, A., Gauglitz, G., Göpel, W., Sensors in Biomolecular Interaction Analysis and Pharmaceutical Drug Screening, in *Sensors Update*, Vol. 3, Baltes, H., Göpel, W., Hesse, J. (eds.); Weinheim: Wiley-VCH, 1998, pp. 573–646.
[45] Zappe, H., Semiconductor Optical Sensors, in: *Sensors Update*, Vol. 5, Baltes, H., Göpel, W., Hesse, J. (eds.); Weinheim: Wiley-VCH, 1999, pp. 1–45.
[46] Spichiger-Keller, U.E., *Chemical Sensors and Biosensors for Medical and Biological Application*; Weinheim: Wiley-VCH, 1998.
[47] Sze, S.M., *Physics of Semiconductor Devices*; New York: Wiley, 1981.
[48] Simpson, M., Sayler, G., Nivens, D., Ripp, S., Paulus, M., Jellison, G., *Trends Biotechnol.* **16** (1998) 332–338.
[49] Simpson, M., Paulus, M., Jellison, G., Sayler, G., Applegate, B., Ripp, S., Nivens, D., in: *Technical Digest of Solid-State Sensor and Actuator Workshop*; Hilton Head, SC, 1998, pp. 354–357.
[50] Simpson, M., Sayler, G., Nivens, D., Ripp, S., Paulus, M., Jellison, G., *Proc. SPIE* **3328** (1998) 202–212.
[51] Yee, G.M., Maluf, N.I., Hing, P.A., Albin, M., Kovacs, G.T., *Sens. Actuators A* **58** (1997) 61–66.
[52] Yee, G.M., Maluf, N.I., Kovacs, G.T., in: *Transducers'99: Digest of Technical Papers*; Sendai, Japan, 1999, pp. 1882–1883.
[53] De Graaf, G., Wolffenbuttel, R.F., *Sens. Actuators A* **67** (1998) 115–119.
[54] Wolffenbuttel, R.F., Silicon Photo Detectors with a Selective Spectral Response, in: *Sensors Update*, Vol. 9, Baltes, H., Göpel, W., Korvink, J.G. (eds.); Weinheim: Wiley-VCH, 2001, pp. 69–101.
[55] Correia, J.H., de Graaf, G., Kong, S.H., Bartek, M., Wolffenbuttel, R.F., *Sens. Actuators A* **82** (2000) 191–197.
[56] Lambrechts, M., Sansen, W., *Biosensors: Microelectrochemical Devices*; Bristol: Institute of Physics Publishing, 1992.
[57] Barsan, N., Weimar, U., *J. Electroceram.* **7** (2001) 143–167.
[58] Simon, I., Barsan, N., Bauer, M., Weimar, U., *Sens. Actuators B* **73** (2001) 1–26.
[59] Geistlinger, H., *Sens. Actuators B* **17** (1993) 47–60.
[60] Heiland, G., Kohl, D., in: *Chemical Sensor Technology*, Vol. 1, Seiyama, T. (ed.); Amsterdam: Elsevier, 1988, pp. 15–38.
[61] Barrettino, D., Graf, M., Zimmermann, M., Hierlemann, A., Baltes, H., Hahn, S., Barsan, N., Weimar, U., in: *Proc. IEEE ISCAS 2002*, Vol. 2; Phoenix, AZ, 2002, pp. 157–160.
[62] Suehle, J.S., Cavicchi, R.E., Gaitan, M., Semancik, S., *IEEE Electron Device Lett.* **14** (1993) 118–120.
[63] Semancik, S., Cavicchi, R.E., *Acc. Chem. Res.* **31** (1998) 279–287.
[64] Heilig, A., Barsan, N., Weimar, U., Schweizer-Berberich, M., Gardner, J.W., Göpel, W., *Sens. Actuators B* **43** (1997) 45–51.
[65] Cavicchi, R.E., Semancik, S., Walton, R.M., Panchapakesan, B., De Voe, D.L., Aquino Class, M., Allen, J.D., Suehle, J.S., *Proc. SPIE* **3857** (1999) 38–49.

[66] Persaud, K.C., Pelosi, P., Sensor Arrays Using Conducting Polymers, in: *Sensors and Sensory Systems for an Electronic Nose*, Gardner, J.W., Bartlett, P.N. (eds.); Dordrecht: Kluwer, 1992, pp. 237–256.
[67] Bartlett, P.N., Gardner, J.W., Odor Sensor for an Electronic Nose, in *Sensors and Sensory Systems for an Electronic Nose*, Gardner, J.W., Bartlett, P.N. (eds.); Dordrecht: Kluwer, 1992, pp. 31–52.
[68] Hatfield, J.V., Neaves, P., Hicks, P.J., Persaud, K., Travers, P., *Sens. Actuators B* **18/19** (1994) 221–228.
[69] Neaves, P., Hatfield, J.V., *Sens. Actuators B* **26/27** (1995) 223–231.
[70] Lonergan, M.C., Severin, E.J., Doleman, B.J., Beaber, S.A., Grubbs, R.H., Lewis, N.S., *Chem. Mater.* **8** (1996) 2298–2312.
[71] Patel, S.V., Jenkins, M.W., Hughes, R.C., Yelton, W.G., Ricco, A.J., *Anal. Chem.*, **72** (2000) 1532–1542.
[72] Cornila, C., Hierlemann, A., Lenggenhager, R., Malcovati, P., Baltes, H., Noetzel, G., Weimar, U., Göpel, W., *Sens. Actuators B* **24/25** (1995) 357–361.
[73] Steiner, F.P., Hierlemann, A., Cornila, C., Noetzel, G., Bächtold, M., Korvink, J.G., Göpel, W., Baltes, H., in: *Transducers '95: Digest of Technical Papers*, Vol. 2; 1995, pp. 814–817.
[74] Sheppard, N.F., Day, D.R., Lee, H.L., Senturia, S.D., *Sens. Actuators* **2** (1982) 263–274.
[75] Senturia, S.D., in: *Transducers '85: Digest of Technical Papers*; 1985, pp. 198–201.
[76] Lundström, I., Shivaraman, S., Svensson, C., Lundkvist, L., *Appl. Phys. Lett.* **26** (1975) 55–57.
[77] Lundström, I., *Sens. Actuators B* **56** (1996) 75–82.
[78] Ekedahl, L.G., Eriksson, M., Lundström, I., *Acc. Chem. Res.* **31** (1998) 249–256.
[79] Rodriguez, J.L., Hughes, R.C., Corbett, W.T., McWhorter, P.J., in: *Int. Electron Devices Meeting 1992: Technical Digest*; San Francisco, 1992, pp. 19.6.1–19.6.4.
[80] Srivastava, A., George, N., Cherukuri, J., *Proc. SPIE* **2642** (1995) 121–129.
[81] *High Accuracy ±1 g to ±5 g Single Axis iMEMS Accelerometer with Analog Input*; Datasheet ADXL 105, Analog Devices, 1999.
[82] Hierold, C., Intelligent CMOS Sensors, in: *Proc. IEEE MEMS 2000*; 2000, pp. 1–6.
[83] Bitko, G., McNeil, A., Frank, R., *Sensors Mag.*, July (2000), pp. 62–67.
[84] Sensonor, Horten, Norway, http://www.sensonor.com.
[85] Intersema Sensoric SA, Bevaix, Switzerland, http://www.intersema.com.
[86] Delphi Corporation, Troy, MI, USA, http://www.delphi.com.
[87] TRW Novasensor, Fremont, California, USA, http://www.novasensor.com.
[88] Robert Bosch GmbH, Reutlingen, Germany, http://www.europractice.bosch.com.
[89] Souloff, B., in: *Advanced Microsystems for Automotive Applications 99*; 1999, pp. 262–270.
[90] Physical Electronics Laboratories, ETH Zürich, Switzerland, http://www.iqe.ethz.ch/pel.
[91] Standard MEMS, Inc., Burlington, MA, USA, http://www.stdmems.com.
[92] Suehle, J.S., Cavicchi, R.E., Gaitan, M., *Electron Device Lett.* **14** (1993) 118–120.
[93] http://mems.sandia.gov.
[94] http://www.dlp.com.
[95] Sensirion AG, Zürich, Switzerland, http://www.sensirion.com.
[96] http://transducers.stanford.edu.
[97] Xie, H., Fedder, G.K., *Sens. Actuators A* **95** (2002) 212–221.
[98] Najafi, K., in: *IEEE Int. Symp. Circuits Systems*; 1987, pp. 233–236.
[99] Senturia, S.D., *Proc. IEEE* **86** (1998) 1611–1626.

[100] Mukherjee, T., Fedder, G.K., *J. VLSI Signal Process.* **21** (1999) 233–249.
[101] Marshall, J.C., Parameswaran, M., Zaghloul, M.E., Gaitan, M., *IEEE Circuits Devices Mag.* **8** (1992) 10–17.
[102] Magic, http://www.research.compaq.com/wrl/projects/magic/magic.htm l.
[103] Ballistic, http://www.eecg.toronto.edu/ ~ gdt/.
[104] Fedder, G.K., Jing, Q., *IEEE Trans. Circuits Syst. II* **46** (1999) 1309–1315.
[105] Sugar, http://www.bsac.eecs.berkeley.edu/ ~ cfm.
[106] Coventor, Cary, NC, USA, http://www.coventor.com.
[107] MEMSCAP, Bernin, France, http://www.memscap.com.
[108] Koll, A., *CMOS Capacitive Chemical Microsystems for Volatile Organic Compounds*; PhD Thesis, No. 13460, ETH Zürich, 1999.
[109] Lange, D., *Cantilever-based Microsystems for Gas Sensing and Atomic Force Microscopy*; PhD Thesis, No. 13984, ETH Zürich, 2000.
[110] Kerness, N., Koll, A., Schaufelbühl, A., Hagleitner, C., Hierlemann, A., Brand, O., Baltes, H., in: *Proc. IEEE MEMS 2000*; 2000, pp. 96–101.
[111] Müller, T., Feichtinger, T., Breitenbach, G., Brandl, M., Brand, O., Baltes, H., in: *Proc. 11th IEEE Workshop on Micro Electro Mechanical Systems*; 1998, pp. 240–245.
[112] Müller, T., *An Industrial CMOS Process Family for Integrated Silicon Sensors*; PhD Thesis, No. 13463, ETH Zürich, 1999.
[113] Brand, O., Baltes, H., in: *Proc. Micro System Technologies 2001*; 2001, pp. 37–42.
[114] Brand, O., Baltes, H., Waelti, M., in: *IEEE European Symp. on Reliability of Electron Devices, Failure Physics and Analysis (ESREF) 2000 and Microelectronics Reliability*, Vol. 40; 2000, pp. 1255–1262.
[115] International Technology Roadmap for Semiconductors, http://public.itrs.net.
[116] Burns, M., Roberts, G.W., *An Introduction to Mixed-signal IC Test and Measurement*; Oxford: Oxford University Press, 2001.
[117] Bushnell, M.L., Agrawal, V.D., *Essentials of Electronic Testing for Digital, Memory and Mixed-signal VLSI Circuits*; Dordrecht: Kluwer, 2000.
[118] Crouch, A.L., *Design-for-Test for Digital ICs and Embedded Core Systems*; Englewood Cliffs, NJ: Prentice-Hall, 1999.
[119] Philips Semiconductors, http://www.semiconductors.philips.com/buses/i2c/facts.
[120] Hilbiber, D.F., in: *Digest of Technical Papers ISSCC 1964*; 1964, pp. 32–33.
[121] Widlar, R.J., *IEEE J. Solid State Circuits* **6**, Feb. (1971), 2–7.
[122] Blauschild, R.A., Tucci, P., Muller, R.S., Meyer, R.G., *Digest of Technical Papers ISSCC 1978*; 1978, pp. 50–51.
[123] Williams, J., *EDN*, April (1991) 180–181.
[124] Meijer, G.C.M., in: *Analog Circuit Design: Low-noise, Low-power, Low-voltage; Mixed-mode Design with CAD Tools; Voltage, Current and Time References*, Huijsing, J.H., van de Plasche, R.J., Sansen, W. (eds.); Dordrecht: Kluwer, 1996, pp. 243–268.
[125] Bakker, A., *High Accuracy CMOS Smart Temperature Sensors*; PhD Thesis, TU Delft, 2000.
[126] Song, B.S., Gray, P.R., *IEEE J. Solid State Circuits* **18** (1983) 634–643.
[127] Tuthill, M., *IEEE J. Solid State Circuits* **33** (1998) 1117–1122.
[128] Tsividis, Y.P., *IEEE J. Solid State Circuits* **15** (1980) 1076–1083.
[129] Freire, R.C.S., Daher, S., Deep, G.S., *IEEE Trans. Instrum. Meas.* **43/2**, April (1994) 127–132.
[130] Wang, G., Meijer, G.C.M., *Sens. Actuators A* **87** (2000) 81–89.

[131] Hagleitner, C., *CMOS Single-chip Gas Detection System Comprising Capacitive, Calorimetric and Mass-sensitive Microsensors*; PhD Thesis, No. 14511, ETH Zürich, 2002.
[132] Bludau, W., Onton, A., *J. Appl. Phys.* **45** (1974) 1846–1848.
[133] Hartung, J., Hansson, L.Å. Weber, J., in: *Proc. 20th Int. Conf. on the Physics of Semiconductors*; 1990, pp. 1875–1878.
[134] Meijer, G.C.M., *Integrated Circuits and Components for Bandgap References and Temperature Transducers*; PhD Thesis, TU Delft, 1982.
[135] Gilbert, B., in: *Analog Circuit Design: Low-noise, Low-power, Low-voltage; Mixed-mode Design with CAD Tools; Voltage, Current and Time References*, Huijsing, J.H., van de Plasche, R.J., Sansen, W. (eds.); Dordrecht: Kluwer, 1996, pp. 269–352.
[136] Leung, K.N., Mok, P.K.T., in: *Proc. 27th European Solid-State Circuits Conference (ESSCIRC)*; September 2001, pp. 88–91.
[137] Kawahito, S., Koll, A., Hagleitner, C., Baltes, H., Tadokoro, Y., *Trans. IEE Jpn.* **119-E**, March (1999) 138–142.
[138] Chu, S., Burrus, C.S., *IEEE Trans.Circuits Syst.* **31** (1984) 913–924.
[139] Brodersen, R.W., Moscowitz, H.S., *VLSI Signal Processing III*; New York: IEEE Press, 1988, pp. 523–534.
[140] Candy, J.C., *IEEE Trans. Commun.* **34** (1986) 72–76.
[141] Meijer, G.C.M., Schmale, P.C., van Zalinge, K., *IEEE J. Solid-State Circuits* **17** (1982) 1139–1143.
[142] Creemer, J.F., French, P.J., in: *Proc. Transducers'01*; 2001, pp. 256–259.

List of Symbols and Abbreviations

Symbol	Designation
A	base-emitter junction area
$f_{\Sigma\Delta}$	sampling frequency of $\Sigma\Delta$ modulator
$I_E[T]$	emitter current
I_R	reference current
$I_s[T]$	current
k	Boltzmann constant
K	constant
L	constant
n	constant
N	number of unit transistors
N_B	Gummel number
$n_i[T]$	intrinsic carrier concentration
q	electron charge
T	absolute temperature
V_{BE}	base-emitter voltage
V_E	Early voltage
$V_G[T]$	silicon bandgap voltage

Symbol	Designation
V_{G0R}	extrapolated bandgap voltage
V_{in}	input voltage
V_{ref}	reference voltage
$\beta[T]$	current amplification factor
η	curvature parameter
$\bar{\mu}[T]$	effective mobility of minor carriers in base

Abbreviation	Explanation
AS^3	application-specific sensor system
BBIC	bioluminescent bioreporter integrated circuit
CCD	charge-coupled device
CMOS	complementary metal oxide semiconductor
DC	direct current
DRC	design rule check
EDA	electronic design automation
EDP	ethylenediamine pyrocatechol
FEM	finite element method
FIR	finite-impulse response
HDL	hardware description language
IC	integrated circuit
IT	information technology
KOH	potassium hydroxide
LCD	liquid crystal display
LED	light-emitting diode
LEL	low explosion limit
LVS	layout versus schematic
MEMS	microelectromechanical system
MOS	metal oxide semiconductor
MPW	multi-project wafer
PCB	printed circuit board
PTAT	proportional to absolute temperature
RIE	reactive ion etching
VHDL	VHSIC hardware desorption language
VHSIC	very high-speed integrated circuit
VOC	volatile organic compound

PART 2

Sensor Applications

2.1 Earthquake Sensor

T. YANADA, H. FURUKAWA, S. ICHIDA and K. TAKUBO,
Yamatake Corporation, Kanagawa, Japan
Y. SHIMIZU and K. KOGANEMARU, Tokyo Gas Co., Ltd., Tokyo, Japan
T. SUZUKI, Toyo University, Saitama, Japan

Abstract

A new concept earthquake sensor was developed utilizing spectral intensity (*SI*) as a measurement value. The acceleration sensing part and signal conditioning part are all integrated in one body as a transducer. A real-time *SI* calculation algorithm was newly developed and proven by actual simulated earthquake waveforms with a three-axis exciter bench. A liquefaction judgment algorithm was also developed based on acceleration signals and integrated into the *SI* sensor. Thus, the first all-in-one *SI* sensor with liquefaction judgment was realized. With this sensor, a new earthquake disaster prevention system for gas supply, Supreme, was developed and introduced into operation around the area of Tokyo Bay, Japan.

Keywords: earthquake; spectral intensity; accelerometer; liquefaction; real-time algorithm

Contents

2.1.1 Introduction

An intense earthquake occurred in Kobe, Japan, in 1995 and caused large losses of buildings that collapsed or burned, and many human lives. Since then, it has been desired to establish quickly the type and extent of damage on the occurrence of an earthquake and install apparatus or instrumentation that can automatically stop the operation of facilities which would cause a secondary disaster in the case of collapse of the building or other severe damage.

Tokyo Gas Co., Ltd. has given attention to the *SI* (spectral intensity) value as an earthquake damage index for earthquake damage and ascertained its effectiveness as an index for earthquake damage. The *SI* value is an index that is strongly correlated to the building and other structural damage calculated from the ground acceleration signals due to earthquake ground motion.

Tokyo Gas Co., Ltd. has installed automatic gas shut-off valves on the gas supply pipelines at 3700 places in and around Tokyo since 1986 in order to minimize earthquake disasters and has built a system to prevent secondary disasters due to the breakage of gas supply pipes by shutting off these governors on the occurrence of an earthquake by means of an earthquake sensor (henceforth called an *SI* sensor). This system uses two conventional types of *SI* sensors. One is a mechanical earthquake switch using a pendulum mechanism and is used for shutting off all the governors. The other is a servo-type acceleration sensor capable of delivering the *SI* value and is installed at about 330 locations to keep track of damage conditions. However, both types were large in size. The mechanical type could deliver a shutdown signal only, so that no detailed damage conditions could be tracked. The servo-type acceleration sensor was expensive and the output *SI* value was low in accuracy because of the low calculation power of the CPU (central processing unit) at that time [1].

Yamatake Corporation and Tokyo Gas started the development of a new earthquake sensor of small size, high function, high reliability and low cost in 1995 with the intention of building a reliable and sophisticated disaster prevention system in addition to offering society a measuring instrument that can control individual facilities on the occasion of a disaster. This development was supported by a small accelerometer pickup implemented owing to the recent progress in MEMS (microelectromechanical systems) technology and the improved performance of the CPU and succeeded in implementing a first product in 1997, which has already achieved many market and field results and gained a favorable reputation.

This chapter outlines the development concept of a new small and inexpensive *SI* sensor in Section 2 and describes the major functional logic in the development to implement the basic concept in Section 3. Section 4 describes actual applications of this developed product.

2.1.2 *SI* Sensor Concepts

2.1.2.1 Background of Development Objective and Product Concept

As described in Section 1, two different types of seismographs have been used in conventional earthquake sensors, one for shutdown and the other for measurement. The earthquake sensor we aimed at in this development was required to have both of these two functions, with small size and low cost.

As a conventional seismograph for measurement of earthquake disaster levels, an earthquake-measuring instrument using a servo-type accelerometer has generally been used. Such a seismograph consists of two functional sections: the detection section accommodating an accelerometer, and the conversion and control section. This is shown in Figure 2.1-1.

The detection section accommodates three servo accelerometers, which detect the acceleration along the *x*- and *y*-axes (horizontal plane) and *z*-axis (vertical axis) and deliver analog signals in proportion to the acceleration. The conversion and control section converts the analog acceleration signals sent from the detection section into digital form by an AD converter. The CPU in it does the recording of acceleration waveforms, calculation and indication of the desired earthquake index, and control of output. In shape, the detection section accommodating an accelerometer has a size of about ϕ 250×150 mm, and the conversion and control section about 250×500×150 mm. The conventional seismograph having such a structure not only is expensive but also involves a large amount of installation work. During maintenance, the detection section and conversion and

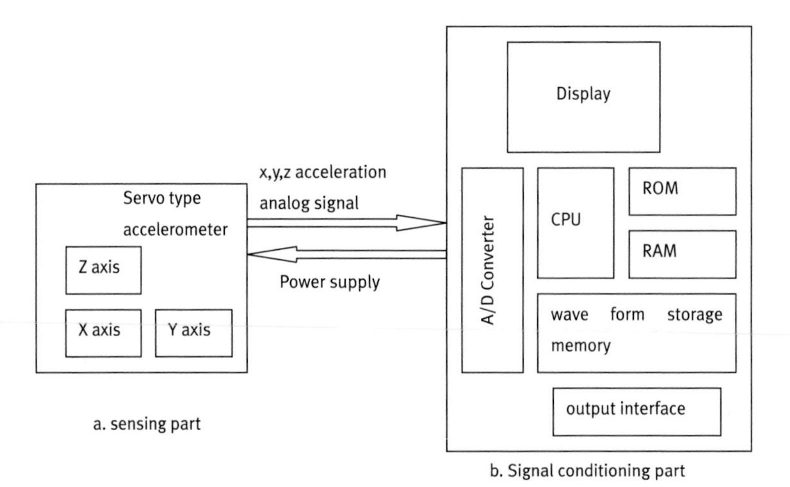

Figure 2.1-1. Existing seismograph consists of sensing part (**a**) and signal conditioning part (**b**). Several kind of signal processing, display and output are done by part (**b**) by receiving analog data from part (**a**).

control section, which are installed at separate locations, must be operated simultaneously. Hence there are several hindrances to using its functions.

This development aimed at implementing an earthquake sensor that overcomes these drawbacks of the conventional type and is easy to handle and install. In such an earthquake sensor, only the indication function will be separated from among the existing functions, and the sensor as a single unit will execute the functions of seismic wave detection and recording and calculation of earthquake index and their output. We also aimed at implementing an earthquake sensor of the all-in-one type in a size about half of the conventional detection section. The development started with deciding a target cost and size and determining the accelerometer pickup, hardware and structure that would satisfy these targets, and employed the procedure of developing an algorithm for the necessary functions working under these restrictions. This section describes the internal structure of the accelerometer pickup, hardware and sensor.

2.1.2.2 Accelerometer

The servo accelerometer was conventionally dominant for the pickup to observe an earthquake of such a magnitude as to incur damage. However, the servo accelerometer is expensive, as much as US$ 1000–5000 per axis, and the size large, as much as about ϕ 50×80 mm in the form of a single unit of the accelerometer. As other options for the acceleration sensor, one can mention the servo type, strain gauge (piezoresistance) type, capacitive type and piezoelectric type according to the detection method.

The performance of accelerometer pickups was compared in terms of their measurable acceleration range and frequency and index of measurable acceleration resolution and based on their detection methods. This comparison is shown in Figures 2.1-2 and 2.1-3. If the detection methods are viewed in terms of measurable acceleration range and measurable frequency, the servo type, strain gage (piezoresistance) type and capacitive type cover low-frequency acceleration signals from static acceleration to 100 Hz as their measurable range while the piezoelectric type covers a measurable range of high frequencies higher than 10 Hz.

The seismic waves to be measured are said to be 50 Hz or <30 Hz and are categorized as low frequencies as vibration sensors. Therefore, we excluded the piezoelectric type from our study because no static acceleration signals could be obtained from its measuring principle, and decided to choose a device appropriate for the purpose from among the strain gage type or capacitive type accelerometer pickups.

As for the piezoresistance and capacitive type accelerometer pickups, MEMS technology has recently been applied to these detection methods and the development of such accelerometer pickups has accelerated rapidly and the scope of application has widened. Their application to automobiles started with their use for activation of air bags and the range of their application has widened to automobile navigation and posture and driving control. The initial MEMS acceleration

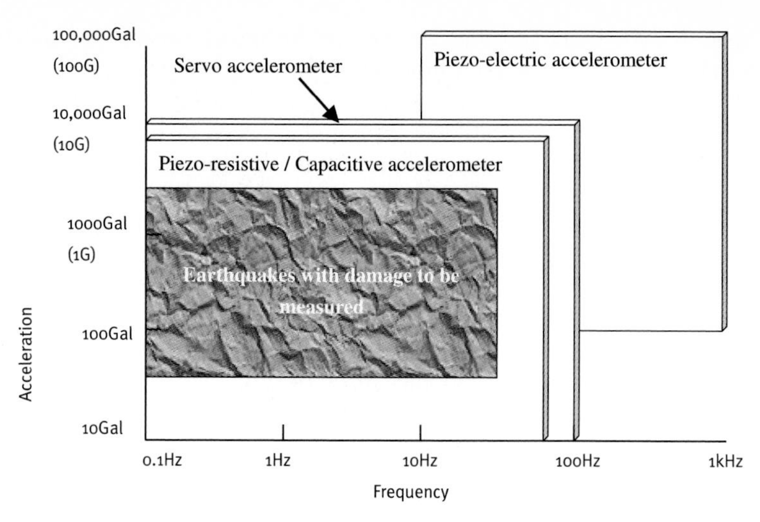

Figure 2.1-2. General vibration sensors and their acceleration and frequency coverage. Earthquakes with damage are in the range of MEMS devices.

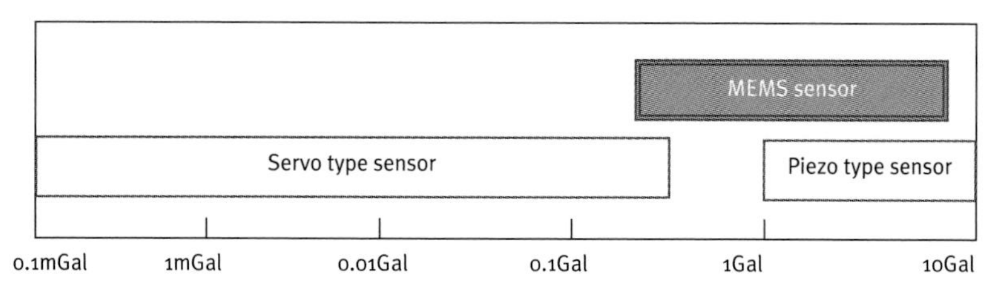

Figure 2.1-3. Typical sensor resolution. Minimum 1 Gal resolution is sufficient for earthquake with damage measurement. MEMS sensors have sufficient resolution.

sensor for air bags had specifications for measurement of high acceleration. As the scope of application has widened to automobile navigation and control, however, inexpensive MEMS accelerometer pickups having specifications for use in the measurement of low acceleration came to the market and they could be used for measurement of earthquakes [2–4].

Although the MEMS acceleration sensor is inferior to the servo accelerometer in resolution, which is about 1 Gal (Gal=cm/s^2), the measurable acceleration range and measurable frequency band cover the range of seismic waves to be measured. If the seismic waves to be measured are limited to reasonable-sized earthquakes, their acceleration is 10 Gal or higher. Therefore, the MEMS accelerometer pickup has sufficient capabilities for measurement of seismic waves that will cause a disaster.

Since the development was aimed at implementing a small, light, and inexpensive earthquake sensor of the all-in-one type, the MEMS accelerometer pickup was judged the most suitable device as an accelerometer pickup. The MEMS accelerometer pickup employed is a micromachined three-axial capacitive silicon

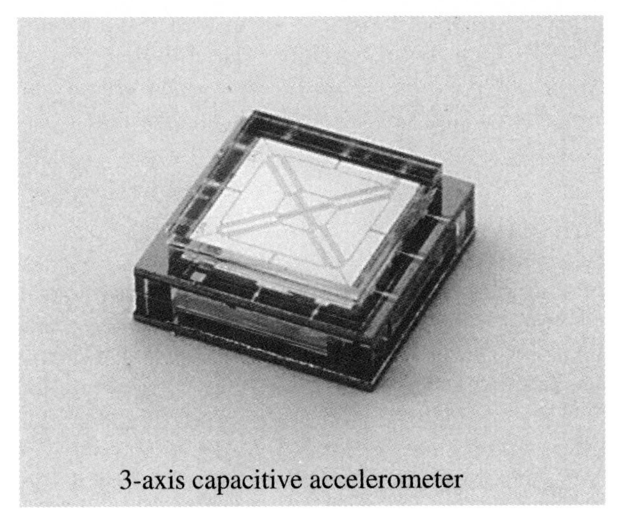

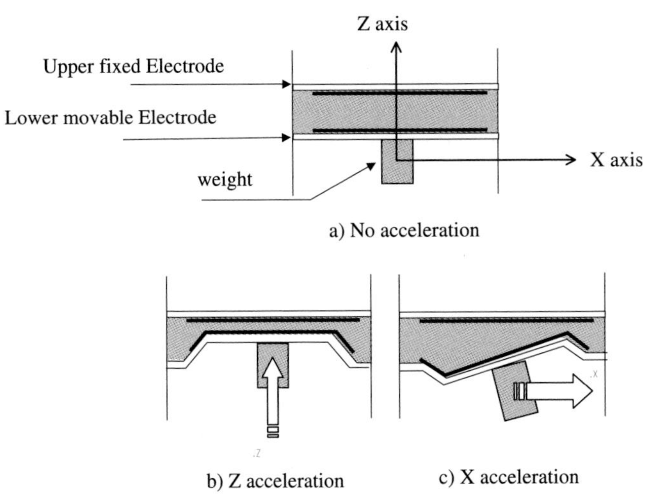

Figure 2.1-4. MEMS capacitive accelerometer.

accelerometer pickup. A photograph and the measuring principle of the sensor chip are shown in Figure 2.1-4. This chip has a weight placed at the center and the detection electrodes are arranged in the direction of each detection axis. Therefore, acceleration detection in three axes is possible in one chip with a die size of 6×6×2 mm [5].

On some capacitive acceleration sensors, an additional electrode is placed to produce an electrostatic attraction. Such a device has the advantage of bringing about a self-diagnosis function to detect a failure of the device by adding an electronic circuit to impose a voltage on that electrode, without adding fabrication processes and in the same structure. The chip that we employed has this additional function and is employed in the developed product as a super small and light accelerometer pickup with a self-diagnosis function.

2.1.2.3 Hardware Configuration

The hardware and internal configuration of the new earthquake sensor are shown in Figures 2.1-5 and 2.1-6.

Three-axial acceleration signals in the *x*-, *y*- and *z*-directions are output from the above-stated MEMS accelerometer pickup. Signals are converted by the analog-to-digital converter into digital information together with the information from the built-in temperature sensor in the pickup and are processed in the CPU. Based on the signals taken in, the CPU makes a compensation on the output in each axis every 10 ms according to the temperature error compensation information obtained in advance and recognizes the compensated signal as an acceleration value. The CPU calculates the earthquake index based on the acceleration data obtained and delivers it to the outside through an output circuit. If a seismic wave is measured at the same time, it is possible to record it in the waveform-recording memory provided inside. It is also possible to change the set value for the output signal and read the waveform data recorded inside via the mounted communication port and by using a dedicated personal computer tool. In this configuration, a drive voltage for the impression of an electrostatic attraction is applied to the accelerometer pickup from the CPU through a digital-to-analog

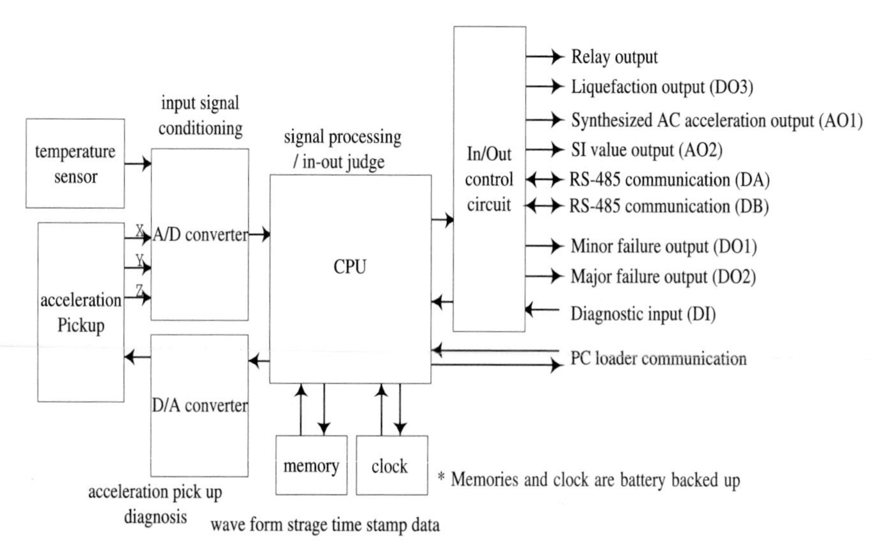

Figure 2.1-5. New type *SI* sensor integrates all functions of sensing part except display unit.

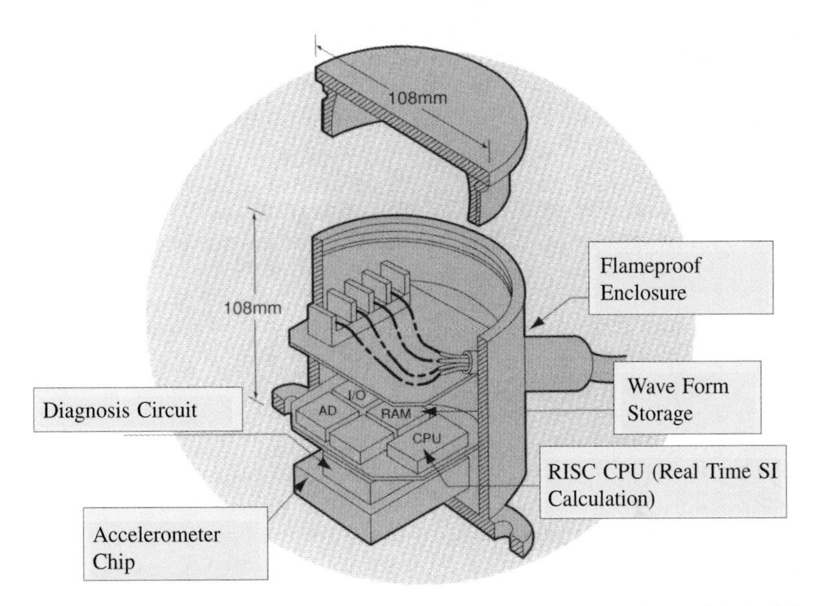

Figure 2.1-6. New *SI* sensor equipped with hardware shown in Figure 2.1-2. All components are placed in a flameproof enclosure which is about half the size of existing seismographs.

converter for the self-diagnosis voltage of the self-diagnosis function. The aim was to mount these electronic circuits on a set of two printed wiring boards and fit it in a flameproof enclosure of about ϕ 100×100 mm.

Thus far, the basic internal configuration and the flow of signal processing have been described. For the development of a sensor that will execute a control function on the occurrence of an earthquake together with waveform recording and other earthquake measuring functions, two restrictions were imposed on the implementation of the functions.

First, if the earthquake exceeds the dangerous level, it is desired that the sensor detect and output this as soon as possible. Therefore, all functional calculations should be executed at every acceleration data sampling at 10 ms intervals and the operation should refresh the output each time.

Second, reliability must be secured as an operating *SI* sensor. For this purpose, all the programs and data used for operations involved in measurement and calculation should be executed only in the built-in ROM and built-in RAM of a stored-program CPU and no external memory device should be used.

To satisfy these two restrictions, the CPU to be used was chosen first. The CPU that we employed was a sophisticated new-generation CPU employed in home game machines, etc. By this means, we succeeded in securing the maximum throughput within the limit of allowable cost. To maximize the throughput of the employed CPU, it is desirable, of course, to operate the CPU at the maximum frequency. However, this sensor has an explosion-proof housing and there is the requirement on it that it be used in an explosive atmosphere. This excludes the use of forced cooling. Therefore, the clock of the CPU is practically limited to about 10 MHz. To complete the measurement and calculation within 10 ms under these conditions, we developed a new algorithm, which would allow fast real-time operation of all the stored logical functions and minimize the consumption of memory resources. As a result, we succeeded in reducing the processing load on the CPU and executing all the measuring and controlling operations with the CPU's built-in memory (ROM and RAM) alone. These and the newly developed algorithm are described in the next main section.

2.1.2.4 *SI* Sensor Specifications

The specifications for the developed new concept *SI* sensor are shown in Tables 2.1-1 and 2.1-2.

In performance, the acceleration measuring section has a measurable range of ±2000 Gal, a sensitivity error of ±2% (±980 Gal, 0–50 °C), a random noise of 5 Gal, and a frequency band (–3 dB) of 50 Hz in the *x*- and *y*-axes and 30 Hz in the *z*-axis.

As for output, it has one controlling relay output, liquefaction detection output, major failure output, and minor failure output as controlling outputs and two industrial standard 4–20 mA outputs (two are freely selectable from among *SI* value, resultant acceleration value, and the measured seismic intensity equivalent) as observation outputs.

As for the earthquake index calculation function, it executes calculations of resultant acceleration, *SI* value, and measured seismic intensity equivalent [6]. In addition, it has a noise-protecting function to distinguish the seismic waves from the measured values that are susceptible to the influence of electromagnetic waves and other noise.

Table 2.1-1. Acceleration measurement part

Item	Specification
Explosion-protection	Exd II BT4X (flameproof, Japan)
Standard acceleration range	±2000 Gal
Acceleration measurement range	±2200 Gal
Acceleration measurement resolution	1 Gal (static acceleration)
Accuracy	±2%
Linearity	±2%
Other axis sensitivity	±3%
Noise	5 Gal (acceleration filter 100 Hz)
Acceleration sampling	10 ms

Table 2.1-2. Functions

Item	Details
Calculation function	Synthesized acceleration calculation *SI* value calculation Seismic intensity equivalent value calculation Liquefaction judgment Noise protect function
Output function	Two 4–20 mA analog output (select from synthesized acceleration/*SI* value/seismic intensity equivalent value) One relay output One liquefaction judgment output Two failure output (minor/major)
Storage function	One waveform data: 10 ms sampling, 120 s data Storage resolution: 1/8 Gal Storage waveform data: 10
Other function	Flameproof enclosure Self-diagnosis function Specialized display unit available (optional)

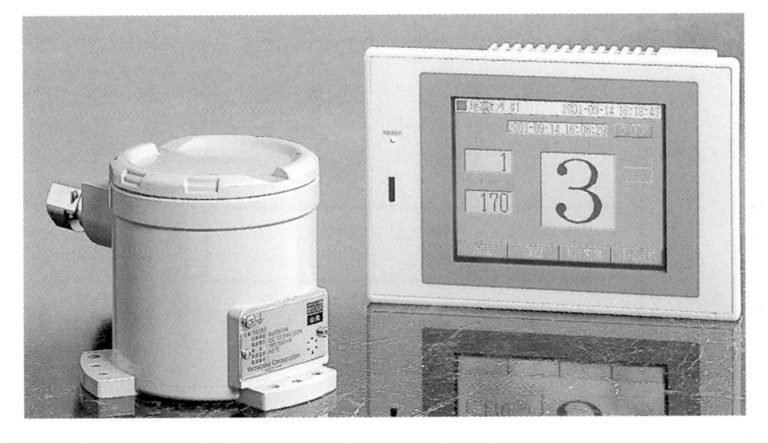

Figure 2.1-7. Photograph of *SI* sensor and optional display unit.

As for the waveform recording function, it can internally store 10 recorded waveforms for 120 s per wave. These functions are accommodated in a flameproof enclosure of ϕ 108×108 mm.

From the beginning of development, the indication function was excluded from the functions to be implemented. However, a monitoring function can be easily realized by connecting this earthquake sensor to an optional display with touch panel via the RS-485 communication port in the sensor (Figure 2.1-7).

2.1.3 *SI* Sensor Algorithms

Several algorithms were developed to attain the objectives described in the previous section: calculation of *SI* value (Section 2.1.3.1), judgment of liquefaction (Section 2.1.3.2) and noise protection (Section 2.1.3.3).

2.1.3.1 Calculation of *SI* Value

The developed earthquake sensor uses the *SI* value, which is one of the estimates of damage due to an earthquake. This section details the features and basic logic of the *SI* value and its application to the actual instrument.

2.1.3.1.1 Relationship of SI Value with Damage

The *SI* value is one of the indices to represent the destructive power of an earthquake against buildings and other structures and was proposed by Housner [7]. The damage to a structure with respect to the intensity of an earthquake is known to be correlated not with the ground acceleration but with the *SI* value, as indicated by Figure 2.1-8.

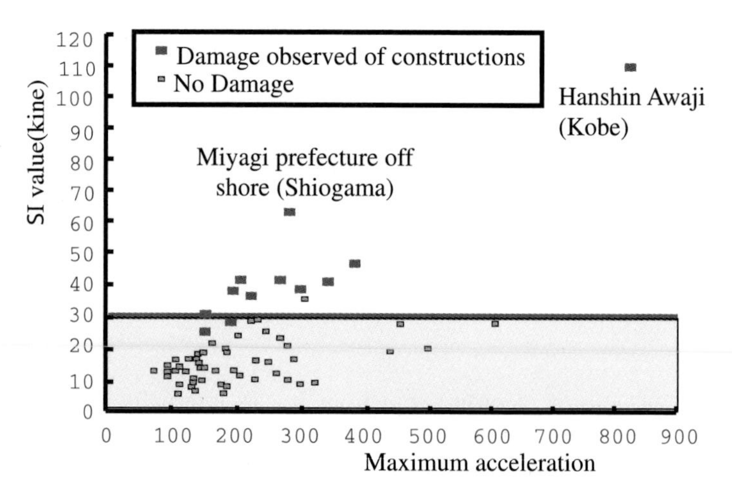

Figure 2.1-8. Construction damage level plotted as correlation between *SI* value and maximum acceleration value of earthquakes. The correlation between construction damage and maximum acceleration is small, but that with *SI* value is strong. The threshold of occurrence of construction damage exists around the 30 kine level.

2.1.3.1.2 *Basic Algorithm for SI Value Calculation [8]*

This section explains the basic theory of the *SI* value, ie, (1) numerical modeling of a structure, (2) maximum velocity response of the structure, and (3) calculation of *SI* value, and finally explains the result of an experiment using earthquake motion-reproducing equipment.

(1) Numerical Modeling of a Structure
In order to derive the damage to a structure by calculation, the structure is treated in the form of a simple physical model.

In the *SI* calculation, the structure is considered to consist of the three elements of a mass, damper and spring, as shown in Figure 2.1-9, and the motion is restricted to a certain direction only. Therefore, this mathematical model is represented by the equation of motion for a damping system of one material point with one degree of freedom [Equation (2.1-1)], where $y(t)$ is the input acceleration [Gal], and the behavior of the structure to the input is represented by $a(t)$, $v(t)$ and $d(t)$, ie, acceleration response [Gal], velocity response [kine], and displacement response [cm], respectively (Gal=cm/s^2, kine=cm/s):

$$ma(t) + cv(t) + kd(t) = -my(t) \qquad (2.1\text{-}1)$$

where m=mass of weight, c=damping constant of damper, and k=spring constant.

Equation (2.1-2) can be derived by rearranging Equation (2.1-1), and this is used for calculation of the *SI* value as the mathematical model for the structure. This model can be replaced by a pendulum model having a damping constant h and a natural vibration period T as shown in Figure 2.1-10.

$$\begin{aligned} &a(t) + 2h\omega v(t) + \omega^2 d(t) = -y(t) \\ &\omega = 2\pi/T \end{aligned} \qquad (2.1\text{-}2)$$

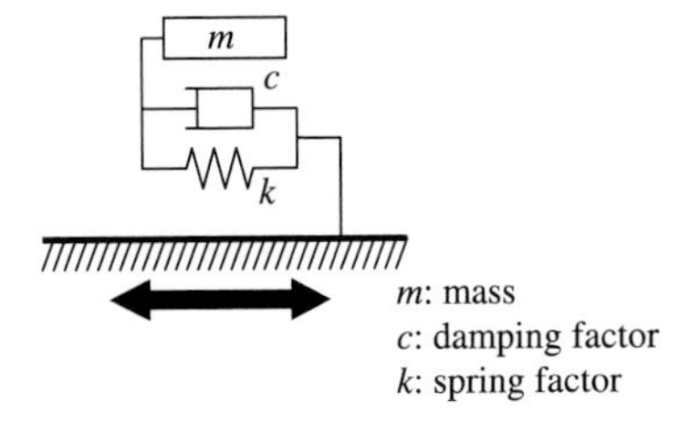

Figure 2.1-9. Physical model 1. A simplified physical model which represents a construction is used for *SI* value calculation. The physical model consists of three factors, mass, damper and spring. It is assumed that movement is one-directional.

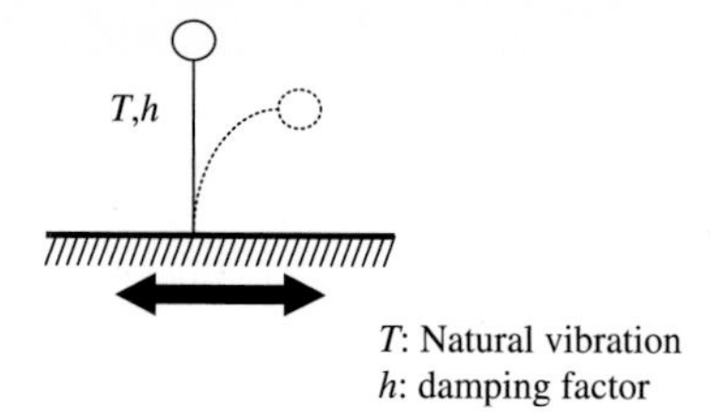

Figure 2.1-10. Physical model 2. This is a conversion of physical model 1 (Figure 2.1-9) to natural vibration T and damping factor h.

(2) Maximum Velocity Response Sv of Structure

By solving Equation (2.1-2), Equation (2.1-3) is derived. This is used to find the velocity response $v(t)$ of the structure to the input acceleration $y(t)$ due to an earthquake. The maximum Sv of $v(t)$ can also be derived [from Equation (2.1-4)].

$$v(t) = \int_0^t y(\tau)\mathrm{e}^{-h\omega(t-\tau)}\left[\cos\omega_\mathrm{d}(t-\tau) - \frac{h}{\sqrt{1-h^2}}\sin\omega_\mathrm{d}(t-\tau)\right]\mathrm{d}\tau \tag{2.1-3}$$

$$\omega = \frac{2\pi}{T}, \quad \omega_\mathrm{d} = \omega\sqrt{1-h^2}$$

$$Sv = |v(\tau)|_{\max} \quad (\tau = 0 \sim t) \tag{2.1-4}$$

A structure has a natural vibration period T and its response $v(t)$ to an input acceleration represents the energy that the earthquake imparts to the structure. Therefore, the maximum (Sv value) of velocity response of the structure represents the maximum energy that the earthquake imparts to the structure. This maximum energy is considered the destructive power to the structure (see Figure 2.1-11).

(3) Calculation of SI Value

A structure has different natural vibration periods, but the principal period T lies between 0.1 and 2.5 s. The Sv in this range can be represented by a spectrogram of maximum velocity response as shown in Figure 2.1-12.

This Sv integrated with respect to T from 0.1 to 2.5 s was defined by Housner as an index to represent the intensity of earthquake motion. Further, this index averaged over the interval T=0.1–2.5 s is the SI value [Equation (2.1-5)].

$$SI = \frac{1}{2.4}\int_{0.1}^{2.5} Sv(T,h)\mathrm{d}T[\mathrm{kine}] \quad (h = 0.2) \tag{2.1-5}$$

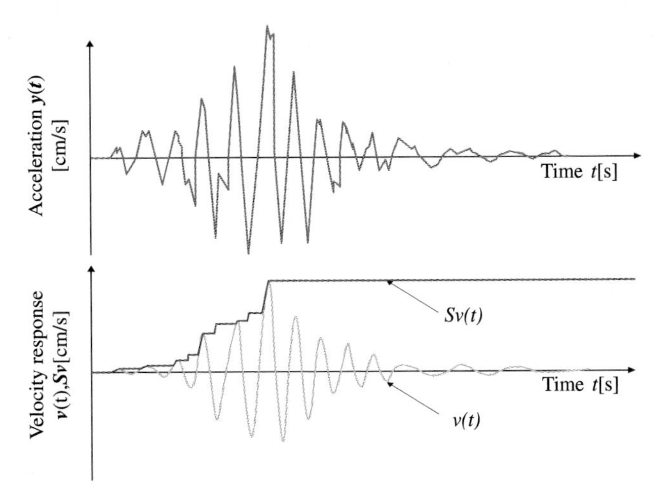

Figure 2.1-11. Acceleration $y(t)$, velocity response $v(t)$, and maximum velocity response $Sv(t)$. Velocity response $v(t)$ and its maximum value Sv can be calculated by applying acceleration to a construction which has natural vibration T.

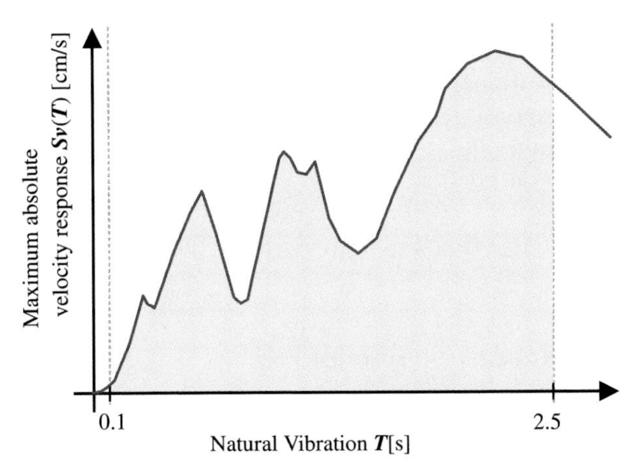

Figure 2.1-12. Sv spectrum. Constructions have several natural vibration T (s) and their major values are between 0.1 and 2.5 s. A spectrum such as that shown here is obtained by obtaining the maximum velocity response $Sv(T)$ of these ranges. The shape of the spectrum varies depending on input acceleration, time, and frequency.

The damping constant h must be chosen according to the structure for which the damage is to be estimated. In this equation, $h=0.2$ is used. This number is set relatively large because the damping factor of the first stage of damage to constructions is adopted.

2.1.3.1.3 Numerical Solution of Velocity Response $v(t)$

If the velocity response $v(t)$ is to be implemented in the earthquake sensor, a numerical solution is employed which is derived by consecutive calculation based on the input signals at sampling intervals of 10 ms. There are various ways to find the numerical solution, and we employed the direct integration method from the point of view of calculation speed of the CPU and memory consumption. The numerical integration equations [Equations (2.1-6) and (2.1-7)] are derived from Equation (2.1-3) by approximation. $v(t)$ can be found by substituting the input acceleration $y(t)$, which varies with time, into these two equations at intervals of Δt.

$$\mathrm{d}(t) = A_{11}(T)\mathrm{d}(t-\Delta t) + A_{12}(T)v(t-\Delta t) + B_{11}(T)y(t-\Delta t) + B_{12}(T)y(t) \tag{2.1-6}$$

$$v(t) = A_{21}(T)\mathrm{d}(t-\Delta t) + A_{22}(T)v(t-\Delta t) + B_{21}(T)y(t-\Delta t) + B_{22}(T)y(t) \tag{2.1-7}$$

where
$y(t)$ = input acceleration [Gal] = [cm/s^2];
$v(t)$ = velocity response [kine] = [cm/s];
$\mathrm{d}(t)$ = displacement response [cm];
t = time of calculation [s];
Δt = calculation interval [s].

The coefficients A_{11}, A_{12}, B_{11}, B_{12}, A_{21}, A_{22}, B_{21}, and B_{22} are defined by Equation (2.1-8):

$$A_{11}(T) = \mathrm{e}^{-h\omega\Delta t}\left(\cos\omega_\mathrm{d}\Delta t + \frac{h\omega}{\omega_\mathrm{d}}\sin\omega_\mathrm{d}\Delta t\right)$$

$$A_{12}(T) = \mathrm{e}^{-h\omega\Delta t}\frac{1}{\omega_\mathrm{d}}\sin\omega_\mathrm{d}\Delta t$$

$$A_{21}(T) = -\mathrm{e}^{-h\omega\Delta t}\frac{\omega^2}{\omega_\mathrm{d}}\sin\omega_\mathrm{d}\Delta t$$

$$A_{22}(T) = \mathrm{e}^{-h\omega\Delta t}\left(\cos\omega_\mathrm{d}\Delta t - \frac{h\omega}{\omega_\mathrm{d}}\sin\omega_\mathrm{d}\Delta t\right)$$

$$B_{11}(T) = e^{-h\omega\Delta t}\left[\left(\frac{1}{\omega^2}+\frac{2h}{\omega^3\Delta t}\right)\cos\omega_d\Delta t-\left(\frac{h}{\omega\omega_d}+\frac{1-2h^2}{\omega^2\omega_d\Delta t}\right)\sin\omega_d\Delta t\right]-\frac{2h}{\omega^3\Delta t}$$

$$B_{12}(T) = e^{-h\omega\Delta t}\left[-\frac{2h}{\omega^3\Delta t}\cos\omega_d\Delta t+\frac{1-2h^2}{\omega^2\omega_d\Delta t}\sin\omega_d\Delta t\right]-\frac{1}{\omega^2}+\frac{2h}{\omega^3\Delta t}$$

$$B_{21}(T) = e^{-h\omega\Delta t}\left[-\frac{1}{\omega^2\Delta t}\cos\omega_d\Delta t+\left(\frac{h}{\omega\omega_d\Delta t}+\frac{1}{\omega_d}\right)\sin\omega_d\Delta t\right]+\frac{1}{\omega^2\Delta t}$$

$$B_{22}(T) = e^{-h\omega\Delta t}\left[\frac{1}{\omega^2\Delta t}\cos\omega_d\Delta t+\frac{h}{\omega\omega_d\Delta t}\sin\omega_d\Delta t\right]-\frac{1}{\omega^2\Delta t} \quad (2.1\text{-}8)$$

2.1.3.1.4 Real-time SI Calculation Algorithm

(1) Problems in SI Calculation Triggered by Calculation Start and Stop Signals (Figure 2.1-13)

As described under the basic principle of *SI* value calculation, the *SI* value is calculated using the maximum *Sv* for each natural vibration $v(t)$ from the beginning to the end of an earthquake. Therefore, if the calculation equation in its original form is to be used in the *SI* sensor, the beginning and the end of the earthquake must be identified.

- With an existing earthquake sensor, *SI* calculation is inactivated while there is no earthquake.
- When the acceleration exceeds a threshold, it is judged to be the occurrence of an earthquake and *SI* calculation is started.

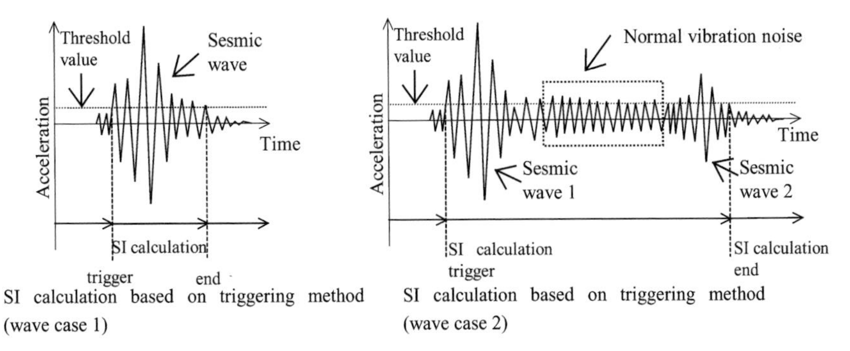

Figure 2.1-13. *SI* calculation based on triggering method. *SI* calculation with a trigger starts when the acceleration exceeds a threshold level, and terminates when the signal level falls below the threshold (wave case 1). However, if the noise level exceeds the acceleration threshold (wave case 2), it may identify two earthquake waves as one.

- If a condition where the acceleration remains less than the threshold continues for a certain time, it is judged to be the end of the earthquake and the *SI* calculation is terminated. *Sv* and other calculation variables are initialized and the state of waiting for the occurrence of the next earthquake is returned.

A problem involved in this method is the existence of a vibration source (automobile, railroad or factory) near the installed location of the earthquake sensor. If such a source emits vibrations that exceed the threshold, it is not assured that the initialization of *Sv* has been accomplished just before the occurrence of an earthquake. In such a case, the correct *SI* value cannot be obtained for the seismic wave that occurred. Even if the vibrations always occurring at the installed location are measured at the installation and the threshold is determined based on the measurement, there may be a necessity for readjustment because of subsequent environmental changes.

To solve this problem, we developed a method of not using a threshold trigger but always performing *SI* calculation.

(2) Basis of Real-time SI Calculation

A seismic wave contains the *P* (preliminary tremor) wave and *S* (principal motion) wave. Both *P* and *S* waves travel from the focus toward the measured point but the former arrives first and the latter later. *SI* values were calculated using various past seismic waves and it was found that the *SI* value is determined by the *S* wave (principal motion) component. Hence the duration of the principal motion was examined and a real-time *SI* calculation method was developed by determining the algorithm and width of the time window for *SI* calculation.

(3) Problems with the Fixed-time Window Method

In a simple method, the time window has a fixed width and is shifted at every calculation interval Δt (called the fixed-time window method). The mechanism of and problems with this method are described below.

An operational example of the fixed-time window method is shown in Figure 2.1-14. The top vibration wave is the velocity response $v(t)$, and the second from the top, $Sv(0)$ through $Sv(t12)$, shows the time windows in which to find *Sv* at each time point. At the time point $t5$, for example, $t2$ through $t5$ are chosen for the time window in $v(t)$, and the maximum of $v(t2)$, $v(t3)$, $v(t4)$, and $v(t5)$ is employed as $Sv(t5)$. When the time length Δt has passed, the time window shifts to $t3$ through $t6$ and $v(t2)$ is discarded and $v(t6)$ is newly added.

A problem with this fixed-time window method is the large memory consumption because of the necessity for memorizing all the $v(t)$ values in the time window. Another problem is the large calculation load on the processor because of the necessity for comparing the $v(t)$ values in the time window to take their maximum. If these functions are to be implemented, the *SI* sensor will become expensive. In Figure 2.1-14, a calculation interval of $\Delta t = 1$ s is assumed. However, if the time interval is set to $\Delta t = 0.01$ s considering that the *SI* value is derived

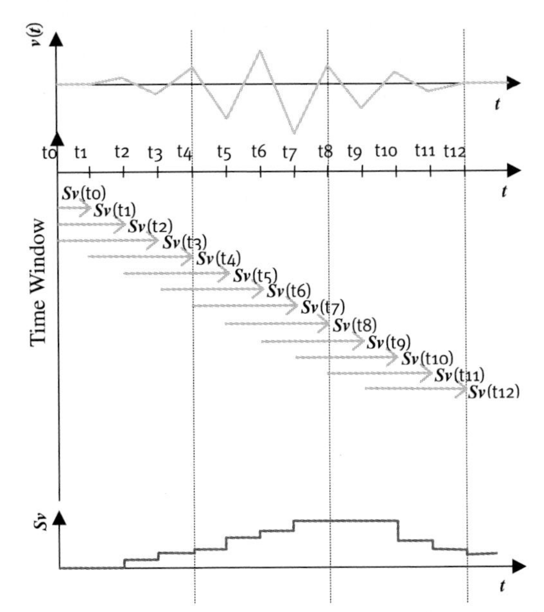

Figure 2.1-14. Real-time *SI* calculation with fixed-time window method. The fixed-time window method features a fixed time window Δt moving at each measure. Every calculated velocity response value $v(t)$ is stored and compared to obtain the maximum velocity response $Sv(t)$.

from actual seismic waves, 100 times as much memory is consumed as in the above example and this will affect the cost and size of the product. To find *Sv* at each time point, it is necessary to find the maximum by comparing all the $v(t)$ values falling in the time window and stored in the memory. This also presents the problem of a heavy calculation load imposed on the CPU.

(4) Solution to the Problem by a Flex-time Window Method

To solve the problems of calculation load and memory consumption in the fixed-time window method, we developed a new flex-time window method. The mechanism of this method is described below.

An operational example of the flex-time window method is shown in Figure 2.1-15. The top vibration wave is the velocity response. The second from the top, $Sv(0)$ through $Sv(t12)$, shows the time window in which to find *Sv* at each time point. The third, *W*0 and *W*1, shows two time windows.

This method uses two time windows, *W*0 and *W*1, and memorizes *Sv* while the object to be memorized was the velocity response $v(t)$ in the fixed-time window method. The preceding time window *W*0 is of a fixed length and the subsequent time window *W*1 is expandable. Each time a new $v(t)$ value is calculated, it is compared with $Sv1$ stored in *W*1 and, if $v(t)$ is larger, updating $v(t) \rightarrow Sv1$ takes place. At certain time intervals, $Sv1$ in *W*1 is copied to *W*0 and $Sv0$ in *W*0 is cleared to zero.

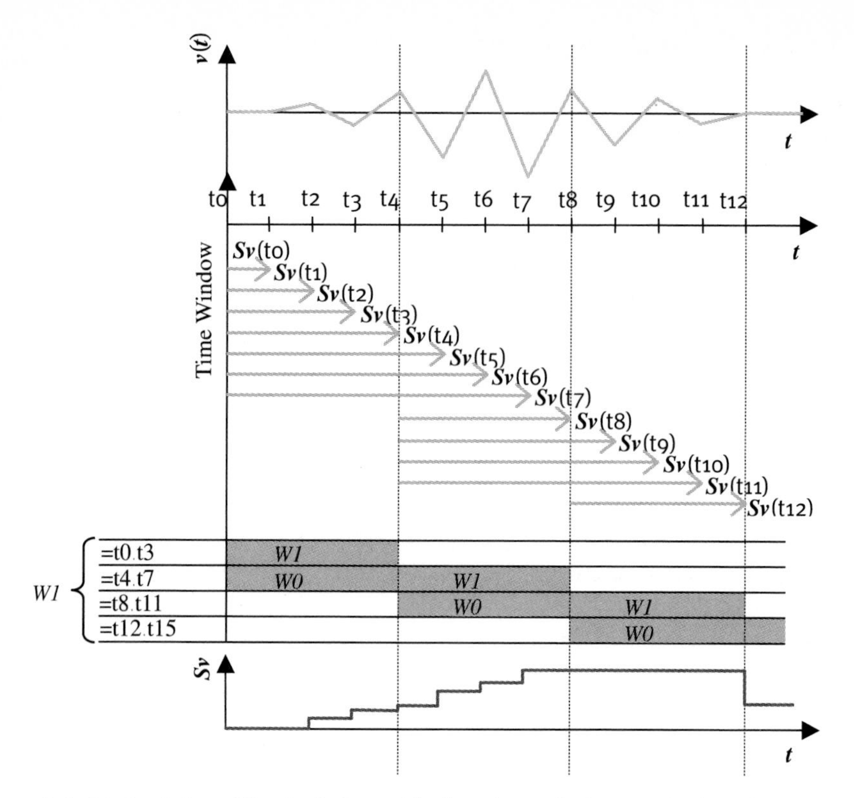

Figure 2.1-15. Real-time *SI* calculation with flex-time window method. Using two time windows *W0*, *W1*, maximum velocity response *Sv*(t) is stored instead of velocity response $v(t)$. Previous time window *W0* is a fixed length and new time window *W1* has a flexible time length.

For example, assume that the calculation starts at the time $t=0$. $v(0)$ is compared with *Sv*1 and, if $v(0)$ is larger, updating takes place. A comparison takes place between *Sv*1 and *Sv*0, and the larger becomes *Sv* to be delivered to the outside. This process repeats itself at every calculation time point. At the calculation time point $t4$, the *Sv* values from $v(t0)$ to $v(t4)$ are memorized in *W*1, and copying of *Sv*1 to *Sv*0 and zero-clearing of *Sv*1 take place. Therefore, at the time point $t5$, $v(t5)$ always updates *Sv*1 in *W*1. For *Sv*($t6$) at the time point $t6$, because *Sv*1 from $v(t5)$ to $v(t6)$ is larger than *Sv*(0) from $v(0)$ to $v(t4)$, *Sv*1 becomes *Sv* to be delivered to the outside. At the time point $t7$, *Sv*0 from $v(0)$ to $v(t4)$ in *W*0 is overwritten by *Sv*1 from $v(t5)$ to $v(t7)$ in *W*1 and *Sv*1 is cleared to zero.

By using the above method, the two memory areas *Sv*0 in *W*0 and *Sv*1 in *W*1 suffice and a substantial reduction is brought about in memory consumption and calculation load compared with the fixed-time window method. Thus, we decided to employ the flex-time window method for real-time *SI* calculation [9].

2.1.3.1.5 SI Value Calculation Parameters

From the required product specifications, the *SI* value calculation should use only the built-in ROM (read-only memory) and RAM (random access memory) of limited capacity in the CPU, and the CPU driving clock is limited to about 10 MHz. Therefore, it is necessary to hold down the calculation processing load and memory consumption.

To process the real-time *SI* calculation inside the *SI* sensor, the parameters below must be optimized. These parameters considerably affect the *SI* value calculation accuracy, calculation processing load and memory consumption of the earthquake sensor.

(a) Width of Calculation Time Window

To find the *SI* value for one seismic wave, the real-time *SI* calculation requires a time window function. With a time window lengthened properly, the calculation accuracy will improve. With too long a time window, however, an accurate calculated value may not be obtained owing to the interference of other seismic waves or vibrations that are not of seismic origin. Note that the calculation processing load and memory consumption are not affected if the time window width is increased or reduced.

(b) Number of Calculation Directions

The basic algorithm for *SI* calculation is a calculation in a certain one-axial direction in the horizontal plane. Because seismic waves vibrate in various directions, the *SI* calculation requires calculations in each of the various directions. The more the directions of calculation, the higher is the calculation accuracy. As the number of directions increases, however, the memory consumption and calculation load also increase.

(c) Number of Natural Vibrations Calculated

The *SI* value is the mean of the maximum velocity response $Sv(T)$ to a natural vibration T, averaged over the interval T=0.1–2.5 s. Therefore, the calculation accuracy improves with the density of the distribution of T calculated with T=0.1–2.5. The higher the density of T calculated, however, the larger are the memory consumption and the calculation load.

2.1.3.1.5.1 Resource Consumption Forecast of Exact SI Value

Here, as the parameters for obtaining the exact *SI* value are used as a standard, the calculation time window is defined as from the beginning to the end of one seismic wave, the number of directions calculated as 180 directions in a horizontal plane, and the number of natural vibrations calculated as 121. In other words, the calculation directions are at intervals of 1°, and the natural vibrations calculated are at intervals of 0.2 s. In the following discussion, the exact *SI* value used as an accuracy standard is the result of calculation of the seismic wave based on these definitions.

Assuming that the calculation function for the exact *SI* value is to be implemented in the *SI* sensor without modification, it is required to mount a DSP (digital signal processor) operating at 300 MHz and a RAM of about 260 kb in the earthquake sensor.

However, these required resources far exceed the specified limits for the *SI* sensor described in Section 1.

To solve this problem, we attempted optimization of resources using a statistical technique. The results are described in the next section. As for the calculation time window width taken up in (a) in the previous section, we determined it using the same technique. The results are also described in the next section.

2.1.3.1.5.2 Determination of Real-time SI Calculation Parameters by Statistical Technique

First, we decided the target calculation accuracy of the *SI* value calculated using the parameters optimized in this section at ±5% of the exact *SI* value. We used many actually observed seismic waveforms as input and calculated the *SI* value. Several *SI* values were obtained for as many sets of set values for the three parameters named above, and their errors from the exact *SI* value obtained for the same input were examined. We employed the technique of determining the minimum parameter value that satisfied the target calculation accuracy from the derived result.

(a) Width of Calculation Time Window

In real-time *SI* calculation, the time window limits the time length to memorize *Sv*, and this time window represents the time width necessary to keep track of the *SI* value of one seismic wave with the required accuracy. We used 60 seismic waves as input acceleration and set the time window width at several lengths. We compared the *SI* values thus obtained with the exact *SI* value by simulation. The results are shown in Table 2.1-3.

Since it was confirmed from the simulation results that a calculation time window width set to 10 s or longer would bring about a calculation error of 1.0%, we decided on a calculation time window width of 10 s. Because the real-time *SI* calculation as described in Section 2.1.3.1.4 consists of two parts, ie, the time window *W0* of fixed length and time window *W1* of variable length, the time window is variable in width from 10 to 20 s.

Table 2.1-3. Time window length and *SI* calculation error by earthquake wave simulation

Time window length (s)	4.0	6.0	7.0	8.0	9.0	10.0	12.0
SI calculation error (%)	2.0	1.5	1.3	1.2	1.1	1.0	0.9

Table 2.1-4. Calculation vector numbers and *SI* calculation error by earthquake wave simulation

SI calculation vector	2	4	5	6	8	9	12	180
SI calculation error (%)	39.9	2.5	1.9	1.6	1.0	1.0	0.9	0.0

(b) Number of Calculation Directions

To decide the direction of *SI* calculation in a horizontal plane, we compared the *SI* calculation results obtained with 60 seismic waves as input with the exact *SI* value by simulation. The results are shown in Table 2.1-4. As a result, it was confirmed that the calculation error could be held down within ±2% with five or more calculation directions. Needless to say, one earthquake sensor measures two orthogonal components of horizontal acceleration. Considering the equiangular division of directions, we decided the number of calculation directions to be eight because the number must be the *n*th power of 2. With eight directions, the calculation error is within 1%.

(c) Number of Natural Vibrations Calculated

We compared *SI* calculation results obtained with 60 seismic waves as input with the exact *SI* value by simulation. The results are shown in Table 2.1-5. From these results, we searched for a set of natural vibrations that would cause a calculation error of 2.2% and decided on seven vibrations arranged at uneven intervals (T=0.1, 0.4, 0.7, 1.0, 1.5, 2.0 and 2.5).

2.1.3.1.6 Summary of SI Value Calculation

Table 2.1-6 summarizes the *SI* value calculation method we finally employed in comparison with the exact *SI* value.

The required resources for exact *SI* calculation were at an unattainable level. We attempted to reduce the resource requirements by minimizing the calculation parameters using a statistical technique within the retainable limits of the required calculation accuracy. As a result, the amount of hardware resources required for this *SI* calculation method was reduced as shown in the table, and an expensive DSP whose addition was initially anticipated became unnecessary. Thus, we succeeded in optimizing the *SI* value calculation method to a practica-

Table 2.1-5. Numbers of natural vibrations and SI calculation error by earthquake wave simulation

Natural vibration	2	4	5	7	7[a]	13	25	49	121
SI calculation (%)	12.9	7.9	4.9	2.4	2.2	2.2	1.1	1.1	1.1

[a] Natural vibration based on uneven interval.

Table 2.1-6. Comparison of required resource by calculation method

SI calculation method	At initial development stage	Adopted
SI calculation trigger	Acceleration threshold	Non-trigger method (real-time *SI* calculation)
Calculation time windows	1	2
Calculation vectors	180	8
Natural vibrations	241	7
Memory	180 kb	~ 0.7 kb
Calculation load	1	0.027

ble calculation processing level with an RISC (restricted instruction set computer) CPU with built-in adders and multipliers alone.

2.1.3.2 Liquefaction Detection

2.1.3.2.1 Necessity for Detection of Ground Liquefaction

The Niigata earthquake occurred in 1964, the Nihonkai-Chubu earthquake in 1983, and the Hyogo-Ken-Nanbu earthquake in 1995. The damage due to these and other past earthquakes was analyzed and it was ascertained that even for the same magnitude of *SI* value the damage to structures becomes severe if ground liquefaction occurs.

Ground liquefaction is a phenomenon that is liable to occur in a landfill, former riverbed, coastal plain or other loose sandy ground saturated with water. When subjected to the shock of an earthquake, the ground, which supported structures until then, abruptly becomes liquid like mud-water. As a result, the ground will lose its ability to support structures. Collapse of structures will occur, and serious damage will also occur to pipelines and other underground structures. Therefore, detection of the ground liquefaction phenomenon is a very important subject in earthquake disaster prevention.

As a conventional method for direct observation of ground liquefaction, the technique of measuring the groundwater pressure has generally been employed [10–12]. This method can detect the occurrence and extent of liquefaction with high accuracy. However, this method requires a large installation space and has high operating costs because it involves ground boring work. Therefore, it is difficult to install the instrument at high density locations.

On the other hand, attempts have been made to judge the occurrence of liquefaction from the features of earthquake acceleration waveforms observed near the points where liquefaction was ascertained, and several methods for this have been proposed [13, 14]. However, they lack the property of being real time, which is indispensable for liquefaction judgment at an actual earthquake disaster.

Therefore, we developed a method to judge the occurrence of liquefaction from the features of acceleration waveforms at an earthquake easily and in real time.

If the earthquake sensors placed on the ground can capture the liquefaction phenomenon in real time from the features of observed acceleration, there is no need to install liquefaction sensors anew and two types of damage estimation, that is, *SI* value and liquefaction detection, can be implemented in one instrument. This makes inexpensive observation of earthquake and disaster prevention information possible.

This section analyzes first the acceleration waveform near the ground where liquefaction occurred and extracts the features of the waveform, and also describes the details of the derivation of an algorithm for simple judgment of liquefaction. In addition, we developed the first real-time liquefaction detection algorithm to be mounted on a low-cost earthquake sensor. This is also detailed below. Next, we verified the operation of the developed liquefaction judgment algorithm by a liquefaction judgment simulation and excitation experiment using past earthquake acceleration waveforms. The results are presented below.

2.1.3.2.2 Analysis of Liquefaction Waveform Characteristics

We collected acceleration waveforms near the location of occurrence of ground liquefaction, divided them into waveforms before the occurrence of liquefaction and waveforms during the occurrence of liquefaction, and extracted their features. Because most of the earthquake acceleration waveforms used for analysis were along two horizontal axes, we decided not to use the vertical acceleration waveforms for liquefaction judgment. Many researchers have investigated the different natures of liquefaction waveforms, and several methods have already been proposed to detect the liquefaction phenomenon from those waveforms. In all these methods, however, processing is impossible with an inexpensive earthquake sensor's internal resources and real-time judgment is difficult. The aim this time was to use parameters that can be processed with an earthquake sensor's limited internal resources and to establish a judgment method that has real-time capabilities. For this purpose, we started our investigation with the extraction of simple feature quantities again for use in judgment from liquefaction waveforms (see Figure 2.1-16).

First, a feature of the waveform before the occurrence of ground liquefaction is that the earthquake must be large enough to cause liquefaction. That is, it was ascertained that the maximum acceleration value and *SI* value must be larger than a certain level (Characteristic 1).

Next, it was also ascertained as features of the acceleration waveform at the occurrence of liquefaction that the acceleration becomes small (Characteristic 2), the acceleration period becomes long (Characteristic 3), and the displacement becomes large (Characteristic 4).

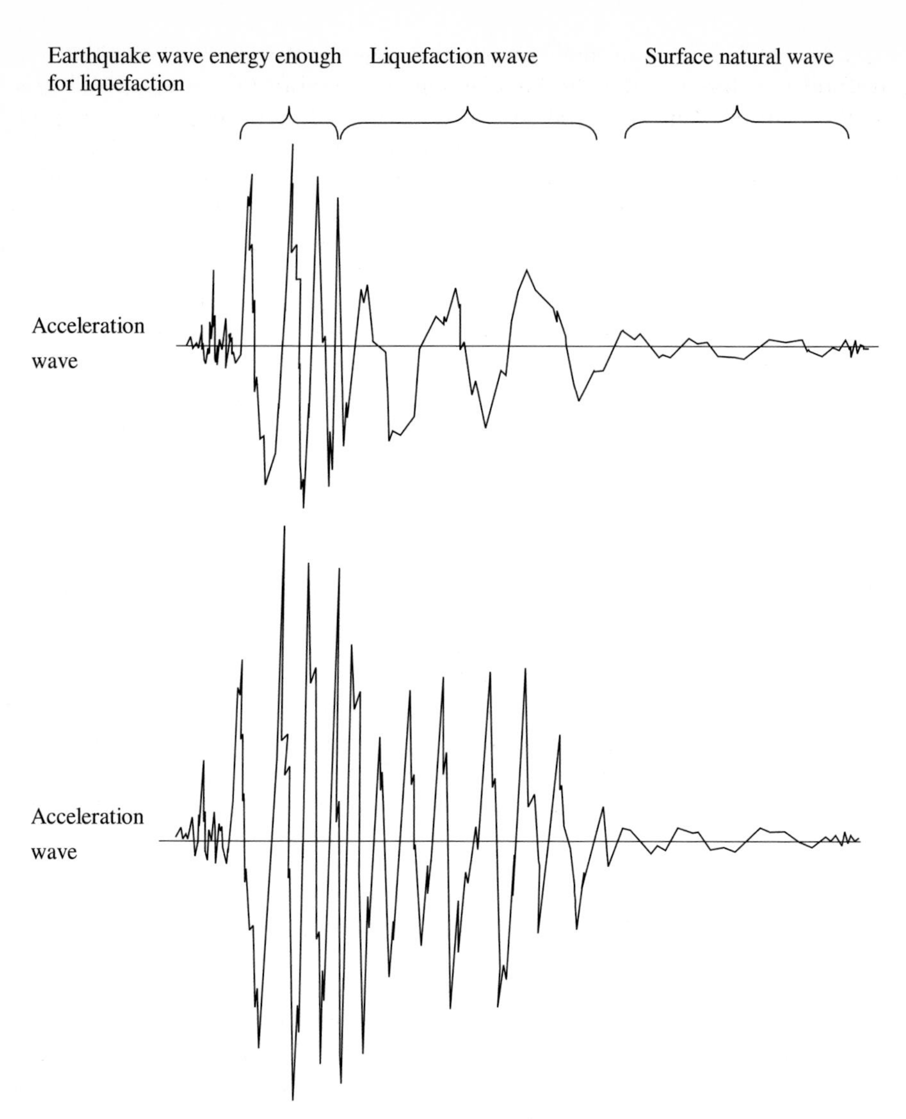

Figure 2.1-16. Two graphs show difference between liquefaction and non-liquefaction earthquake waveforms.

2.1.3.2.3 Extraction of Liquefaction-extracting Parameters

We investigated liquefaction judgment criteria and selected those conditions from the above-stated features of acceleration waveforms that were considered practical in actual use. As a result, we excluded the condition that the acceleration waveform becomes small at the occurrence of liquefaction (discriminating condition according to Characteristic 2) since we searched for a method to judge the occurrence of liquefaction and because this condition cannot judge whether the waveform is before or after the liquefaction. To simplify the measurement of acceleration period (Characteristic 3), we decided to employ the time *T_z* from a zero crossing to the next zero crossing at which the acceleration changes it's sign, or a time length equivalent to a half period. As for the displacement (Characteristic 4), integrating the acceleration simply twice yields the displacement. However, it was considered that errors would accumulate and the accuracy would not be satisfied if this technique was employed. Therefore, as the method for calculating a stable displacement amplitude, the equation proposed by Towhata *et al.* was used [15]. According to research by these authors, the displacement amplitude can be found by finding the displacement amplitude on the assumption of a stationary vibration and multiplying it by an appropriate correction factor. If the correction factor is 2, the estimate of the maximum displacement D_{max} is as follows:

$$D_{max} = 2.0 \frac{SI^2}{PGA} \tag{2.1-9}$$

where *SI* is the *SI* value (kine) and *PGA* is the maximum acceleration (Gal) in a horizontal plane.

From these results, we attempted to use four variables: maximum acceleration, *SI* value, zero-crossing period, and displacement estimate, as the judging parameters to be used for liquefaction detection. These four parameters are called *PGA*, *SI*, *T_z*, and D_{max} hereafter. As a result of analysis of observed earthquake waveforms, it was ascertained that the conditions $SI > 20$ kine, $PGA > 100$ Gal, $D_{max} > 10$ cm, and $T\_z > 1$ s are the parameter values that characterize liquefaction waveforms [16].

2.1.3.2.4 Algorithm Implemented in Actual Instrument

2.1.3.2.4.1 Real-time Liquefaction-judging Algorithm

From the above, we succeeded in ascertaining the four parameters for use in judging the occurrence of liquefaction in a simplified manner from the features of acceleration waveforms at the occurrence of liquefaction. Next, we investigated real-time algorithms to be implemented in an actual earthquake sensor.

2.1.3.2.4.2 Maximum-triggered Algorithm

Because it was necessary to make a judgment within the duration of one earthquake wave, we investigated an algorithm that allowed the liquefaction-judging function for a certain length of time from the detection of the maximum value of *SI* or *PGA*. If this algorithm is used, however, because a wave called the surface wave is propagated along the ground surface and arrives after the principal motion, the influence of this surface wave on liquefaction judgment must be considered.

First, the surface wave has the feature of taking a long-period waveform like the acceleration waveform during the occurrence of liquefaction, and it follows that the T_z value satisfies the liquefaction condition even with a non-liquefaction seismic wave.

PGA and *SI* take large values for a seismic wave that is strong but does not cause liquefaction. As for D_{max}, the same seems possible because it is derived not from actual displacement but from *PGA* and *SI*.

Therefore, the possibility of misjudgment can arise if the time of enabling the liquefaction-judging function is long. Furthermore, if the distance from the earthquake epicenter to the measuring point is short, the interval from the principal motion to the surface wave becomes so short that it is difficult to exclude the influence of the surface wave by adjusting the enabling time for liquefaction judgment.

We verified this using actual earthquake waveforms. As a result, a judgment-enabling time of 10 s was necessary for judging the occurrence of liquefaction. With 10 s, however, a misjudgment that liquefaction occurred was made for the Izu-Ohshima-Kinkai Earthquake at Mishima and other places where no liquefaction occurred.

Thus, the algorithm enabling liquefaction judgment for a certain time using a maximum value as a trigger led to a misjudgment due to the influence of the surface wave. Therefore, it was necessary to develop an algorithm that is not susceptible to the influence of the surface wave.

2.1.3.2.4.3 Judgment Algorithm using a Zero-crossing Time Window

If the acceleration waveform during liquefaction is compared with the acceleration waveform of a surface wave, their periods are certainly of the same length. However, it was ascertained as a distinctive feature that the liquefaction waveform was larger in acceleration value. Because surface waves propagate from the focus vertically first and then along the ground surface before they arrive at the measuring point, they have been damped to a small acceleration at the measuring point. Therefore, if an acceleration of a long period occurs, distinction between the liquefaction waveform and surface wave may be possible by comparing the maximum value of acceleration in that period. We investigated this method.

As the judging parameter of acceleration, we attempted to use not the maximum *PGA* of the whole seismic wave but the maximum acceleration A_{max}_z in the zero-crossing period. Because *PGA* is for judging the scale of an earthquake before the occurrence of liquefaction and has the same meaning as *SI*, we decided to exclude it from the judging conditions.

From the waveform of actual seismic waves, we took the threshold for A_{max}_z judgment at 100 Gal or over. By this means, we could ascertain that the correct judgment was possible for the Izu-Ohshima-Kinkai Earthquake at Mishima and other places where the maximum-triggered algorithm led to a misjudgment.

For consistency in parameter calculation, the maximum in the zero-crossing period was also used for *SI* and D_{max}. Therefore, the four liquefaction-judging parameters were replaced with SI_{max}_z, D_{max}_z, A_{max}_z and T_z.

From the above-stated result, we decided to use the liquefaction-judging algorithm based on the zero-crossing time window as the real-time algorithm to be implemented on an actual instrument.

2.1.3.2.4.4 Influence of Installed Orientation of SI Sensor on Accuracy

For the purpose of reducing the CPU's calculation load and memory consumption, we initially attempted to perform a liquefaction judgment using the acceleration along two axes in a horizontal plane, or the x- and y-axis. As described in Section 2.1.3.1, however, it is known that the acceleration waveform and calculation result of *SI* value differ for different calculation directions. Thus, dependence of the *SI* sensor installation on orientation for liquefaction judgment was studied.

As the ascertaining method, the seismic waves observed along two horizontal axes were subjected to vector projection to reproduce them as seismic waves in different directions and liquefaction judgment took place using the reproduced waveforms. As the number of projection directions, eight directions at 22.5° intervals were employed that had been used in *SI* value calculation.

The ascertainment was done using actual seismic waves. As a result, it was ascertained that, with the liquefaction waveform observed at the Nihonkai-Chubu Earthquake at the Tsugaru Bridge (treated as a liquefaction waveform because there was a sign of liquefaction in the vicinity), correct judgment was impossible if the earthquake sensor was installed in a different direction in the horizontal plane. Therefore, we decided that liquefaction judgment should take place for each of the acceleration waveforms projected in eight directions and one should judge that liquefaction occurred if any one direction satisfied the conditions.

2.1.3.2.4.5 Detailed Calculation of the Four Parameters

This section describes the parameter calculation and detailed operations for judgment decided by the above-stated discussion and taking place in an actual instrument (see Figure 2.1-17).

For *SI*, the maximum values in eight horizontal directions were found by the above-stated *SI* value-calculating algorithm. The maximum values are held at their maximum in the flex-time window (10–20 s), and these are used as the *SI* value. SI_{max_z} is the maximum value held in the liquefaction-judging, zero-crossing time window.

For D_{max_z}, according to Towhata *et al.*, it is necessary to calculate it using the *SI* and *PGA* values in the same axis direction. Therefore, one must find eight values from *SI* and *PGA* in each of the eight axis directions and take the maximum of the eight values as D_{max}. Because *PGA* is held at its maximum in the calculation time window in the same way as *SI*, D_{max} also is held at the same value for 10–20 s. D_{max_z} is the value calculated and held as the maximum of D_{max} in the liquefaction-judging, zero-crossing time window.

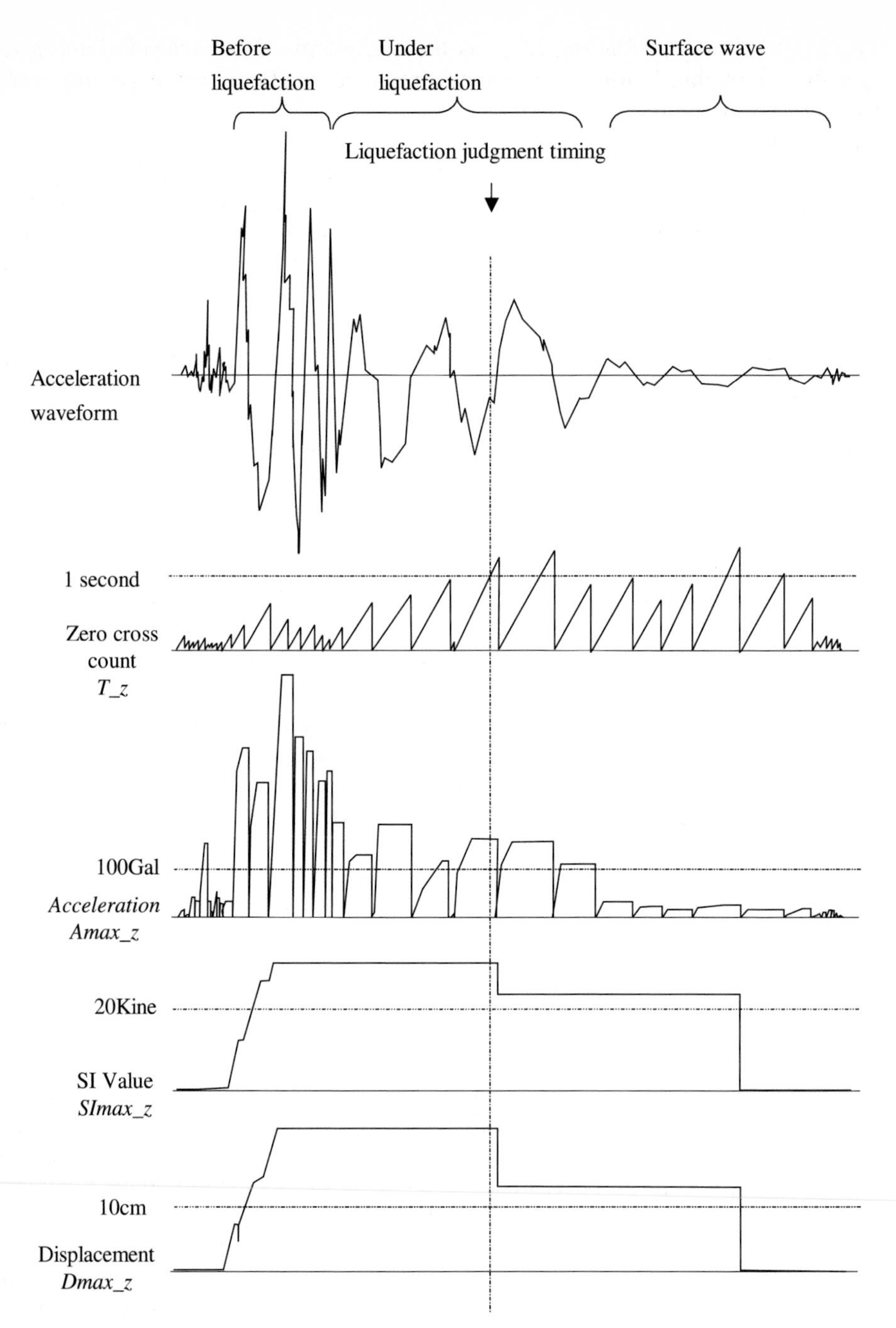

Figure 2.1-17. Actual T_z, $A_{\max}\_z$, $SI_{\max}\_z$ and $D_{\max}\_z$ transition. Liquefaction is judged when all conditions exceed set values.

A_{max_z} is found by finding A_{max} as the maximum of the accelerations in the eight directions and holding it at its maximum in the liquefaction-judging, zero-crossing time window.

T_z is the time count to a zero crossing, which can be found from the acceleration waveforms projected in the eight directions and calculated in real time. As a matter of course, this time count is accumulated in the zero-crossing time window and is cleared to zero when a zero crossing occurs. Here, the judging threshold for zero-crossing period is 2 s. However, because T_z is the time count in the zero-crossing window and is half of the zero-crossing period, its threshold is equivalent to half, or 1 s.

Liquefaction judgment by an actual instrument takes place as follows. When SI_{max_z}, D_{max_z} and A_{max_z}, which are held in each zero-crossing time window, exceed their set thresholds and the time count to a zero crossing exceeds its threshold, this is judged to be the occurrence of liquefaction at that timing.

2.1.3.2.5 Verification of Liquefaction-judging Function

2.1.3.2.5.1 Simulation of Algorithm Implemented in Actual Instrument

We fabricated a simulator that performed the same liquefaction-judging operation as an actual instrument and input observed earthquake waveforms to it to ascertain the accuracy of its judgment results.

A total of 45 waveforms including 19 liquefaction waveforms (○) and 26 non-liquefaction waveforms (×) were input to the simulation. The simulation results are shown in Table 2.1-7. As a result, 40 waveforms, around 90%, were judged correctly and this confirms that the developed algorithm will pose no problems in practical use.

2.1.3.2.5.2 Ascertainment of Actual Instrument Operation by Excitation Experiment [17]

The acceleration input used for the judgment verification by simulation was waveforms captured by different measuring seismographs. Therefore, it is necessary to ascertain the influence of errors in acceleration measurement by the actual instrument. An acceleration pickup and acceleration-measuring circuit are mounted on the actual instrument. To ascertain the influence of these and their installed conditions and installed direction, we conducted an excitation experiment. As the liquefaction judgment results used for the acceptability judgment, waveforms measured by an accurate accelerometer (servo accelerometer used for ordinary precise earthquake observation) and recorded during excitation were used. These waveforms were input to a simulation on a personal computer and the simulated liquefaction judgment results were used for the acceptability judgment. For the experiment, a three-axis exciter bench at the Pipeline Technical Center of Tokyo Gas Co., Ltd. was used. The instrument that we developed and the servo accelerometers were mounted on the exciter bench and the liquefaction judgment results obtained by both were compared. Four units of the developed

Table 2.1-7. Liquefaction judgment results: the developed algorithm met around 90% of judgment compared with actual field investigation (○, liquefaction; ×, non-liquefaction)

Category	Earthquake names	Observed points	Actual field investigation	Simulation results based on the developed algorithm	Judgment: met/not met
Liquefaction or possible liquefaction	Niigata	Kawagshi Town	○	○	Met
	Tokachi-Oki	Aomori Harbor	○	○	Met
	Nihonkai-Chubu	Hachirou Lagoon	○	○	Met
	Nihonkai-Chubu	Tugaru Bridge	○	○	Met
	Nihonkai-Chubu	Akita Harbor	○	×	Not met
	Superstition Hills	Wildlife	○	○	Met
	Loma Prieta	Treasure Island	○	○	Met
	Kushiro-Oki	Kushiro Harbor	○	○	Met
	Hokkaido-Nansei-Oki	Hakodate Harbor	○	×	Not met
	Hyougo-Ken-Nanbu	Amagasaki (Kansai Electric Power Co., Inc.)	○	○	Met
	Hyougo-Ken-Nanbu	Amagasaki Bridge (Ministry of Construction)	○	○	Met
	Hyougo-Ken-Nanbu	Amagasaki Harbor	○	○	Met
	Hyougo-Ken-Nanbu	Rokko Island	○	○	Met
	Hyougo-Ken-Nanbu	Higashi-Kobe Bridge	○	○	Met
	Hyougo-Ken-Nanbu	Kobe Harbor Construction Office	○	○	Met
	Hyougo-Ken-Nanbu	Kobe Port Island	○	○	Met
	Hyougo-Ken-Nanbu	Japan Railways Takatori Station	○	×	Not met
	Taiwan Chichi	Yuanlin	○	○	Met
	Tottori-Ken-Seibu	Sakai Harbor	○	○	Met
No liquefaction	El Centro	El Centro	×	×	Met
	Kern County	Taft	×	×	Met
	Kern County	Taft	×	×	Met
	Hyuga-Nada	Itajima Bridge	×	×	Met
	Tokachi-Oki	Muroran	×	×	Met
	Tokachi-Oki	Hachinohe	×	×	Met
	Izu-Oshima-Kinkai	Mishima	×	×	Met
	Miyagi-Ken-Oki	Kaihoku Bridge	×	×	Met

Table 2.1-7 (continued)

Category	Earthquake names	Observed points	Actual field investigation	Simulation results based on the developed algorithm	Judgment: met/not met
	Miyagi-Ken-Oki	Shiogama	×	×	Met
	Miyagi-Ken-Oki	Tarumi	×	×	Met
	Miyagi-Ken-Oki	Sendai	×	×	Met
	Miyagi-Ken-Oki	Tohoku University	×	×	Met
	Miyagi-Ken-Oki	Aomori	×	×	Met
	Michoacan, Mexico	SC&T	×	○	Not met
	Hokkaido-Touhou-Oki	Kushiro	×	×	Met
	Hokkaido-Touhou-Oki	Nemuro	×	×	Met
	Sanriku-Haruka-Oki	Aomori	×	×	Met
	Sanriku-Haruka-Oki	Hochinohe	×	×	Met
	Hyougo-Ken-Nanbu	JMA Hikone	×	×	Met
	Hyougo-Ken-Nanbu	JMA Kobe	×	×	Met
	Hyougo-Ken-Nanbu	Shijo	×	×	Met
	Hyougo-Ken-Nanbu	Senri	×	×	Met
	Hyougo-Ken-Nanbu	Fukiai	×	×	Met
	Hyougo-Ken-Nanbu	Fushimi	×	×	Met
	Taiwan Chichi	Shihkang National Elementary School	×	○	Not met
	Tottori-Ken-Seibu	K-net Yonago	×	×	Met

instrument were placed in directions of 0° (two units), 11.25° and 45°, with respect to the x-axis of the exciter bench to ascertain the direction of acceleration measurement and eight directions of calculation, and this arrangement was used to ascertain the liquefaction judgment results.

Simulated waveforms for which liquefaction judgments are critical were selected for experiments that were excitable by the three-axis exciter bench (Table 2.1-8). Thus, six waveforms which are influential in liquefaction and 15 waveforms which are not influential in liquefaction were used. From the results, we

Table 2.1-8. Actual earthquake excitation experiments proved that liquefaction judgment function with different orientations met standard sensor judgment. Actual measured earthquake waves were put into a three-axis exciter bench (○ liquefaction judgment)

No.	Input earthquake waveforms		Judgment				
	Earthquake name	Observed points	Reference servo type 0°	Sensor A 0°	Sensor B 11.25°	Sensor C 45°	Sensor D 0°
1	Hougo-Ken-Nanbu	Amagasaki Harbor	○	○	○	○	○
2	Wildlife	Wildlife	○	○	○	○	○
3	Nihonkai-Chubu	Aomori					
4	Izu-Oshima-Kinkai	Mishima					
5	Nihonkai-Chubu	Hachiro Lagoon	○	○	○	○	○
6	Sanriku-Haruka-Oki	Aomori					
7	Miyagiken-Oki	Tohoku University					
8	Hyougo-Ken-Nanbu	Port Island (surface)	○	○	○	○	○
9	Hyougo-Ken-Nanbu	Port Island (12 m below ground)					
10	Hyuga-Nada	Kouchi					
11	Tokachi-Oki	Hachinohe					
12	Niigata	Kawagishi Town	○	○	○	○	○
13	Nihonkai-Chubu	Tugaru Bridge	○	○	○	○	○
14	Hyougo-Ken-Nanbu	Kobe marine meteorological					
15	Hyougo-Ken-Nanbu	Senri					
16	Hyougo-Ken-Nanbu	Port Island (28 m below ground)					
17	Hyougo-Ken-Nanbu	Port Island (79 m below ground)					
18	Miyagiken-Oki	Sendai					
19	El Centro	El Centro					
20	Tokyo-Wan	Meguro					
21	Miyagiken-Oki	Shiogama					

could ascertain that the liquefaction-judging function, which was ascertained by simulation, would cause no problems in the judging operation also in an actual instrument.

2.1.3.2.6 Summary

We analyzed earthquake liquefaction waveforms recorded in the past and decided basic parameters by extracting features from the liquefaction waveforms. Further, we devised a mechanism for parameter correction and judgment operation. By this means, it was possible to provide a low-cost earthquake sensor with the capability of judging the occurrence of liquefaction with high accuracy.

According to the simulation results, liquefaction was incorrectly judged to have occurred for the waveform of the Taiwan Chichi Earthquake at Shihkang National Elementary School and other waveforms. These are waveforms with long periods and large amplitudes containing the influence of fault movement. However, since the damage is severe near a fault in the same manner as liquefaction, such information can be effective from the point of view of disaster prevention.

2.1.3.3 Noise Protection

Sensors are used in various environments and are generally disturbed by their surroundings. Researchers and developers are making efforts to ensure the selective output of the required measured physical quantity alone by excluding these disturbances. However, there is still no sensor in existence that is not influenced by any disturbance. Therefore, to implement a sensor and offer it to society as an instrument that matches social requirements, it must be fabricated as a product that excludes the influence of these disturbances to the necessary level.

In this development also, the influences of radiation fields on the acceleration signal were observed in laboratory and field tests of a prototype model. This section describes the function added to our developed product to exclude noise-induced malfunction.

2.1.3.3.1 Background of Noise Protection Function

The *SI* value, the instrumental equivalent of seismic intensity and liquefaction judgment form indices to estimate earthquake damage, is calculated based on the acceleration signals output from acceleration sensors. Therefore, if the acceleration signals are influenced by noise, an erroneous acceleration value, *SI* value, and instrumental equivalent of seismic intensity and liquefaction judgment will be output.

In the course of this sensor's development, a prototype model, which incorporated a real-time seismic index-calculating algorithm, was fabricated and subjected to various environmental, installation, and operation tests in the field. It was found that the acceleration signal output was susceptible to electromagnetic waves. In implementing the sensor as a commercial product, measures had to be taken to exclude this influence and ensure that no misjudgment would occur in the earthquake sensor output.

The methods to deal with the noise superimposed on sensor output are broadly divided into two groups. One is the method of modifying the sensor structure or electronic circuit to improve the immunity to noise to the required level, thereby preventing misjudgment from occurring. The other is the method of discriminating and removing noise from the noise-containing signal on the side of signal processing, thereby preventing misjudgment from occurring.

In the present development, the first method, or the noise immunity-improving method, was not pursued. In actual fact, no modification was made to the acceleration signal-extracting circuit of the MEMS acceleration sensor, and the changes and modifications for noise reduction were limited to the power system and signal processing section that were outside the sensor. The second approach was employed as the method to prevent the earthquake sensor from misjudging. A function to discriminate and suppress noise on the output was developed and this function was implemented in the CPU as logical operations. There are several reasons why this approach was employed, and the major ones are given below.

First, the MEMS sensor body and acceleration signal extracting circuit had already been on sale as a module for other applications. The MEMS sensor has the features of low cost, small size, and high performance. Of these, low cost is mainly due to mass production of the order of several hundred thousand units per year in the automobile, amusement and other markets. If the structure or circuit of a sensor module were modified as a measure to counter noise, it would have been necessary to customize a MEMS sensor module for exclusive use in an earthquake sensor. The extent of production of earthquake sensors is of the order of several thousand units per year at most. Therefore, customizing leads to much hindrance in implementing a MEMS sensor at low cost and deviates from the basic development concept of our earthquake sensor.

The second reason is the nature of the electromagnetic noise source itself, which caused the disturbances observed this time. Noise was observed in the field in a prototype model test and the cause of noise was established to be due to the use of radio waves of illegal intensity. It is regrettable that, although few, there are people in Japan who do not observe the law in this respect. As detailed in Section 2.1.4, the product developed this time is to be used in an unmanned facility or equipment and is required to maintain its function for approximately 10 years. Even if the performance of noise immunity is improved in the radio bands currently open to the public and measures are taken to prevent the product from misjudgment in the current field test, there is still the risk that such measures will lose their effect if the radio regulations are amended and the radio bands open to the public are changed in the future. It has regrettably been ascertained that the transmitter power of illegal radio users is increasing with the progress of science and technology. With these as the background, it must be adequately anticipated that the bands and intensity of noise sources will change during the forthcoming 10 years of operation. That is, even if appropriate measures are taken in the circuit to counter electromagnetic noise in the band and at the intensity currently observed, this cannot ensure that misjudgment will be excluded in the future.

Hence this development was aimed at implementing a product at low cost and maintaining stable performance for 10 years, and a function was devised to discriminate noise from the signal and not to reflect the calculation result in the controlling output of the earthquake sensor if noise is recognized. Details are described in the following sections.

2.1.3.3.2 Ratio Noise Protection

2.1.3.3.2.1 Basic Logic for the Extraction of Distinctive Quantities and Noise Discrimination

The prototype model was subjected to intense electromagnetic radiation in a radio anechoic chamber and acceleration was measured. The results are shown in Figure 2.1-18. Figure 2.1-19 shows a waveform observed by the prototype model installed in an actual field. It seems to be influenced by radio electromagnetic radiation of an illegal intensity. Although these waveforms are different in shape, a common characteristic is seen in both. This is because all the acceleration signals drift to one side while irradiated with electromagnetic waves. The waveform observed in the field is triangular in shape, and it is surmised that the object emitting radio waves of illegal intensity is a moving object (automobile) and the triangular shape is due to the behavior of this object approaching and leaving the prototype sensor while emitting radio waves. In contrast, seismic waves basically produce an output that oscillates to and from. This is distinctively different from the noise-induced output, which is observed on one side only.

Based on this distinctive behavior, a basic logic was employed to discriminate between an acceleration output due to seismic waves and output due to noise. This is explained using a Figure 2.1-20. Figure 2.1-20 a shows a typical acceleration sensor output observed when injected with electromagnetic noise from a

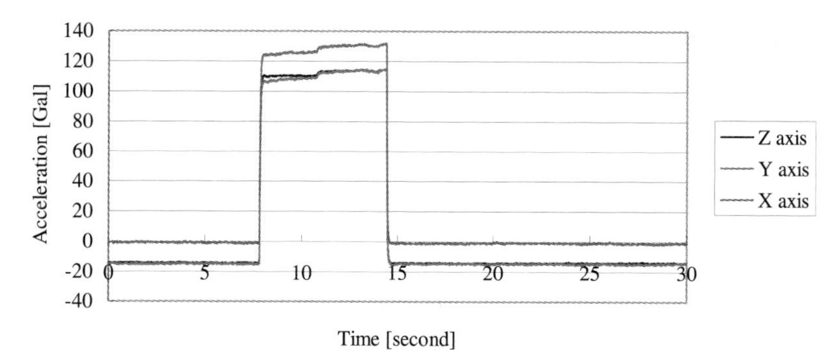

Figure 2.1-18. Noise waveforms observed at test laboratory.

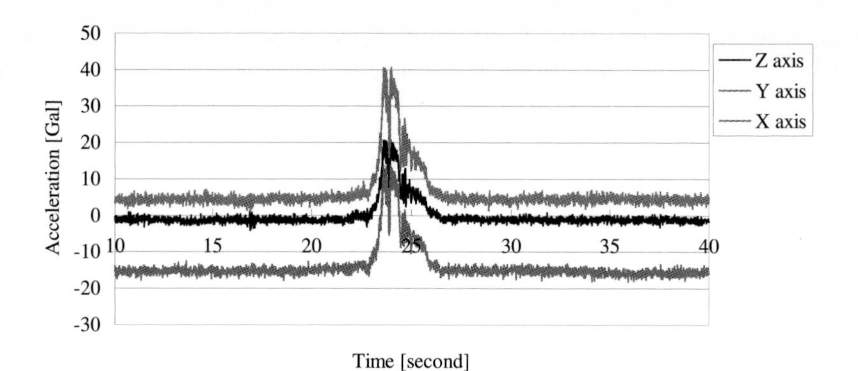

Figure 2.1-19. Noise waveform from radio transmittance of illegal intensity observed in the field.

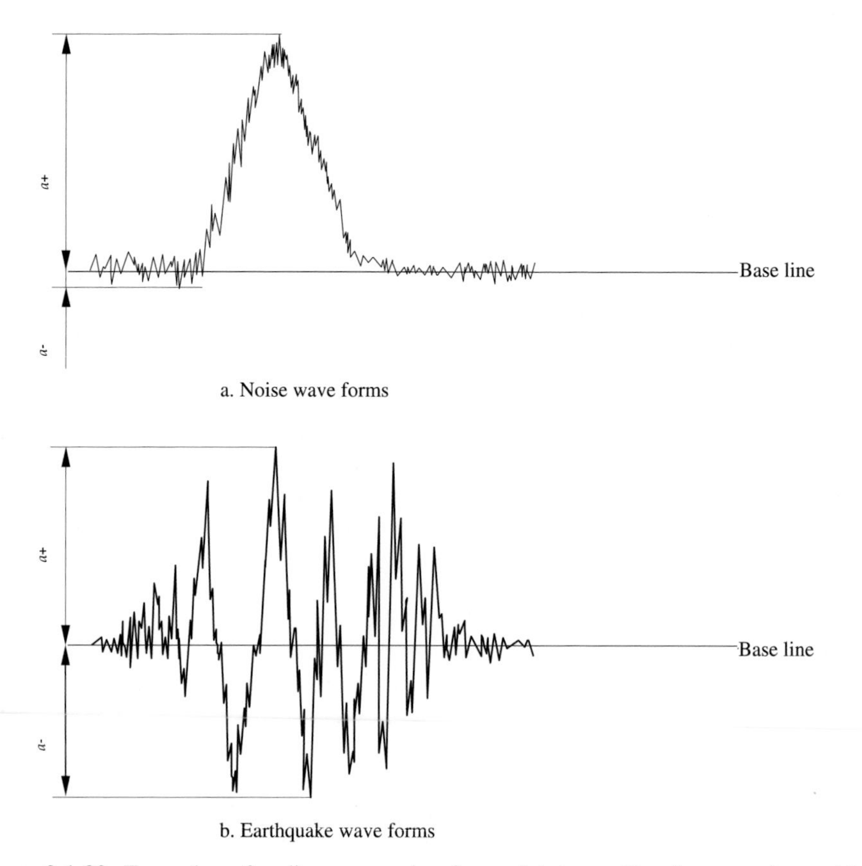

Figure 2.1-20. Examples of radio wave noise forms (**a**) transmitter in a moving vehicle and (**b**) earthquake waveforms measured by an accelerometer. Distinction is judged based on difference in wave amplitudes. Ratio of plus (a+) and minus (a–) is small in earthquakes, but larger for noise waves.

moving vehicle. Figure 2.1-20b shows an acceleration sensor output observed when injected with actual seismic waves. The ratio in absolute value of the positive quantity $a+$ of the sensor output with respect to the baseline to the negative quantity $a-$ is distinctively different between noise-induced output and actual earthquake-induced output. Attention was paid to this fact and it is used as the basic logic for noise discrimination. That is, this method discriminates between noise and actual earthquakes based on whether or not the ratio of positive acceleration output to negative output (hereafter called the noise ratio) lies within a certain threshold r ($r > 1$), as indicated by the equation

$$1/r < |a+|/|a-| < r \tag{2.1-10}$$

Judgment: not due to noise if true;
due to noise if false.

2.1.3.3.2.2 Application of Basic Logic

The basic concept was explained in Section 2.1.3.3.2.1. However, some problems arose when the operation of the earthquake sensor was considered as an actual instrument.

First, as is obvious from Figure 2.1-20 used for the explanation, the acceleration sensor output patterns are input to the logic, which compares and judges the input as one whole noise wave or as one whole seismic wave. In actual measurement and calculation processing, the acceleration sensor output is input as an ever-changing quantity from the left side of the output waveform in Figure 2.1-20. In an actual instrument, it is necessary to apply the logic so as to make a real-time judgment on this ever-changing acceleration sensor output and yet to prevent a misjudgment.

Second, fluctuating acceleration sensor output variations can occur in an ordinary still state owing to some causes other than electromagnetic noise. Measures must be taken to prevent frequent misjudgment as to noise in the ratio judgment according to Equation (2.1-10). Such a phenomenon is supposed in a situation where the acceleration sensor output fluctuates due to a steep temperature change.

Third, the description in Section 2.1.3.3.2.1 used the ratio in absolute value of the positive maximum $a+$ to the negative maximum $a-$ of the accelerator sensor output entered in the earthquake sensor. If the entered maximum is stored and used as it is, a ratio is determined as a discriminating value once noise is introduced. Therefore, any seismic wave will be judged as noise and not reflected in the output if it does not exceed the ratio so determined for some noise previously injected. In contrast, once a seismic wave is entered, any subsequent noise that does not exceed the noise ratio determined for that seismic wave will not be correctly judged as noise. As an extreme example, take a noise wave that produces $a+ = 2000$ Gal and a ratio $r = 5$. If a subsequent seismic wave entered does not exceed $a- = -400$ Gal, it will be judged as noise and the earthquake judging function will not work. In contrast, if an earthquake of $a+ = 400$ Gal and

$a-$=–400 Gal has been measured in the past, it is impossible to judge any noise input as noise except that exceeding 2000 Gal.

A function was incorporated in an actual instrument in order to avoid these problems and still utilize the basic logic. The operation of the function is described below.

(a) Real-time Judgment

Real-time noise judgment using the acceleration sensor output value entered every moment was desired and an attempt was made to secure real-time judgment and prevent a misjudgment by making a noise judgment in two stages in an actual instrument. A description is given below using Figure 2.1-21.

First, a ratio judgment is performed in real time according to the basic logic. An explanation of the operation is given for the case where a noise waveform is entered and where an earthquake waveform is entered. Figure 2.1-21a shows the case of a noise waveform. The quantity of acceleration is entered to the CPU every moment as an acceleration input from the left of the figure. At the time point $ta\text{–}x$ in the figure, $a+$ takes the value $a(ax+)$ and $a-$ takes the value $a(ax-)$. At this point of time, the ratio of $|a(ax+)|$ to $|a(ax-)|$ is within the noise ratio r, and the noise is not judged as noise. As time passes and when the time point $tb\text{–}y$ arrives, $a+$ is refreshed to $a(ay+)$ but $a-$ remains $a(ax-)$. At this time, the ratio of $|a(ay+)|$ to $|a(ax-)|$ exceeds the noise ratio r, and the noise is judged as noise.

In a similar manner, Figure 2.1-21b applies to the case of an actual seismic wave. At the time point $tb\text{–}x$, $a+$ takes the value $a(bx+)$ and $a-$ takes the value $a(bx-)$. At this point of time, the ratio of $|a(bx+)|$ to $a(bx-)|$ is within the noise ratio r, and the earthquake is not judged as noise. As time passes and when the time point $tb\text{–}y$ arrives, $a+$ is refreshed to $a(by+)$ but $a-$ remains $a(bx-)$. At this time, the ratio of $|a(by+)|$ to $|a(bx-)|$ exceeds the discriminating ratio r, and the earthquake is judged as noise. That is to say, if the basic logic is applied to a real-time system without modification, a seismic wave may be recognized as noise depending on the oscillating state of the seismic wave. This condition was thought to have the possibility of occurrence in the preliminary shock section in particular where a shock starts from a no-shaking state. However, when time advances to $tb\text{–}z$, $a-$ is refreshed to $a(bz-)$. At this time point, the ratio of $|a(by+)|$ to $|a(bz-)|$ is within the noise ratio r and the earthquake will be correctly judged as a seismic wave. Thus, the following logic operation was added to avoid this phenomenon, which is apt to occur in the preliminary shock portion of earthquake motion.

This logic 'sets a judging time width Tr, measures the time length of the condition of the ratio of input values exceeding a noise ratio and judges it as noise protection if this condition continues longer than the set time length Tr. The judging time width Tr is set to a value which contains one cycle of reciprocation of earthquake motion.'

With this added logic operation, a noise judgment is made for the noise waveform in Figure 2.1-21a at the time point $ta\text{–}j$ after the time length Tr from the time point $ta\text{–}y$ when the input data exceeds the noise ratio. For the actual earthquake

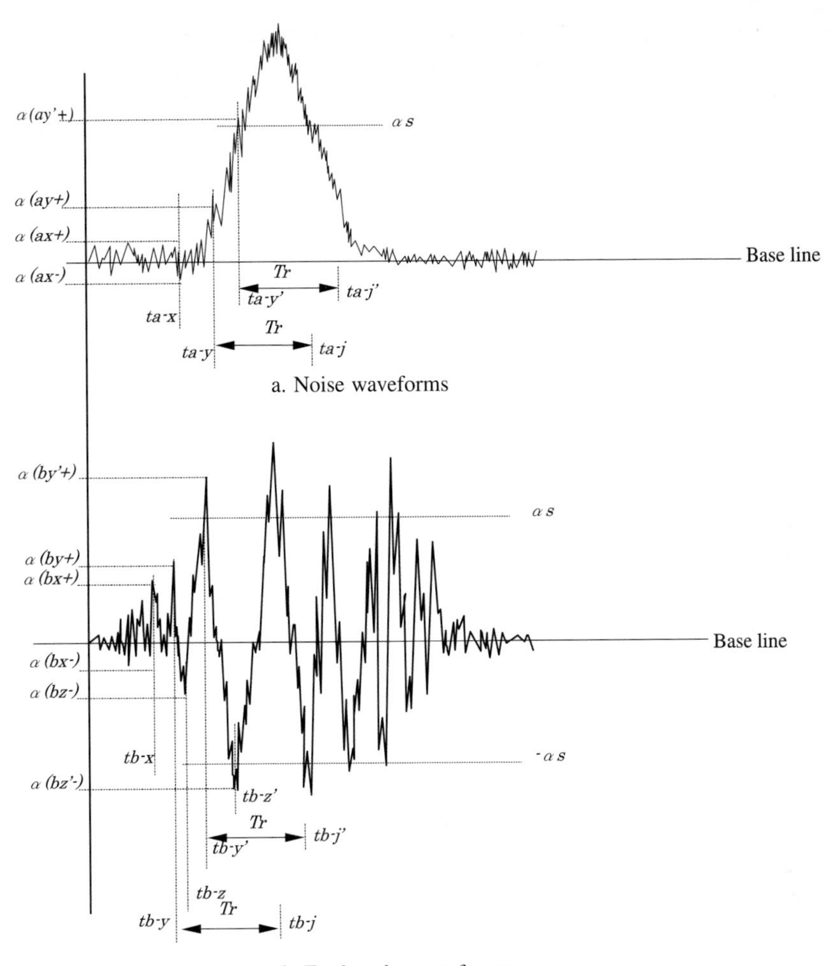

Figure 2.1-21. Initial earthquake detection logic is added to avoid misjudgment as noise. This is for actual *SI* sensor application at the field. Threshold *as* (*as*>0) and judgment time length *Tr* are set.

motion in Figure 2.1-21 b, the input data exceeds the noise ratio at the time point *tb–y*, but the ratio of positive and negative acceleration peaks falls below the noise ratio *r* at a time point *tb–z* before the time point *tb–j* when the time length *Tr* has passed from *tb–y*. Therefore, the earthquake is not judged as noise.

Hereafter, the condition where the input does not exceed the noise ratio is called the measuring state, the condition where the noise ratio is exceeded and a judgment on whether that condition continues for the time length *Tr* is waited for is called the safety state and the condition, where the noise ratio is exceeded but judged as noise because that condition had continued longer than the time length *Tr*, is called the protection state.

(b) Fluctuating Output Component of Acceleration Sensor
As mentioned under the second problem, the acceleration ratio may exceed the noise-judging ratio and a false recognition of noise may result owing to the fluctuating component in the acceleration sensor output. To avoid this, a threshold as ($as>0$) was set for the input acceleration subjected to ratio judgment. That is, the ratio judgment takes place only if the condition $|a+| > as$ or $|a-| > as$ holds for the input to be judged, and a no-noise condition is unconditionally assumed if this condition does not hold. This operation is explained using Figure 2.1-21 a.

With this added function, the time when the input data exceeds the noise ratio is $ta\text{–}y$ for the noise waveform in Figure 2.1-21 a, but the noise ratio judgment does not take place because $|a+|$ does not exceed as at this time point. The noise ratio judgment takes place first at the time point $ta\text{–}y'$ when the input acceleration exceeds as, and a safety state is entered if the noise condition is satisfied. After that, a protection state is entered at the time point $ta\text{–}j'$ when the time length during which the condition is satisfied has continued for Tr.

For the seismic wave in Figure 2.1-21b, the input data exceeds the noise ratio at the time point $tb\text{–}y$, but the noise ratio judgment does not take place at this time point because $|a+|$ does not exceed as. The noise ratio judgment takes place first at the time point $tb\text{–}y'$ when the input acceleration exceeds as, and a safety state is entered if the noise condition is satisfied. However, the input is not judged as noise because the ratio of positive and negative acceleration peaks ($|a(by'+)|$ and $|a(bz'-)|$) falls below the noise ratio r at a time point $tb\text{–}z'$ before the judgment time width Tr has passed.

(c) Influence of Past Acceleration Input History
As mentioned under the third problem, the past input components must be excluded. A window of a certain time length was established in the same manner as the Sv forgetting the algorithm for SI and the method for preventing the noise detecting function from becoming ineffective was employed, by allowing the function to forget the past window information ($a+$ and $a-$ of that window) as time went by.

In Figure 2.1-21 b, the noise waveform of Figure 2.1-21 a is injected and the values $a+$ and $a-$ used for the noise ratio are cleared to zero at certain time intervals Tw. This prevents a subsequently observed seismic wave of Figure 2.1-21 b from being misjudged as noise.

2.1.3.3.2.3 Operation Logic and Output When Incorporated in an Actual Instrument

(1) Time-dependent Forgetting Logic Incorporated in an Actual Instrument
Of the logic described in Section 2.1.3.3.2.2, (a) and (b) can be incorporated without modification. For the forgetting time of (c), however, this logic was incorporated in the form of using the time window of the SI value calculation algorithm for the purpose of reducing the complexity of internal operations and be-

cause there were already no CPU resources (ROM, RAM and throughput) to spare for it at the time of incorporation, although it was a logic that could be handled basically independent of the *SI* value calculation algorithm.

The basic forgetting method is described below using Figure 2.1-22. The basic interval window used for forgetting is 10 s in the same manner as in the *SI* algorithm (the basic interval window is called *Tw* hereafter) and the number of windows used for the noise judgment is three (the reason why the protection logic uses three windows while the *SI* algorithm uses only two windows will be explained later).

Of these windows being used, the oldest in the time history is called the back window (*Wb*), the intermediate one is called the center window (*Wc*) and the one in the current process of receiving the acceleration sensor output in real time is called the front window (*Wf*). Let $\alpha(b+)$, $\alpha(c+)$ and $\alpha(f+)$ be the maximum of absolute value of positive acceleration in the respective windows, and let $\alpha(+)$ be the maximum of these three. Likewise, let $\alpha(b-)$, $\alpha(c-)$ and $\alpha(f-)$ be the maximum of absolute value of negative acceleration and let $\alpha(-)$ be the maximum of these three. A noise ratio judgment was made using $\alpha(+)$ and $\alpha(-)$ according to Equation (2.1-10).

At time *t*4–1, noise begins to be injected. At this time, the accelerations for judgment are $\alpha(+)=\alpha(041+)$ and $\alpha(-)=\alpha(03-)$, and they exceed the judgment ratio *r*. However, because the input does not exceed the acceleration threshold *as*, no noise judgment takes place as a consequence and the instrument continues its measuring state. When the time arrives at the time point *t*4–2, the noise input exceeds *as* and the noise ratio condition is satisfied. The instrument enters a safety state in which it waits for a time continuation of *Tr*. After that, the condition of a time continuation of *Tr* is satisfied at the time point *t*4–3, and the instrument enters a protection state at this time point. After that, time passes and the front window *Wf* comes to *W*08. At this time point, the acceleration memorized in the window *W*05 that has exceeded *as* is forgotten from the memory of the back window and the noise ratio judgment becomes ineffective. However, the operation of the instrument does not immediately recover from the protection state. The recovery from the protection state is restricted until the acceleration input due to past noise ceases to affect the seismic intensity calculation result. In the actual instrument, the waiting time for restoration is set as the number *np* of the forgetting windows, and restoration from the protection state takes place at the point of time when the number of times of forgetting exceeds *np* after the three windows used for noise judgment have all entered a no-noise state. In the case of $np=1$, in the example of Figure 2.1-22, the protection state is terminated at the time point *t*9–1. For this protection-terminating condition, it may be considered that the logic of ‘until the acceleration input stabilizes below a certain threshold’ is incorporated. However, this was not employed this time because this would have made the logic complex and the load on the CPU was already marginal.

This operation was applied to earthquake waveforms and it was found that, although basically a normal operation took place, such a pattern might arise at the end of the principal motion of an earthquake that caused the earthquake

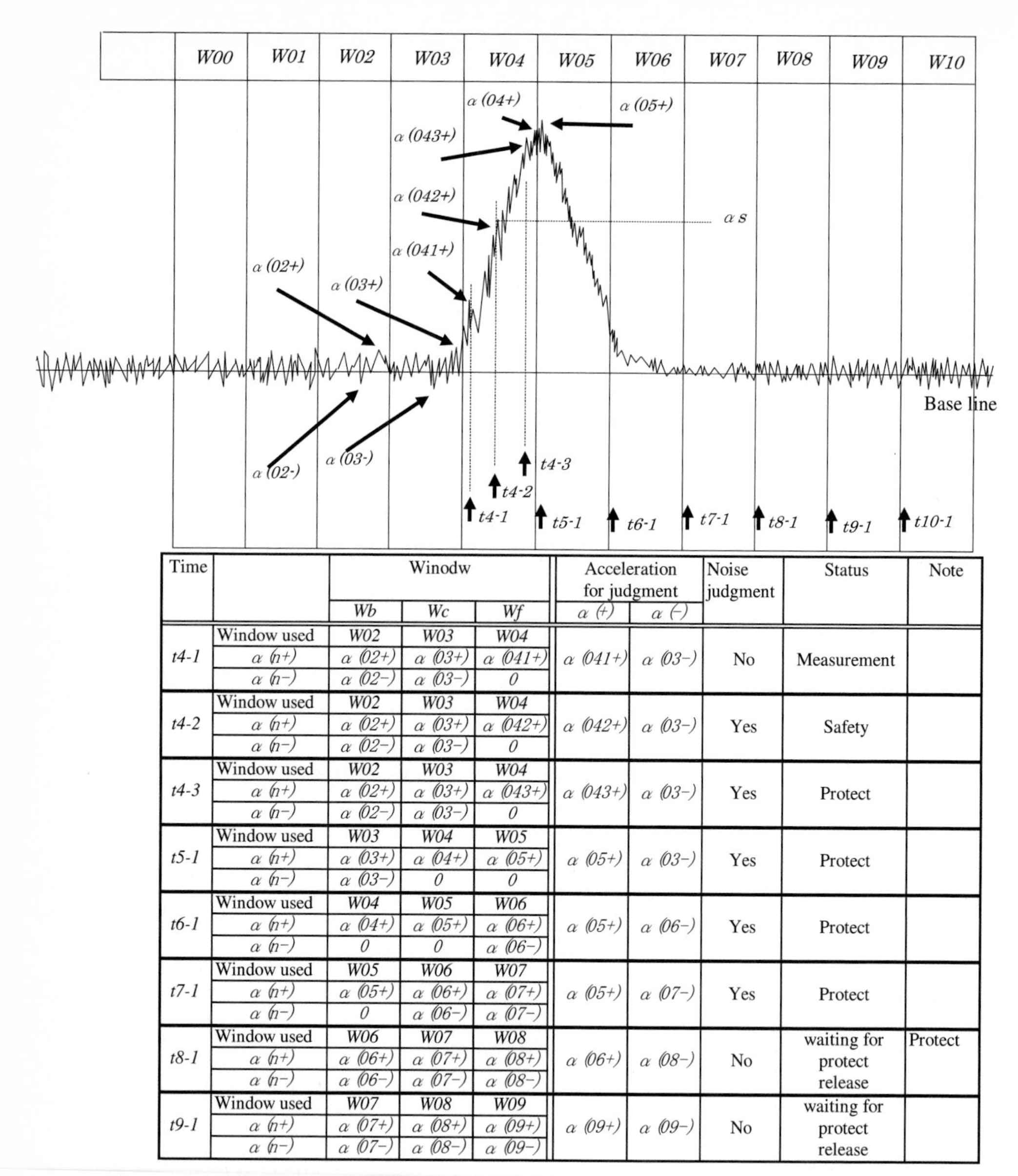

Time		Winodw			Acceleration for judgment		Noise judgment	Status	Note
		Wb	*Wc*	*Wf*	α (+)	α (−)			
t4-1	Window used	*W02*	*W03*	*W04*	α (041+)	α (03−)	No	Measurement	
	α (n+)	α (02+)	α (03+)	α (041+)					
	α (n−)	α (02−)	α (03−)	0					
t4-2	Window used	*W02*	*W03*	*W04*	α (042+)	α (03−)	Yes	Safety	
	α (n+)	α (02+)	α (03+)	α (042+)					
	α (n−)	α (02−)	α (03−)	0					
t4-3	Window used	*W02*	*W03*	*W04*	α (043+)	α (03−)	Yes	Protect	
	α (n+)	α (02+)	α (03+)	α (043+)					
	α (n−)	α (02−)	α (03−)	0					
t5-1	Window used	*W03*	*W04*	*W05*	α (05+)	α (03−)	Yes	Protect	
	α (n+)	α (03+)	α (04+)	α (05+)					
	α (n−)	α (03−)	0	0					
t6-1	Window used	*W04*	*W05*	*W06*	α (05+)	α (06−)	Yes	Protect	
	α (n+)	α (04+)	α (05+)	α (06+)					
	α (n−)	0	0	α (06−)					
t7-1	Window used	*W05*	*W06*	*W07*	α (05+)	α (07−)	Yes	Protect	
	α (n+)	α (05+)	α (06+)	α (07+)					
	α (n−)	0	α (06−)	α (07−)					
t8-1	Window used	*W06*	*W07*	*W08*	α (06+)	α (08−)	No	waiting for protect release	Protect
	α (n+)	α (06+)	α (07+)	α (08+)					
	α (n−)	α (06−)	α (07−)	α (08−)					
t9-1	Window used	*W07*	*W08*	*W09*	α (09+)	α (09−)	No	waiting for protect release	
	α (n+)	α (07+)	α (08+)	α (09+)					
	α (n−)	α (07−)	α (08−)	α (09−)					

Figure 2.1-22. For noise detection, similar to the *Sv* forget algorithm of *SI*, constant time windows are prepared and past window data (*a*+, *a*– of those windows) are abandoned. Thus, continuous noise detection is avoided.

ground motion to be recognized as noise at fault. This is shown in Figure 2.1-23. The earthquake reaches the end of the principal motion in the window *W*06. Assume that the division of the forgetting windows is located as shown in this figure. As shown in Figure 2.1-23, at the time point *t*8–1, when time changes over

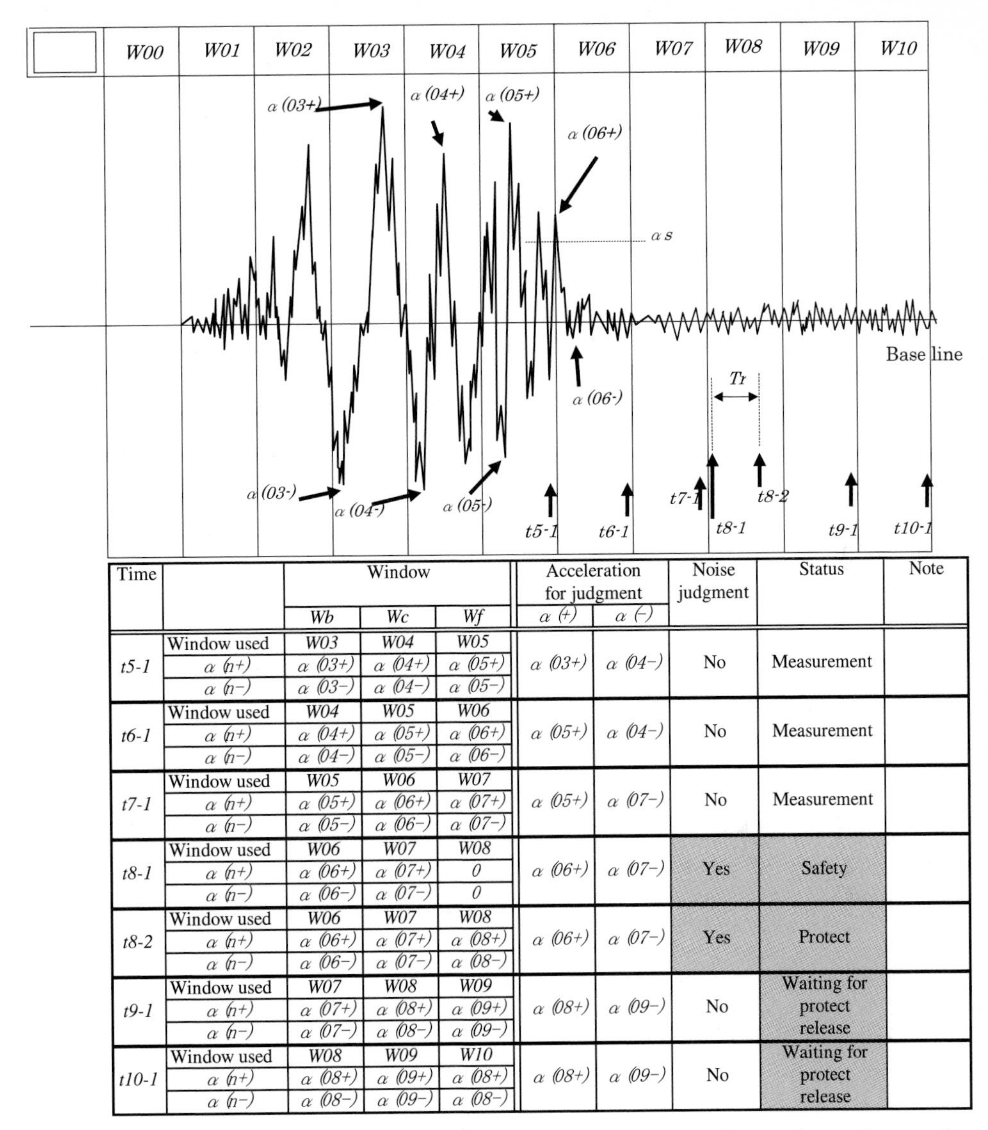

Time		Window			Acceleration for judgment		Noise judgment	Status	Note
		Wb	*Wc*	*Wf*	$\alpha(+)$	$\alpha(-)$			
t5-1	Window used	*W03*	*W04*	*W05*	$\alpha(03+)$	$\alpha(04-)$	No	Measurement	
	$\alpha(n+)$	$\alpha(03+)$	$\alpha(04+)$	$\alpha(05+)$					
	$\alpha(n-)$	$\alpha(03-)$	$\alpha(04-)$	$\alpha(05-)$					
t6-1	Window used	*W04*	*W05*	*W06*	$\alpha(05+)$	$\alpha(04-)$	No	Measurement	
	$\alpha(n+)$	$\alpha(04+)$	$\alpha(05+)$	$\alpha(06+)$					
	$\alpha(n-)$	$\alpha(04-)$	$\alpha(05-)$	$\alpha(06-)$					
t7-1	Window used	*W05*	*W06*	*W07*	$\alpha(05+)$	$\alpha(07-)$	No	Measurement	
	$\alpha(n+)$	$\alpha(05+)$	$\alpha(06+)$	$\alpha(07+)$					
	$\alpha(n-)$	$\alpha(05-)$	$\alpha(06-)$	$\alpha(07-)$					
t8-1	Window used	*W06*	*W07*	*W08*	$\alpha(06+)$	$\alpha(07-)$	Yes	Safety	
	$\alpha(n+)$	$\alpha(06+)$	$\alpha(07+)$	0					
	$\alpha(n-)$	$\alpha(06-)$	$\alpha(07-)$	0					
t8-2	Window used	*W06*	*W07*	*W08*	$\alpha(06+)$	$\alpha(07-)$	Yes	Protect	
	$\alpha(n+)$	$\alpha(06+)$	$\alpha(07+)$	$\alpha(08+)$					
	$\alpha(n-)$	$\alpha(06-)$	$\alpha(07-)$	$\alpha(08-)$					
t9-1	Window used	*W07*	*W08*	*W09*	$\alpha(08+)$	$\alpha(09-)$	No	Waiting for protect release	
	$\alpha(n+)$	$\alpha(07+)$	$\alpha(08+)$	$\alpha(09+)$					
	$\alpha(n-)$	$\alpha(07-)$	$\alpha(08-)$	$\alpha(09-)$					
t10-1	Window used	*W08*	*W09*	*W10*	$\alpha(08+)$	$\alpha(09-)$	No	Waiting for protect release	
	$\alpha(n+)$	$\alpha(08+)$	$\alpha(09+)$	$\alpha(08+)$					
	$\alpha(n-)$	$\alpha(08-)$	$\alpha(09-)$	$\alpha(08-)$					

Figure 2.1-23. Example of acknowledging earthquake wave as noise at the end part of earthquake major waves.

from window *W*07 to *W*08, the earthquake is judged as noise and a safety state is entered because the acceleration for the noise ratio judgment is calculated from the data of windows *W*07 and *W*08. Because the earthquake ground motion terminates after this, the instrument enters a protection state at the time point *t*8–2 after the time length *Tr*. Because the possibility of this condition occurring in an actual earthquake motion could not be denied, the following condition was added to solve this problem.

Figure 2.1-24 shows the additional condition. The occurrence of the noise ratio judgment in each window is used as the information quantity. In window *W*05, for example, the judgment of presence or absence of noise made based on the accelerations $\alpha(05+)$ and $\alpha(05-)$ in the window is used, and absence is represented by 0 and presence by 1. At the timing of window forgetting, let the window-based judgment result pattern of the current back, center and front windows be 0, 1 and 0, respectively. That is, the judgment result of the center window is 1 at forgetting. The acceleration data in this window is masked as not to be used for noise judgment and the data in the back window are used. This information is added to the column of the window-based judgment in the table. Let the operation be ascertained according to this logic. Under the same conditions as in Figure 2.1-23, the window-based judgment pattern is 0/1/0 at the time point *t*7–1. Therefore, at the forgetting timing when a transition occurs from the time point *t*7–1 to *t*8–1, the acceleration data in window *W*06 are masked and the noise judgment at the time point *t*8–1 is made by calculating the data in the center window at that time point (*W*07 in this case) and the acceleration data at the current time point *t*8–1, and the judgment result of no noise is obtained. The noise protection function uses three windows whereas the *SI* value algorithm uses two windows. The reason for this is to work at this additional condition.

(2) Output Operation

The noise protection function is incorporated as an operation to continue calculation of the seismic index from the acceleration output and restrict the output if the noise condition is satisfied.

An output operation logic block diagram of the earthquake sensor is shown in Figure 2.1-25. This noise protection function is inserted between output value processing and output peak hold processing based on calculated values. The processing varies according to whether it is in a measuring state, safety state, or protection state. However, this function is designed not to interfere at all with the output value processing or peak hold processing.

An example of actual output operation is given below to explain the output restriction operation in each state.

The output operation for actual earthquake ground motion is shown in Figure 2.1-26 for the case of the presence and absence of this logic. In this earthquake ground motion example, the instrument enters a safety state at the time point *tb–y′* in the first half. If this noise protection function is not incorporated, an analog output (the discussion here takes it as the *SI* value output) is delivered as delineated by the dotted line, and the ON/OFF output enters an ON state because the set value is exceeded at the time point *tst*. If this logic is incorporated, the analog output and ON/OFF output are both refreshed as delineated by the continuous line at the time point *tb–z′* when the safety state changes over to a measuring state. Therefore, there is a possibility that the time output of (*tb–z′*)–(*tb–y′*) is delayed with respect to the actual calculated value in the first half of the earthquake ground motion by incorporating this function.

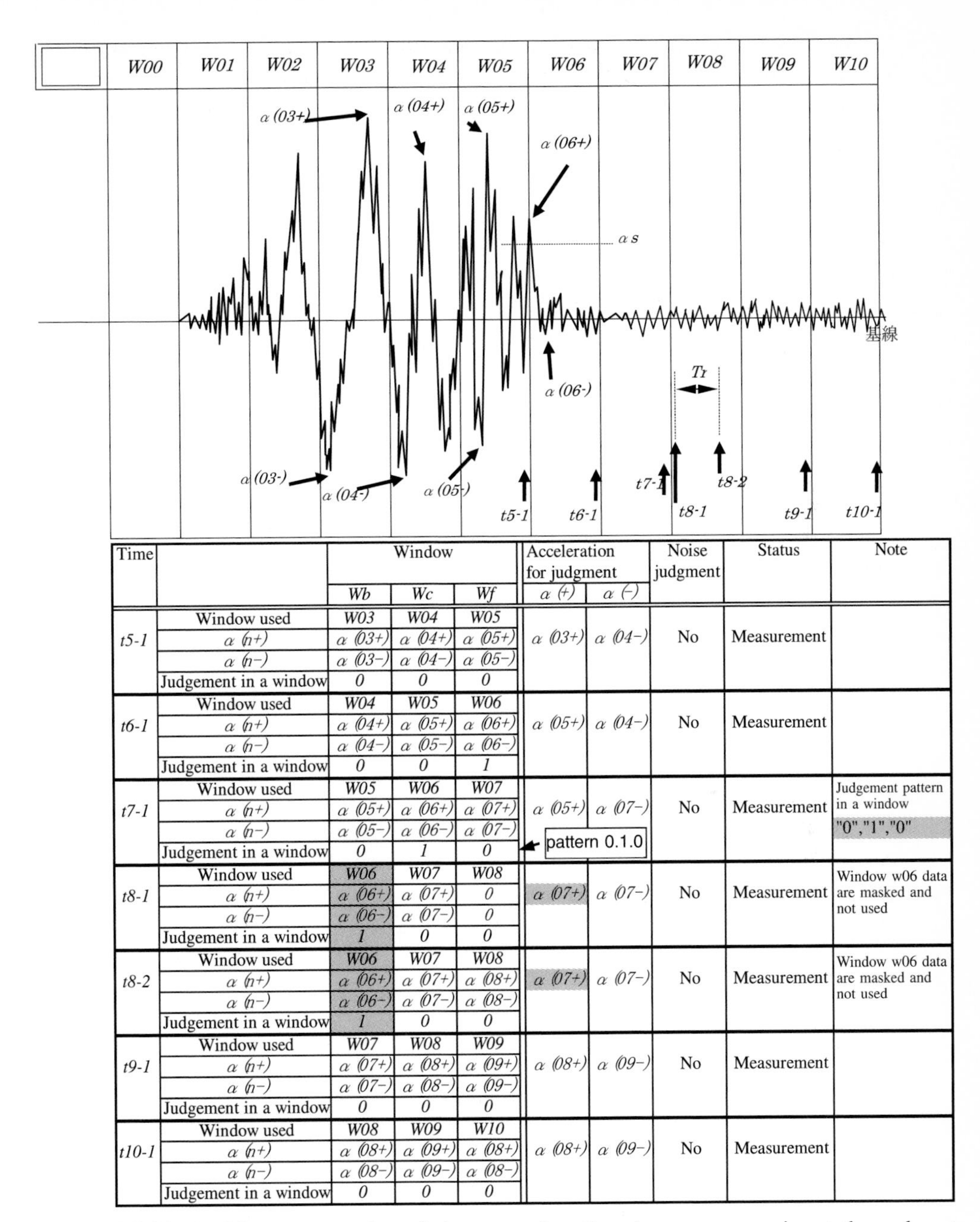

Time		Window			Acceleration for judgment		Noise judgment	Status	Note
		Wb	Wc	Wf	α (+)	α (−)			
t5-1	Window used	W03	W04	W05	α (03+)	α (04−)	No	Measurement	
	α (n+)	α (03+)	α (04+)	α (05+)					
	α (n−)	α (03−)	α (04−)	α (05−)					
	Judgement in a window	0	0	0					
t6-1	Window used	W04	W05	W06	α (05+)	α (04−)	No	Measurement	
	α (n+)	α (04+)	α (05+)	α (06+)					
	α (n−)	α (04−)	α (05−)	α (06−)					
	Judgement in a window	0	0	1					
t7-1	Window used	W05	W06	W07	α (05+)	α (07−)	No	Measurement	Judgement pattern in a window "0","1","0"
	α (n+)	α (05+)	α (06+)	α (07+)					
	α (n−)	α (05−)	α (06−)	α (07−)					
	Judgement in a window	0	1	0	pattern 0.1.0				
t8-1	Window used	W06	W07	W08	α (07+)	α (07−)	No	Measurement	Window w06 data are masked and not used
	α (n+)	α (06+)	α (07+)	0					
	α (n−)	α (06−)	α (07−)	0					
	Judgement in a window	1	0	0					
t8-2	Window used	W06	W07	W08	α (07+)	α (07−)	No	Measurement	Window w06 data are masked and not used
	α (n+)	α (06+)	α (07+)	α (08+)					
	α (n−)	α (06−)	α (07−)	α (08−)					
	Judgement in a window	1	0	0					
t9-1	Window used	W07	W08	W09	α (08+)	α (09−)	No	Measurement	
	α (n+)	α (07+)	α (08+)	α (09+)					
	α (n−)	α (07−)	α (08−)	α (09−)					
	Judgement in a window	0	0	0					
t10-1	Window used	W08	W09	W10	α (08+)	α (09−)	No	Measurement	
	α (n+)	α (08+)	α (09+)	α (08+)					
	α (n−)	α (08−)	α (09−)	α (08−)					
	Judgement in a window	0	0	0					

Figure 2.1-24. Avoiding wrong acknowledgement of earthquake waves as noise at the end part of earthquake major waves.

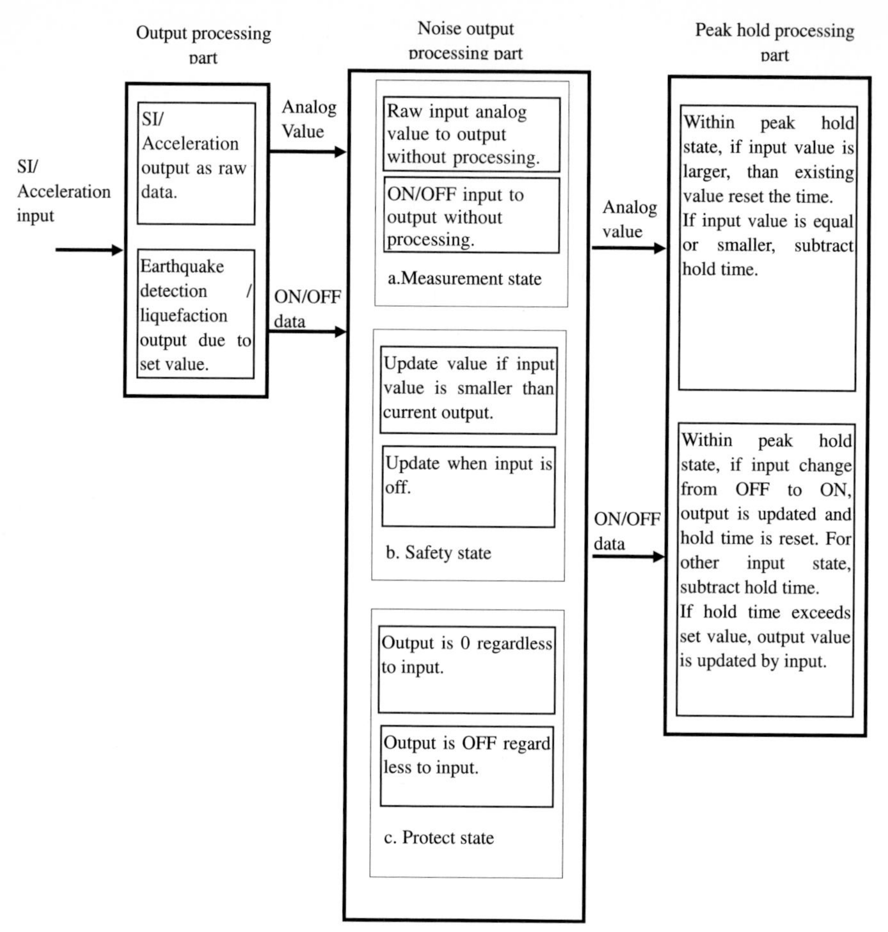

Figure 2.1-25. Noise protection function is inserted between output processing step and output peak hold step. Process is changed due to the status of measurement, safety, and protect state.

2.1.3.3.2.4 Determination of Parameters

Thus far, this section has explained the noise protection function. The parameters to be determined actually to operate this function consist of the following:

(a) noise ratio (*r*): ratio of positive and negative accelerations to discriminate between noise and an earthquake;

(b) function threshold (*as*): threshold to prevent a misjudgment due to output fluctuation of the acceleration sensor;

(c) judging time width (*Tr*): waiting time to a judgment as noise.

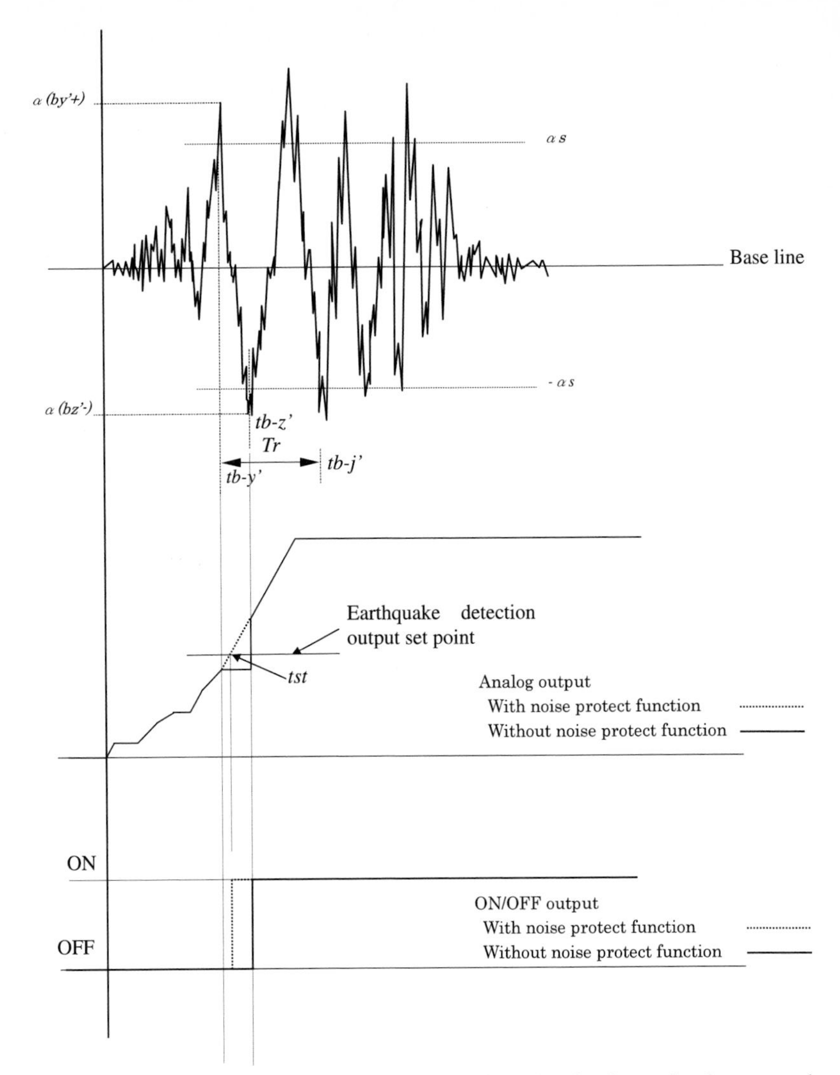

Figure 2.1-26. Output signal difference of real earthquakes is shown in the case when the noise protect function is used or not. Output operation is only different in the safety state from time $tb-y'$ to $tb-z'$.

For the noise ratio r, the ratio of positive and negative accelerations of actual earthquake waveforms was calculated by deriving them from 26 seismic waves with the result that the ratio was less than 3. From this, it was determined that $r=5$.

For the function threshold *as*, because an *SI* value of 30 kine was used for the controlling threshold in the major applications described in the next section, it was determined that $as=50$ Gal, at which a value of 15 kine is not exceeded when a noise component of a triangular waveform is input.

For the judging time width *Tr*, because the zero-crossing period was 2 s in the long-period portion of earthquake ground motion as described under liquefaction, the basic period of noise detection was assumed as twice that period, or 4 s. From this, it was determined that the judging time width *Tr* at $Tr = 2$ s assuming the positive and negative peak intervals, or half-period.

As described at the beginning of this section, there is a fair possibility that the noise environment will change in the future. Therefore, it was decided to place these parameters at a location where they could be altered inside the *SI* sensor. In the case where circumstances change, requiring the parameter threshold to be altered, the threshold setting can be changed through wired communication.

2.1.3.3.3 Summary

The acceleration sensor employed for this development exhibits a certain behavior in a radiation field. A feature was found in this behavior and a method was developed to discriminate noise from a seismic wave in real time on the side of logical processing in order to restrict the output in case of noise. This functional operation was successfully incorporated within the limits of inexpensive earthquake sensor resources as a function to discriminate between a seismic wave and noise by utilizing the time window technique used for the *SI* value calculation.

Because these measures to counter electromagnetic noise are taken on the signal processing logic side, they are thought to function effectively for a practical service period of 10 years after installation (15 years if the life of the last installed sensor is included), even if changes take place in the radio environmental conditions during that period.

2.1.4 Application

As described in the previous sections, this *SI* sensor has two built-in functions: one is to measure acceleration and calculate and deliver multiple seismic indices in real time and the other is a protect function to prevent the control output from causing misjudgment owing to noise, etc. These make it a highly functional control earthquake sensor. This section considers the system of Tokyo Gas Co., Ltd. as an example and presents an application in which this *SI* sensor is actually used. The application is intended to shut off gas pipelines at the occurrence of a disaster and estimate damage with high accuracy.

2.1.4.1 Shut-off of Gas Pipelines at the Occurrence of a Disaster

Tokyo Gas Co., Ltd. has pipelines of 48 000 km total length in an area of 3100 km^2 in the metropolitan area of Tokyo and supplies city gas from factories to consumers through those pipelines. Natural gas is refined at the factory and then is sent out in a high-pressure state. The high-pressure city gas goes through governors successively to undergo pressure reduction from high to medium pressure and further to low pressure before reaching common households.

Of these pipelines at different pressures, the high- and medium-pressure pipelines form a system that can suspend gas transportation as follows (see Figure 2.1-27). If an earthquake occurs and it is judged that gas transportation should be suspended on the high- or medium-pressure pipelines, the headquarters of Tokyo Gas Co., Ltd. uses radio links to control the shut-off valves installed at several places on these pipelines. On the other hand, the low-pressure pipelines are set up like a mesh over the consuming district and they are the largest of all types of pipelines in total length. To shut off the supply through the low-pressure pipelines, gas transportation will be suspended at the governor that reduces pressure from medium to low. The governors are installed at 3700 locations in the service area. If all the governors are to be shut off by a radio system, the system will be expensive because of the facilities to be installed and their maintenance costs. This choice is not practical.

Therefore, we chose a system to shut off the low-pressure governor automatically by earthquake sensors installed at the same location in order to achieve security on the occurrence of a disaster at low cost. The newly developed earth-

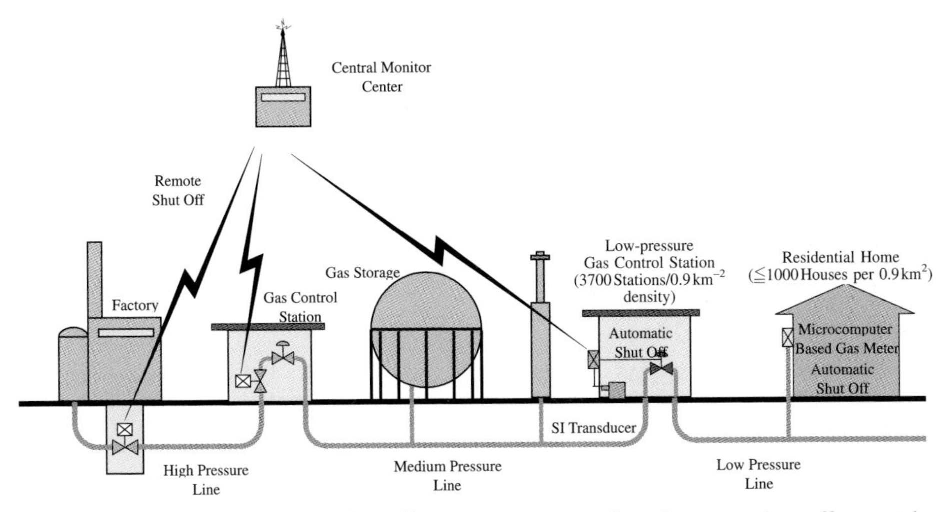

Figure 2.1-27. High-pressure and medium-pressure control stations are shut off remotely from the central control center in case of heavy earthquakes. 3700 low-pressure control stations are shut off automatically by *SI* sensors and *SI* value and liquefaction information will be transmitted to the center.

quake sensor is used for shut-off control of this low-pressure governor. So far, a mechanical *SI* sensor has been used for automatic shut-off. However, replacing it with the new *SI* sensor has made control based on accurate *SI* value measurement results possible. The new earthquake sensor is intelligent with a CPU mounted on it. Therefore, if it becomes necessary to change the shut-off threshold owing to facility improvements in the future, this sensor can flexibly adapt itself to such a change.

2.1.4.2 Use in a High-density Earthquake Disaster Prevention System

Tokyo Gas Co., Ltd. has so far installed measuring *SI* sensors (servo type) at about 300 locations, liquefaction sensors (water level sensors) at 20 locations and based seismographs at five locations in its service area and observed earthquake information from these instruments using private microwave communication network. It also observes information on pressure, flow rate and shut-off of the low-pressure governor at about 300 locations using the same private communication network. These two types of information are entered into an 80000-mesh database with a mesh size of 250×175 m that was prepared beforehand based on information on ground conditions and the like. Tokyo Gas developed an earthquake sensing and alerting system, SIGNAL, to estimate the damage to the pipelines by combining the information obtained thus far with the database and has operated the system since 1994 (see Figure 2.1-28) [18, 19].

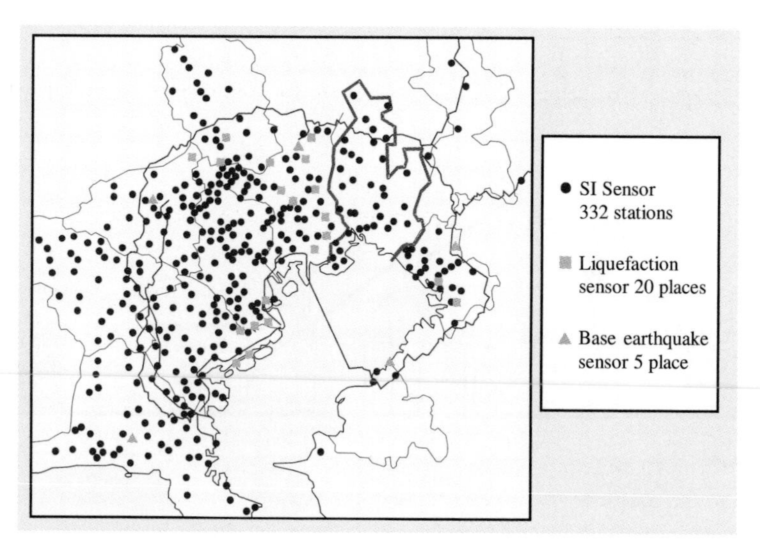

Figure 2.1-28. Existing earthquake information monitoring stations with existing earthquake disaster prevention system SIGNAL. *SI* value measurement spots are 332 locations and 20 locations for liquefaction monitoring in overall 3700 km^2.

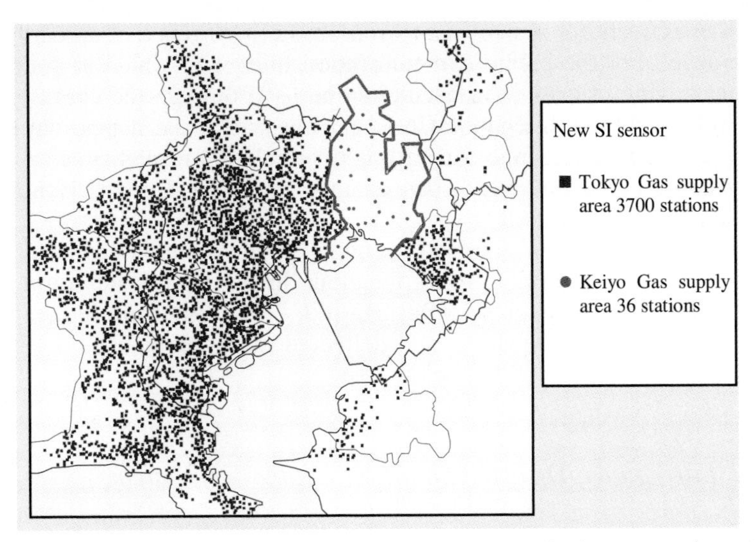

Figure 2.1-29. New *SI* sensor placement. *SI* value and liquefaction measuring points are 3700 locations in overall 3100 km^2. This constructs a high density earthquake monitoring network for every 0.9 km^2 and high-grade earthquake damage prediction with Supreme will become possible.

The installation of the new earthquake sensor on the automatic earthquake-sensing and shut-off device described in the previous section will be completed for about 1800 locations by March 2002 and at all 3700 locations by 2007 (see Figure 2.1-29). Replacing the conventional mechanical *SI* sensor with this new *SI* sensor will make it possible to hold manifold information such as *SI* values, acceleration values, and liquefaction detection at the 3700 governor-installed locations on the occurrence of an earthquake. This means that measured earthquake information will be available in an area of every 1 km^2 and the world's highest density earthquake observation network will have been established. Tokyo Gas Co., Ltd. began construction of the "Supreme" earthquake disaster prevention system in January 1998 using high-density earthquake information obtained from the sensor and bore data at more than 60000 locations. This system is outlined below.

2.1.4.2.1 Configuration of Earthquake Information Acquisition System

On the low-pressure governor at 3700 locations, a telemeter device had been installed and connected to public communication lines even before the development of this earthquake sensor to record pressure and other information at the governor. The earthquake information from the new earthquake sensor, such as *SI* value, Gal value, liquefaction detection, and governor shut-off, is taken in this telemeter device and sent to the situation room at the headquarters together with gas pressure information on the occurrence of an earthquake. In constructing this disaster prevention system, we partly improved the recording telemeter device to make it able to acquire data effectively for a short time even on the occurrence of an earthquake of disastrous level. For the public communication lines with which to connect, telephone lines having priority at disasters can be used through the courtesy of NTT (Nippon Telegraph and Telephone Co.), and this will make it possible to collect data efficiently for a short time. According to simulation results, 80% of alert data can be gathered in 25 min even when damage at the level of the Hanshin-Awaji earthquake class has occurred.

Earthquake information obtained through public communication lines, the earthquake information at about 300 locations linked by private communication network, and liquefaction sensor information at 20 locations are integrated in the situation room at the headquarters and delivered to the earthquake disaster situation estimating system described below.

2.1.4.2.2 Earthquake Disaster Situation Estimating System

The Supreme system (Figure 2.1-30) can interpolate and estimate the following earthquake disaster situation from a database prepared beforehand for the 1.4 million meshes of 50×50 m based on the gathered earthquake information, topological information, geological information, boring data, etc.

(a) *SI* value information spatially interpolated in the area;

(b) spatial distribution information estimated from *SI* value, etc.;

(c) low-pressure gas pipeline damage information.

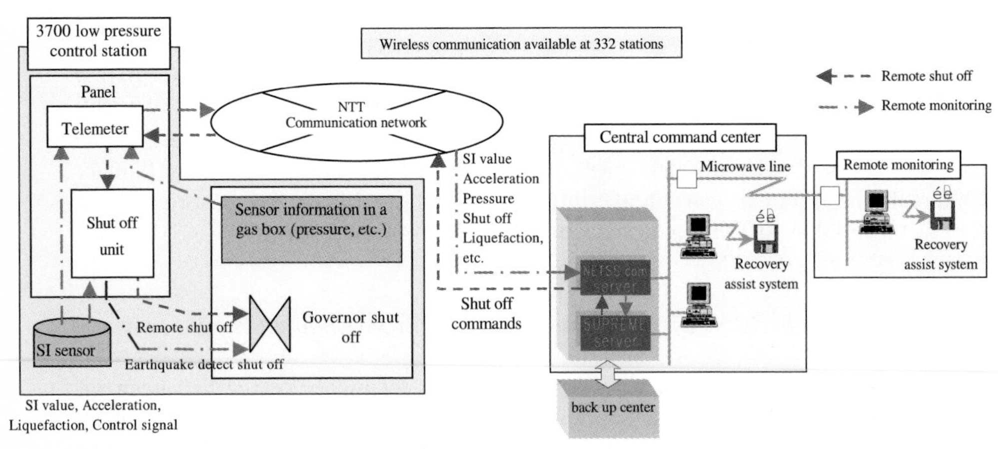

Figure 2.1-30. Schematic diagram of the newly developed earthquake disaster prevention system Supreme. Earthquake information is sent (dashed line) from low-pressure control stations to the central disaster prevention center, the Supreme server predicts damage level based on collected information, and the center determines areas to shut off gas supply. Shut off commands will be sent to stations where automatic shut off is not activated and then the gas supply is shut off at areas in critical condition.

These estimation algorithms are incorporated in Supreme based on study results attained by researchers in various fields after the commissioning of SIGNAL [20].

With this system in commission, it is possible to determine the area of gas supply suspension to prevent secondary disasters and carry out operations of the subsequent restoration program accurately and swiftly, based on these pieces of precisely estimated information supplied in real time at the occurrence of an earthquake [21].

2.1.4.3 Summary

The new earthquake sensor will be installed at 3700 locations in the metropolitan area of Tokyo to measure various kinds of earthquake information and will function as the world's highest density earthquake observation network. Accurate damage estimation has become possible based on this densely gathered earthquake information. This system started operation in July 2001 and is expected to improve substantially the earthquake disaster prevention level of the natural gas supply network in the metropolitan area.

2.1.5 Conclusion

As an instrument that is effective for earthquake disaster prevention, we have developed an earthquake sensor using a combination of a MEMS acceleration sensor and a high-performance CPU. Although small and inexpensive, multiple functions surpassing conventional seismographs are implemented in this earthquake sensor. We are proud because it would have been difficult to satisfy this target with the mere combination of readily available electronic devices and an algorithm.

As described in Section 2.1.1, we established the target of being small, inexpensive, and yet reliable as an earthquake sensor to be used in large numbers, from the start of the development.

Section 2.1.2 describes the concept and internal configuration introduced to satisfy the required specifications of small size, low cost, and high reliability, among others. We planned from the start to employ an existing mass produced MEMS acceleration sensor and CPU. Within these limitations, we aimed at developing a high-performance earthquake sensor that would surpass the function of ordinary seismographs demanded in the market. When we attempt to implement a target function, we are inevitably confronted with the problem of insufficient capability of various components. We are apt to attach too much priority to the function, and the product as the result of development often assumes a very different form from the initial target. This time, we continuously carried out development without losing sight of the imposed limitations and succeeded in attaining the targeted specifications and product cost as a result.

Section 2.1.3 presents the *SI* value calculation algorithm developed to attain this target. With this newly developed algorithm, it became possible as a novel function to calculate *SI* values with high precision in real time inside the small earthquake sensor itself.

Next, based on the results of various studies, this section describes the liquefaction-detecting algorithm that can simply detect the occurrence or absence of a liquefaction phenomenon in real time from the *SI* value derived by this calculation. According to verification results, this algorithm is fully suitable for practical use, and will provide effective information for earthquake disaster prevention in the coastal zone on the Bay of Tokyo in particular where liquefaction-prone ground is present.

At the end of Section 2.1.3, we explained the noise protect function that was devised for effective use of the MEMS sensor. As described in the text, MEMS sensors are devices from which one can draw the features of small size and low cost on the premise that they are mass produced. For effective use of a MEMS sensor, efforts are required to minimize its customization to the application. Since the device employed this time is of the electrostatic type, the wave detection and rectification circuit in the processing circuitry of the device is influenced by intrusion of electromagnetic waves. It seems that an error output is induced on one side by electromagnetic waves. However, we succeeded in suppressing misjudgement of the control function by making good use of this feature. Even if a MEMS sensor is used, the same technique cannot be used as a matter of course if the sensor is, for example, of the piezoresistance type. Like this development, it is important to employ an appropriate way of dealing with each problem by understanding the behavior of the device used.

Section 2.1.4 presents an application for this sensor. With the small, inexpensive, and highly functional earthquake sensor developed, it has become possible to implement a dense earthquake observation network. By making dense measurement feasible, it has become possible to obtain important information in emergency situations, that is, damage forecast at an earthquake disaster with high accuracy. For this implementation, the density of measuring points was augmented by installation of this sensor. Further, a large contribution is accredited to a ground amplification factor forecast method achieved by various studies and the results of research on forecast of damage to the infrastructure.

To reiterate, MEMS sensors have the features of small size, low cost, and high performance. To draw these features to the maximum, however, peripheral technology to draw the detection output from the MEMS sensor to the maximum is also very important to the performance required by the intended application. We made an effort to implement all the logical functions in a single-chip CPU in order to impart the features of small size and low cost to the product body without modification. For this purpose, we developed the new algorithm presented here and used an assembler language having the highest CPU resource efficiency for logical description. For the implementation, we used as many as 60000 steps of instruction. This time we succeeded in developing a new product without losing the features of MEMS sensors by comprehensive development including peripheral technology. We hope that this earthquake sensor will be useful for the prevention of earthquake-induced secondary disasters not only in Tokyo but also throughout the world.

2.1.6 Acknowledgments

We would like to express our sincere appreciation to all those who provided important guidance and advice during the new *SI* sensor development.

Mr. Tokue, Mr. Takahashi, and Mr. Torayashiki of Silicon Sensing Products Ltd. helped in selecting an optimized accelerometer for signal pick-up and gave useful advice on the measurement method. Proof of newly developed real-time *SI* algorithm was done under the guidance of Dr. Tsuneo Katayama, Director-General of the National Research Institute for Earth Science and Disaster Prevention. Professor Yasuda, College of Science and Engineering, Tokyo Denki University, provided extensive guidance for developing the world's first real-time liquefaction algorithm based on acceleration waveforms. Associate Professor Yamazaki, University of Tokyo, provided extensive advice on the earthquake disaster prevention system Supreme utilizing the new *SI* sensor. Mr. Shimizu, Mr. Kanamori, and Mr. Nagata, Yamatake Corporation, provided marketing, engineering, and managing guidance for the new *SI* sensor development.

2.1.7 References

[1] Shimizu, Y., in: *Proceedings of the Second Real-time Earthquake Disaster Prevention Symposium*, Tokyo, JSCE (Japan Society of Civil Engineers), 2000, pp. 127–134.
[2] Roylance, L.M., Angell, J.B., in: *Microsensors*; New York: IEEE Press, 1991, pp. 352–358.
[3] Spangler, C.L., Kemp, J.C., in: *Transducers '95, Proceedings of the 8th International Conference on Solid-State Sensors and Actuators*, Volume 1; Lausanne: Elsevier Science S.A., 1996, pp. 523–529.
[4] Roessig, T.A., Howe, R.T., Pisano, A.P., Smith, J.H., in: *Transducers '97, Digest of Technical Papers*; Piscataway: IEEE, 1997, pp. 859–862.
[5] Torayashiki, O., Takahashi, A., Tokue, R., *Trans. IEEE Jpn.* **116-E** (1966) 272–275 (title translated; only available in Japanese, *Shin Jishindo no Supekutora Kaiseki Nyuumon*).
[6] Tong, H., Yamazaki, F., in: *Seisan Kenkyu*; IIS, Univ. Tokyo, 1996, pp. 547–550.
[7] Housner, G.W., in: *Journal of the Engineering Mechanics Division*; USA, ASCE, 1959, pp. 109–129.
[8] Osaki, Y., *New Spectral Analysis of Earthquake Motion*; Tokyo: Kajima Press, 1994.

[9] Takubo, K., Yanada, T., Furukawa, H., Nagata, M., Koganemaru, K., Nakayama, W., Shimizu, Y., Araki, Y., Takahashi, A., in: *7th IEEE International Conference on Emerging Technologies and Factory Automation, ETFA'99*; Piscataway, 1999, Vol. 1, pp. 379–384.
[10] Youd, T.L., Holzer, T.L., *J. Geotech. Eng.* **120** (1994) 975–995.
[11] Ishihara, K., Anazawa, Y., Kuwano, J., *Soils Foundations* **27** (1987) 13–30.
[12] Shimizu, Y., Yasuda, S., Morimoto, I., Orence, R., in: *Soils and Foundations*, Vol. 42, No. 1, pp. 35–52, Japanese Geotechnical Society, Japan.
[13] Miyajima, M., Nozu, S., Kitaura, M., Yamamoto, M., *Pap. Annu. Conf. Jpn. Soc. Civil Eng.* No. 647, vol. 1 51 (2000) 405–414.
[14] Kostadinov, M., Yamazaki, F., *Earthquake Eng. Struct. Dyn.* (2001) 173–193.
[15] Towhata, I., Park, J.K., Orense, R.P., Kano, H., *Soils Foundations* **36** (1996) 29–44.
[16] Nakayama, W., Shimizu, Y., Suzuki, T., in: *Program and Abstracts, Seismological Society of Japan, Fall Meeting*, B69; 1998.
[17] Furukawa, H., Yanada, T., Takubo, K., Yoshikawa, Y., Ichida, S., Nagata, M., Shimizu, Y., Koganemaru, K., Nakayama, W., in: *Proceedings of 38th SICE Annual Conference Domestic Session Papers*, Vol. 1, SICE, Tokyo, 1999, pp. 1–2.
[18] Yoshikawa, Y., Kano, H., Yamazaki, F., Katayama, T., Akasaka, N., in: *Proceedings of 4th US Conference on Lifeline Earthquake Engineering*, USA, ASCE 1995, pp. 160–167.
[19] Shimizu, Y., *J. Soc. Instrum. Control Eng.* **36** (1997) 41–44.
[20] Hosokawa, N., Watanabe, T., Shimizu, Y., Koganemaru, K., Ogawa, Y., Nakahama, S., Isoyama, R., in: *Proceedings of the 26th JSCE Earthquake Engineering Symposium*; JSCE, Tokyo, 2001, Vol. 2, pp. 1333–1336.
[21] Shimizu, Y., in: *Proceedings of the First Real-time Earthquake Disaster Prevention Symposium*; JSCE, Tokyo, 1999, pp. 13–18.

List of Symbols and Abbreviations

Symbol	Designation
c	damping constant of damper
d(t)	acceleration response
D_{max}	maximum displacement
h	damping constant
k	spring constant
m	mass
np	number of forgetting windows
PGA	maximum acceleration on a horizontal plane
r	discriminating noise ratio
SI	seismic intensity
Sv	maximum velocity response
t	calculation time
Δt	calculation interval

Symbol	Designation
T	vibration period
T_z	zero crossing period
Tr	judging time width
Tw	time interval
$v(t)$	velocity response
Wb	back window
Wc	center window
Wf	front window
$y(t)$	input acceleration
a	acceleration
as	function threshold

Abbreviation	Explanation
CPU	central processing unit
DSP	digital signal processor
Gal	cm/s^2
kine	cm/s
MEMS	microelectromechanical systems
RISC	restricted instruction set computer

2.2 Microelectronic Bonding Processes Monitored by Integrated Sensors

M. MAYER, ESEC SA, Cham, Switzerland
J. SCHWIZER, ETH ZÜRICH, ZÜRICH, SWITZERLAND

Abstract

Real-time monitoring methods are beneficial for the control, optimization, and failure analysis of microelectronic packaging processes. The focus of this chapter is on the real-time monitoring of two widely used processes, soft solder die bonding and thermosonic ball bonding. The methods presented here use uniquely developed microsensors integrated on custom-made test chips using commercial double-metal CMOS processes.

For the soft solder die bonding process, nine aluminum-based resistive temperature detectors were integrated on various locations on a test chip. The temperature was monitored during the placement of the test chip on the leadframe. The experimental results describe the wetting and spreading effects of the molten solder under the drip. Together with numerical results obtained with a transient thermal finite element model, the beginning of wetting can be quantified. For the soft solder PbSn10, wetting occurred around 30 ms after touch down. This method hence can be used to determine wetting times under different process conditions. The wetting time determines an upper limit of the throughput of the process.

For the thermosonic ball bonding process, temperature and force sensors were integrated on test chips. The sensors are aluminum resistors and piezoresistive p^+- and n^+-diffused resistors placed around bond pads. A bond quality parameter can be derived from the temperature signal variation during bonding. The piezoresistive sensors permit the measurement of forces in the x-, y-, and z-directions simultaneously. The ultrasonic tangential force signal reached a typical value of 0.12 N on a 50 μm diameter contact zone and revealed significant process characteristics. Four process phases are identified which are necessary for a successful ball bond formation. These phases are assigned to initial stiction, sliding, bond growth, and deformation effects.

Keywords: integrated sensor; microsensor; die attach; wire bonding; monitoring

Contents

2.2.1 Introduction

Integrated electronic components such as circuits, optical components, and micro systems reach the consumer only after being mechanically protected and electrically interconnected. Electronic packaging currently adds substantially to the cost of most electronic components.

Just like the entire field of microelectronics, research and development in electronic packaging is driven by factors such as cost, size, reliability, and performance [1–3]. As the production processes evolve towards higher precision and throughput, new challenges emerge, and problems solved for one process generation again become challenges for the next generation of technologies [4].

How to meet these challenges by introducing a microsensor-based approach is reported in this chapter. The family of integrated microsensors presented here enables real-time in situ monitoring methods for two microelectronic bonding processes – solder-based die bonding and of ultrasonic wire bonding. Out of the various other electronic packaging processes, the two bonding processes are identified as most promising for the application of integrated process monitors. The two processes lend themselves to a practical realization of the necessary tasks, including:

- the design of integrated microsensors compatible with commercial complementary metal oxide semiconductor (CMOS) processes, assisted by finite element (FE) analysis;
- the engineering of devices for electrical connection to the sensors inside the processes;
- the separation of thermal and mechanical signals by appropriate selection of sensor material, sensor design, experimental conditions, and evaluation techniques; and
- the development of evaluation software to cope efficiently with the large amount of experimental data.

The next section introduces the two bonding processes and the two mainly used microsensor types – temperature sensors and piezoresistors. Different possible integrated temperature sensors are outlined as well as some basic theory of resistive temperature detectors and integrated piezoresistors. Sections 2.2.3 and 2.2.4 show how the tasks mentioned above are addressed for the two bonding processes. Finally, Section 2.2.5 lists the conclusions and gives an outlook on future research.

2.2.2 Process Monitoring

Process monitoring is a vital routine in semiconductor wafer fabrication. Test structures are included on each wafer to measure critical dimensions and electrical parameters. Measurements of critical dimensions and impurity density are carried out off-line in between process steps. Laser-based techniques such as ellipsometry and ultrasonic laser sonar are used for thickness measurements of transparent and opaque films [5]. Besides these well-known methods, research on real-time process control techniques is advancing rapidly. For reactive ion etching (RIE), a closed-loop etching rate control has been reported using impedance measurements of the plasma [6]. Recently, a micromachined resonant sensor has been used to monitor RIE more directly as reported in [7], a reference containing an overview of in situ monitoring techniques for RIE. The resonant sensor is integrated on a test wafer. Real-time etch rate is monitored by measuring the decrease in the sensor fundamental frequency during the removal of material by the etching process.

Similarly to wafer fabrication processes, microelectronic packaging processes are also monitored and evaluated using a considerable variety of techniques [8, 9]. A brief summary is given in the following subsection.

2.2.2.1 Microelectronic Package Evaluation

A natural distinction can be made between off-line and on-line packaging process characterization methods. *Off-line methods* usually consist of an evaluation of the package after the process step. The result is an indirect measure allowing qualification of the process. As an example, *destructive testing*, simultaneous force measurement, and subsequent visual inspection [9, 10] are applied off-line to characterize bond connections such as die bonds, wire bonds, and solder joints. Special equipment is used during the destructive test to measure the force needed to break the bond between two materials. Another example is *reliability testing*, which is extensively used and usually destructive. In reliability testing, the devices usually are subjected to temperature cycling or exposed to hot and humid atmospheres. Such test procedures are described in a number of standards. Subsequently, the devices are opened for visual inspection. Off-line process monitoring generally relies on spot checking or *sampling*, as used for process qualification, process control, and production lot acceptance tests. In the worst case, this may imply the rejection of entire production lots if the measured values are out of specification. This might be followed by a recalibration of the process.

Off-line evaluations of stress and temperature fields within the package are useful for the assessment of package materials and geometries [11]. For this purpose, various acoustic and electromagnetic waves are used to nondestructively screen the product. Visible light is used in Moiré interferometry [12], laser interferometry, and laser speckle interferometry to quantify surface deformations.

Scanning laser-reflectance thermometry is reported to resolve the die-attach thermal resistance [13]. X-rays are used to inspect the interior of molded packages [9]. Ultrasound is used in scanning acoustic microscopy (SAM) to detect voids at interfaces between the chip and packaging materials. Various off-line molded package characterization techniques are summarized in [14].

In contrast to off-line methods, *on-line* or *real-time methods* are directly applicable during the process. If they are compatible with the production, they permit an automated feed-back to the machine and thus can be used for a closed-loop process control technique. Automatic production stops are possible if a process characteristic shifts out of a specified range. *Machine vision* [15] is an on-line technique extensively used for quality monitoring and accurate component placement. Real-time *impedance spectroscopy* of screen printing solder paste used in surface-mount technology has been reported [16]. *Sensors* are extensively used for the real-time monitoring of pressure, temperature, and other important process parameters. The in situ temperature during lead-on-chip die attach was measured using a miniaturized resistive temperature detector (RTD) attached to the chip [17]. Quartz pressure transducers and thermocouples have been used for real-time investigations in transfer molding of microchips [18]. These pressure and temperature sensors were inserted in the mold cavity and used for measurements during mold compound injection. Recently, Hall sensors were used to determine die-pad tilting during mold compound injection [19, 20]. In this work, the chip was substituted by a magnet, the orientation of which was detected using Hall sensors. Another sensor for real-time monitoring was applied during assembly on printed-circuit boards [21]. It measured the microelectronic package deformation during infrared reflow soldering using cantilever beam probes touching the package surface. The deformation signal exhibited characteristic peaks when popcorning of the package occurred. An on-line process control method for wedge bonders consists of monitoring the wire deformation during the bond [22]. Ultrasound is stopped as soon as the specified deformation is obtained. More robust processes are feasible using this technology.

In addition to these macroscopic transducers, microsensors [23, 24] have found fascinating applications in the field of microelectronic packaging monitoring. Microsensors usually are sensors integrated on a semiconductor chip using very large-scale integration (VLSI) technologies. VLSI technologies normally are used to produce integrated circuits. Microsensors (*integrated sensors*) are characterized by their small dimensions – typically in the range of a few hundred microns. In the following, a brief summary of such devices is given.

2.2.2.2 Packaging Characterization by Microsensors

Microsensors permit nonintrusive in situ measurements [25–37]. Test chips with integrated microsensors for plastic package analysis were reported in considerable detail [11, 25]. Such microsensors measure thermal stress [26, 27], corrosion [28–31], and moisture [25, 31]. Various test chips are commercially avail-

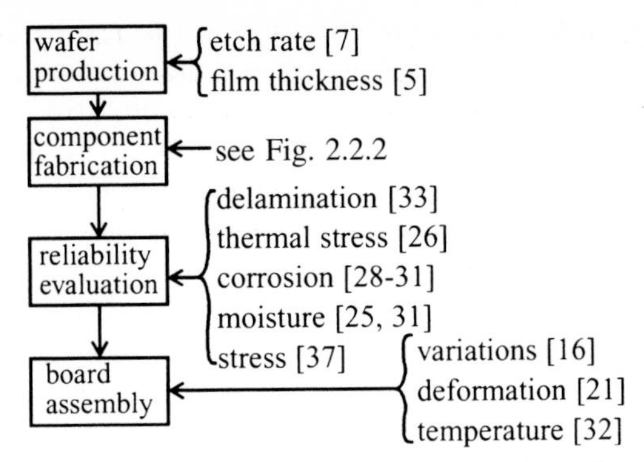

Figure 2.2.1. Microelectronic manufacturing process flow [2] and references to monitoring techniques.

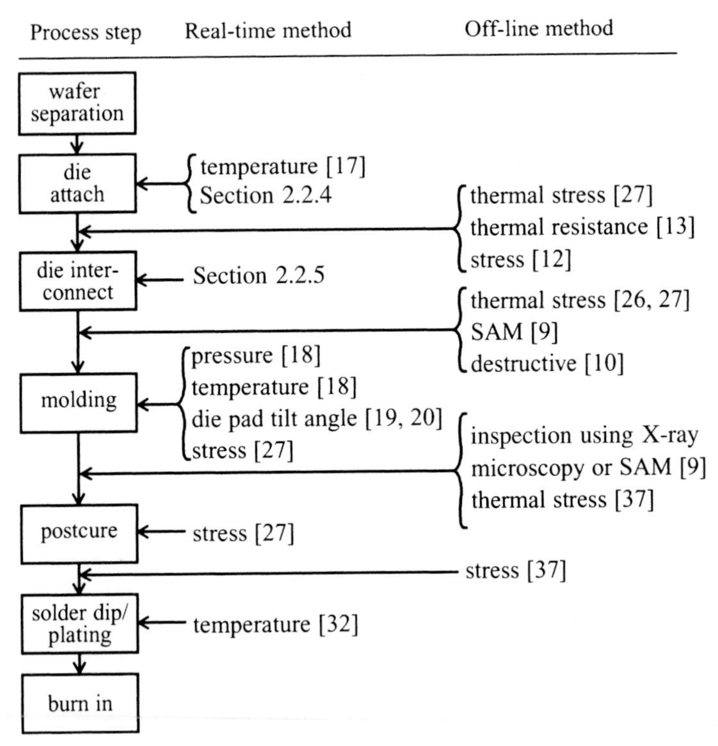

Figure 2.2.2. The packaging process flow [43] typical for dual in-line packages (DIP) and quad-flat-packs (QFP) and references to monitoring techniques.

able [38–40]. On these chips, *p–n* junctions [32] and aluminum resistors [30, 41] serve as temperature sensors, while piezoresistors are used as stress detectors [11, 35–37]. Piezoresistive sensors were used to investigate thermal stress on the chip [11]. They were also used to determine chip stresses after molding, postcure, and thermocycles, respectively [37]. Reference [42] describes the application of integrated strain gages to the on-line measurement of the creep behavior of a soft solder alloy after die attach. The experimental results were used to calibrate a finite element (FE) model, which in turn can be used to predict the creep behavior for different device geometries, materials, and process variables. More recently, in situ stresses were measured before and after die attach and during curing of the liquid encapsulant of a chip on board (COB) process [27]. It was shown that die attach stresses were small compared with the encapsulation stresses. In addition, the results showed that the dominant part of mold compound shrinkage occurred during the first few minutes at the cure temperature. The final stress essentially develops during the cool-down period. In another example, the temperature distribution in plastic packages was measured during the vapor-phase and infrared reflow processes, as well as during a dip-coating process [32]. The thermal stress during temperature cycles has been measured using piezoresistive microsensors [26]. Viscoelasticity of the molding compound was identified as the reason of cyclic stress increase in the package.

All of these devices and techniques contribute to ensuring and improving the high-quality standards of the microelectronic manufacturing process, from wafer production via component fabrication and reliability evaluation to board assembly, as shown in Figures 2.2.1 and 2.2.2 [43].

2.2.3 Bonding Processes and Microsensors

The joining of materials plays a crucial role in microelectronic packaging. In particular, critical issues are reliable adhesion of semiconductors on substrates and reliable electrical contacts to integrated circuits by wire bonds or flip chip technologies. The enhancement of bond interface quality focuses on materials research and manufacturing process research. For effective process research, appropriate monitoring techniques are required.

In this section, two microelectronic bonding processes are addressed. Soft solder die bonding and thermosonic wire bonding are discussed in Sections 2.2.3.1 and 2.2.3.2, respectively. The discussion includes process flows, qualification techniques, current challenges, and in situ real-time measurement techniques. Section 2.2.3.3 discusses the fabrication and cross sections of the integrated sensors developed for process monitoring. Characteristic quantities of integrated temperature sensors and some theory of the piezoresistive effect in resistors on (100) silicon are summarized. Basic microsensor design considerations are discussed.

2.2.3.1 Soft Solder Die Bonding Process

2.2.3.1.1 Process Considerations

Soft solder die attach is a common method to bond power semiconductor devices on leadframes [44,45]. An automatic soft solder die bonder attaches thousands of chips per hour. The bonding cycle comprises several distinct operations. First, solder is dispensed on to a hot leadframe. Simultaneously, the chip is picked up from the diced wafer by a rubber tool. The rubber tool is mounted on a bondhead which transports the chip from the wafer to the leadframe, and places it on the liquid solder film. This step is schematically shown in Figure 2.2.3. Machine components include a hot-plate to heat the leadframe, a steel adapter to mount the rubber tool to the bondhead, and a spring ensuring repeatable normal forces during bonding.

Critical parameters during chip soldering are the temperature, the type and quality of surfaces, and the type and amount of solder. Additional parameters include the bondhead force (bonding force), the bond time, and the overtravel distance. The bondhead force assures intimate contact between chip and solder. The bond time determines how long the rubber tool presses the chip on to the molten solder. Finally, the overtravel is the distance travelled by the bondhead beyond the point of first contact, ie, the incursion into the spring, as shown in Figure 2.2.3 b.

The thermal properties of soft solders make them preferable over organic bonding materials. This is particularly true for power applications. Soft solder-bonded devices can be found in mobile phones. Soft solders are less expensive than silver-filled epoxies. Reliable die bonding requires a low void density and a controlled thickness of the bonding layers. To fulfil this requirement, highly uniform solder materials and surfaces are used [46]. Many solder alloys are available, each requiring the process to be tuned for the specific solder/substrate material combination.

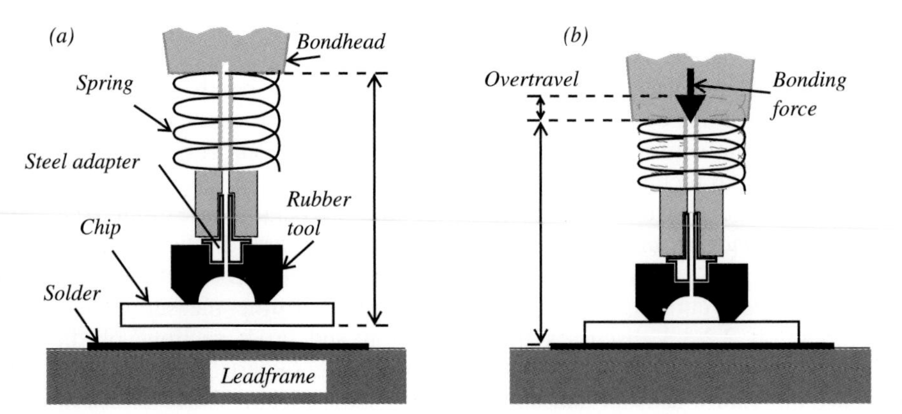

Figure 2.2.3. Schematic view of die bonding using soft solder layer: (**a**) approaching and (**b**) bonding. The chip is held to the rubber tool by a differential pressure.

2.2.3.1.2 Quality Assessment

Evaluation of the bonding layer and process control is usually done off-line and by spot checks. Common methods include SAM, x-ray analysis, and destructive shear testing. Microsensors on a test chip to measure the temperatures at different locations on the chip surface have been reported [47]. Nine *p–n* diodes were used as temperature sensors. They were located in the chip center, at the middle of the sides, and in the corners. The test chip also contained six heating resistors. After die attachment, the chip was heated with at least 30 W. The temperature at the chip surface during heating was reported and qualitatively correlated with the solder layer void distribution.

2.2.3.1.3 Challenges

A considerable effort is focused on speeding up die attachment processes to increase their throughput without loss of quality. This needs a good understanding of the process itself. In view of their off-line nature, the evaluation methods mentioned in Section 2.2.3.1.2 require large numbers of measurements before conclusions about the process can be drawn. Novel real-time in situ monitors, however, promise to give direct access to the process. Such direct monitoring has the potential to accelerate the optimization of critical process parameters. A key parameter to investigate is the process temperature as it determines the efficiency of surface wetting needed for successful bonding. Direct in situ measurements of the temperature were performed for a laser soldering process using infrared thermography [48]. The reported procedure required a large chip area guaranteeing a sufficient signal level. The method was demonstrated for soldering ceramic chip capacitors on to copper-plated printed circuit boards (PCBs) using an Nd:YAG laser. The onset of wetting was monitored using infrared sensors. No discussion is given in [48] on how to apply the method to standard soft solder die bonding where the visual path to the die surface is blocked by the bondhead. This drawback makes a laser-based temperature monitoring method less suited for monitoring soft solder die bonding processes. Intrusive thin-film temperature sensors overcome such difficulties if they can be contacted during bonding, as has been reported for a lead-on-chip die bonding process [17].

Integrated sensors exhibit several advantages compared with intrusive sensors. They offer a higher spatial resolution. At the same time they require less process modification because they are part of the chip. Section 2.2.4 reports how such sensors were developed and used to gain information about the soft solder die attachment process.

2.2.3.2 Thermosonic Ball Bonding Process

2.2.3.2.1 Process Flow

Wire bonding is a technique to establish electrical contact between integrated circuits and substrates [49–51]. This is achieved by bonding thin gold or aluminum wires between the two components. The resulting connections are called *wire bonds*. In *thermosonic* wire bonding, ultrasonic energy and heat is used to accelerate and stabilize the bonding process. The following paragraphs briefly review the mechanical components needed for the process, the wire bonding cycle, details of the evaluation of process parameters during bonding, and recent advances in real-time wire bonding process monitoring.

Instruments needed for thermosonic wire bonding are shown schematically in Figure 2.2.4 a. The system includes a heater stage (hot-plate), a tool (capillary) fixed to the tip of an amplifying horn, a piezoelectric stack fixed on the rear end of the horn, an electrical flame-off (EFO) electrode, and wire clamps. *Transducer* is the jargon term for the system comprising the piezoelectric stack and horn. In addition to these indispensable parts, some wire bonders also include an internal normal force measurement system. An example based on piezoceramics is shown in Figure 2.2.4 b.

The basic cycle of a thermosonic wire bond is shown in Figure 2.2.5. In (a), a capillary tip with a wire and an electrode are shown schematically. A high d.c. voltage is applied between wire and electrode. This produces an arc between wire and EFO electrode, causing the tail to melt and subsequently solidify into a ball. In (b), the capillary guides the gold wire with ball termination towards the bonding pad. In (c), the ball is deformed by the touch down impact and bonded to the pad by the interaction of bonding force, heat, and ultrasonic energy from the transducer. In (d), the capillary retreats from the pad. Subsequently, the capillary follows a well defined trajectory, shown in (e), and shapes the wire into a loop, before reaching the second (wedge) bonding site. In (f), the capillary deforms the wire and bonds it to the substrate, again using the combination of bonding force, heat, and ultrasound. The capillary moves up with open wire clamps, as shown in (g). In (h), a piece of wire called the *tail* is formed by closing the clamps and tearing off the wire, which breaks at its weakest spot close to the second bond. This tail supplies the material needed for subsequent ball formation.

Bonding is influenced by machine parameters such as initial ball diameter, bonding force, ultrasonic energy and time, and chip temperature [52]. These input parameters are shown schematically in Figure 2.2.6. In standard processes, the bonding and ultrasound level are controlled as shown schematically in Figure 2.2.7 a and b, respectively. The impact of the ball on the pad produces a high force, causing the ball to undergo a strong initial deformation. After impact, the bonding force is usually controlled at a lower level. Now, a constant a.c. voltage is biased to the transducer to produce the ultrasonic vibrations. The ultrasonic power rises exponentially with a typical rise time of a few milliseconds. After about 10–12 ms, the bond has formed, and the ultrasound is switched off. Details about the ball bond formation [49, 53, 54] and effects of each input parameter are now briefly described.

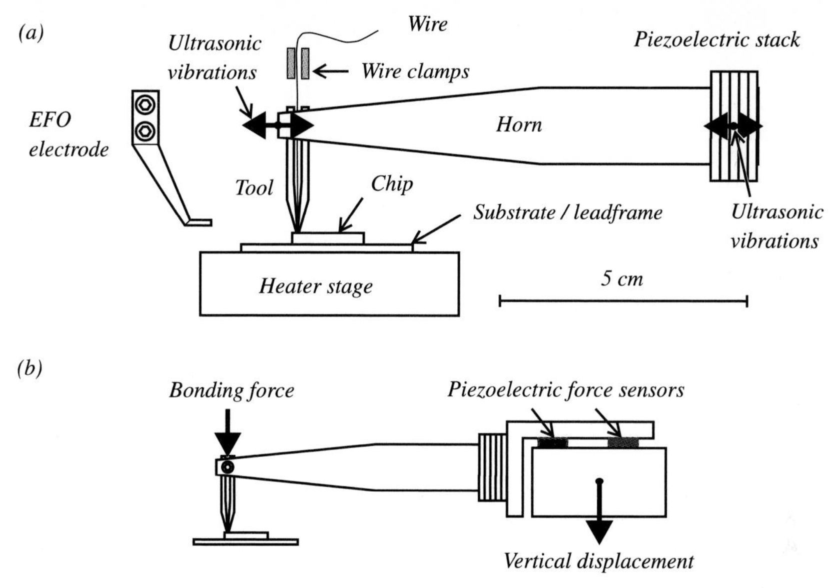

Figure 2.2.4. (**a**) Schematic setup for thermosonic wire bonding. The substrate is pressed to the stage by clamping the plate (not shown) and underpressure valve (not shown). (**b**) Location of force sensors.

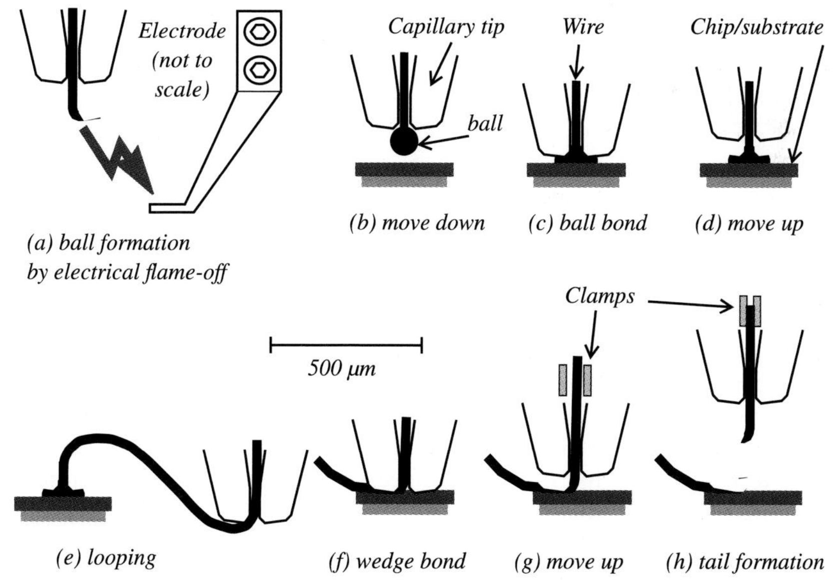

Figure 2.2.5. Consecutive stages (**a**)–(**h**) of bonding cycle. Electrode not to scale.

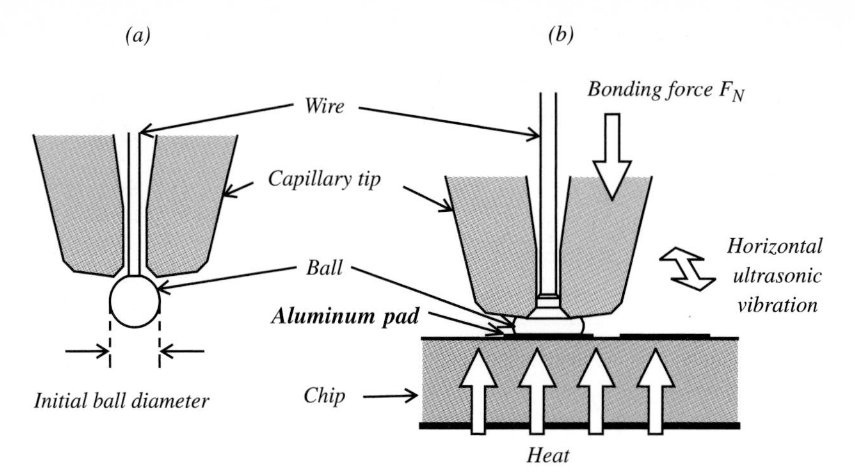

Figure 2.2.6. Schematic illustration of (**a**) initial (free-air) ball and (**b**) thermosonic ball bonding process. Large arrows indicate process variables.

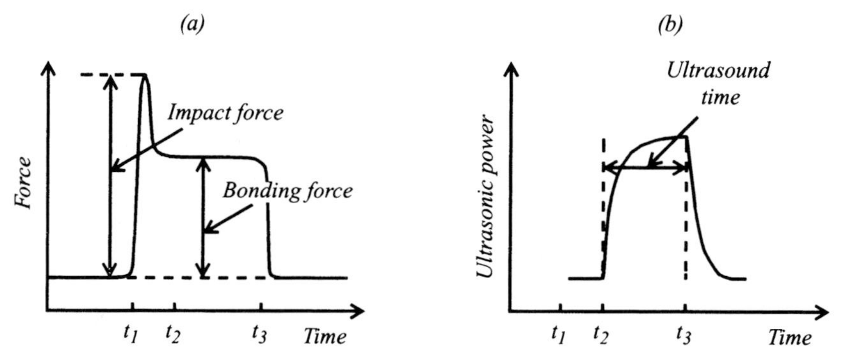

Figure 2.2.7. Schematic variations of (**a**) bonding force and (**b**) ultrasonic power during a ball bond.

The perpendicular bonding force ensures the intimate contact between the bonding partners. Friction energy is dissipated at the interface by the ultrasonic vibrations. The first benefit of this interfacial motion is the removal of the native aluminum oxide from the pad, followed closely by the formation of the intermetallic bond. By dehydrating the chip surface, an elevated chip temperature causes a more uniform surface quality, increasing the stability of the process. Both temperature and ultrasound dissipation enhance the interdiffusion of aluminum and gold. Excessive ultrasonic energies can produce breakage (*cratering*) of the silicon under the pad. Excessive bonding forces decrease the interfacial friction by damping the oscillation. Excessive chip temperatures, ultrasonic energies, and bonding forces all individually contribute to excessive ball deformation, resulting in shorts between adjacent balls.

2.2.3.2.2 Quality Assessment

Ball bond quality is assessed by off-line measurements of characteristics such as the height and diameter of the deformed ball, its shear strength, and the percentage of intermetallic compound at the bond interface [10, 55]. Definitions of the geometric ball parameters are shown schematically in Figure 2.2.8 a. The distance between top edge and pad (ball height), and the ball diameter, taken at the top edge, are usually determined using an optical microscope [10]. Shear force values are obtained using a shear tester, as shown in Figure 2.2.8 b. A shear tool removes the ball from the pad by a horizontal motion and simultaneously measures the force needed. The shear strength of a ball is obtained by dividing shear force by the area defined by the diameter of the deformed ball [10]. The shear strength is a better indicator of bond interface quality than the shear force [56]. The intermetallic compound at the ball bottom is shown schematically in Figure 2.2.9. The amount of intermetallic compound is determined by first removing the balls from the pads by etching away the aluminum pad. This is done using, eg, a potassium hydroxide (KOH) solution [10]. Next, the ball bottom is in-

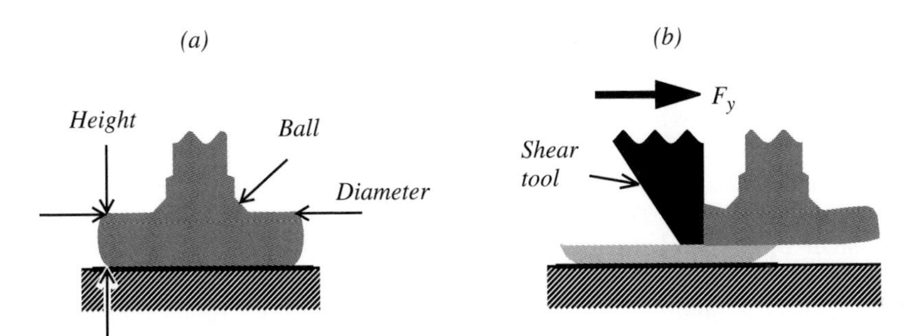

Figure 2.2.8. Schematic illustrations of ball bond characterization methods: (**a**) geometry assessment using microscope and (**b**) shear force measurement using shear tester.

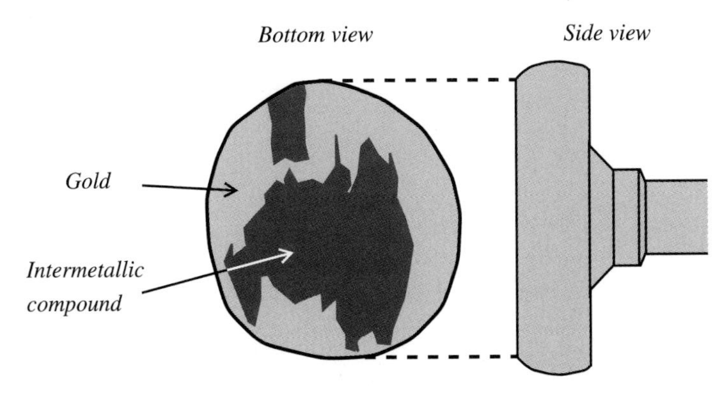

Figure 2.2.9. Schematic illustrations of bonded ball bottom and side. Intermetallic compound, dark color; gold, light color

spected under a microscope and the intermetallic compound fraction of the total area is estimated. The quality of the bond is closely related to the amount of intermetallic compound. However, experimental values are tedious to obtain. Therefore, process characteristics such as geometry and shear strength are more commonly used for bond qualification.

The characteristics of a stable process are within specified lower and upper limits. Optimal input parameters can be found by statistical design of experiment (DOE) [57–61]. A DOE comprises the investigation of process characteristics for combinations of input parameters. Process windows defined by ball geometry and shear strength, and found using DOEs are reported in [62, 63].

2.2.3.2.3 Wire Bonding Importance and Recent Efforts

Wire bonding has been used to connect most integrated circuits to the external world. Wire bonding is a flexible and inexpensive technique producing highly reliable interconnects. Wire bonders evolve rapidly to meet ever more severe requirements such as bonding temperatures below 150 °C, ultra-fine ball pitches, and shorter cycle time. Technologies such as high-frequency bonding [62], lightweight transducers, and novel substrate cleaning technologies [64] fulfil these increased requirements. Nevertheless, process optimization relying on conventional off-line methods becomes more and more demanding and time-consuming. At present, wire bonding optimization is the most time-consuming task when setting up a production line [65]. Research on novel optimization methodologies and tools is therefore justified. A strategy to decrease process optimization time is to apply physical models [4]. Ideally, model functions promise to reduce the number of measurements needed to determine a process window. Unfortunately, there is a general lack of quantitative models describing how the process characteristics depend on the various input parameters. The reason possibly is that a realistic physical description of the welding mechanisms during thermosonic bonding of gold balls to aluminum pads involves several disciplines. Continuum mechanics describe stress fields during initial deformation and during bond formation. Materials science describes how cold deformation and ultrasound dissipation change mechanical material constants. Tribology deals with frictional effects, including abrasive and welding action between ball and pad. Material science provides knowledge about diffusion mechanisms and intermetallic phase formation, necessary for the assessment of bond reliability.

Numerical and experimental research efforts in these fields with respect to wire bonding are summarized in the following.

Two-dimensional FE analyses of the initial deformation have been reported for thermosonic ball bonding [66, 67] and thermocompression wedge bonding [68]. The influence of strain rate-dependent yield stress of gold on the initial deformation behavior have been investigated [67]. Ultrasonic shearing force and its effect on the stress field have not been taken into account. In particular, the authors are not aware of theoretical models or simulations of tribological effects during thermosonic bond growth. Early experimental research efforts focused on the monitoring of feedback signals of the ultrasonic transducer [69]. Such signals were used to qualify bonds heuristically. A more direct method using a piezoelectric sensor attached to the transducer horn was demonstrated [70]. This sensor permitted measurements of harmonics of the ultrasonic vibration during ball bonding. Bond quality was determined by interpreting the envelope function of an observed harmonic. Recently, the same approach was used with a wedge bonder [71]. A correlation between the second harmonic and the shear force was reported. Likely, this heuristic method is useful in production for on-line quality monitoring of ultrasonic wedge bonds. However, it is not applicable for ball bonds because there, the second harmonic is absent owing to the rotational symmetry. With a laser vibrometer, it is possible to shift the measurement point from the horn to the tool tip and even on to the wire [72, 73]. However, in those investigations the variation of higher harmonics with time was not specified. Experiments where the laser beam was focused on a ball or capillary tip under production conditions or for a fine pitch process have not been reported. Recently, a modified ball was used to monitor in situ temperature during bonding [74]. An additional nickel wire was connected to the ball to form a gold–nickel thermocouple. The in situ temperature during bonding correlated with the degree of pad surface contamination. A custom-made instrument and capillary were used. In view of these recent efforts, a more detailed investigation of different welding process phases and their dependence on bonding parameters would help to build quantitative physical models.

Reference [75] reports on real-time in situ signals for temperature and stress using microsensor integrated below the bonding pad. Real-time microsensor signals are expected to be highly appropriate for thermosonic ball bonding process research. No other method promises the same spatial resolution and vicinity to the process. An enhanced understanding of the bond formation can be obtained by real-time measurements with test chips. In addition, test chips can help to improve the reliability of a whole package [25]. Such microsensors were further used to determine a bonding force process window [76] or to get information about the bond growth during the dissipation of ultrasound [77, 78]. Recently, several small piezoresistive sensors were integrated along a line below a bonding pad and read out during bonding a ball [79, 80]. The spatial resolution of these measurements was 20 μm. The results are useful for understanding the stress distribution at the contact zone during bonding.

Microsensors have been used for off-line qualification of electronic packages [25, 33–37] and for on-line monitoring of electronic packaging processes [79, 80]. Microsensors have been used to investigate ball bonds [81]. A *p–n* junction was placed below the bonding pad and reverse-biased after bonding on these pads. Excessive leakage currents correlated with mechanical damage induced by nonoptimized bonding. To learn more about the bonding process, it is advantageous to use microsensors as real-time monitors of the various mechanisms of the ball bonding process.

No other method promises the same spatial resolution and vicinity to the process. Therefore, real-time microsensor signals are expected to be highly appropriate for thermosonic ball bonding process research. Section 2.2.5 reports on the development of such sensors and their application the characterization of thermosonic ball bonding processes. In the following section, the fabrication of microsensor test chips is discussed.

2.2.3.3 Microsensor Basics

To monitor die attachment and ball bonding processes, the authors designed microsensors integrated on test chips. Details about the fabrication including schematic cross sections of microsensors are given in the next section. Basic principles of resistive temperature sensors and piezoresistors are described in Sections 2.2.3.3.2 and 2.2.3.3.3, respectively. Section 2.2.3.3.4 presents resistive sensor design considerations including the basics of four-wire and Wheatstone bridge measurement configurations.

2.2.3.3.1 Fabrication

Most test chips reported in this chapter were fabricated using the standard single poly, double metal application specific CMOS [82] processes *alp2lv* [83, 84] and *alp1mv* of EM Microelectronic-Marin SA, Switzerland. The gate lengths of these processes are 2 and 1 μm, respectively. Microstructures were designed using the lower aluminum metallization and n^+- and p^+-type source/drain diffusions as resistive sensing materials, measuring temperature and stress. Sensors fabricated using aluminum consist of lines with the specified minimum linewidth and a meander shape. This permits resistance densities of about $10\,\text{k}\Omega/\text{mm}^2$. Two possible structures based on aluminum are shown as cross sections in Figure 2.2.10 a and b. Structure (a) comprises several metal lines. They are parallel to each other and shown perpendicular to the view plane. The metal lines are sandwiched between two dielectric layers on the top and two on the bottom. For the *alp1mv* process, the dielectric layers from top to bottom are an oxynitride passivation, an undoped silicate glass–spin on glass–undoped silicate glass (USG–SOG–USG) intermetal oxide layer sandwich, a borophosphosilicate glass (BPSG) contact oxide, and a thermal field oxide next to the p-type bulk silicon. Structure (b) shows a wire bonding pad. It consists of the two metal layers and openings in the passivation and intermetal oxide layers. Sensing aluminum lines are integrated at the periphery of the pad. Two structures based on diffusions are shown in Figure 2.2.10 c and d. Both structures are covered by a bonding pad. Structure (c) comprises n^+-type resistor lines diffused in the bulk silicon. Structure (d) consists of a broad p^+-type resistor line diffused in an n-well.

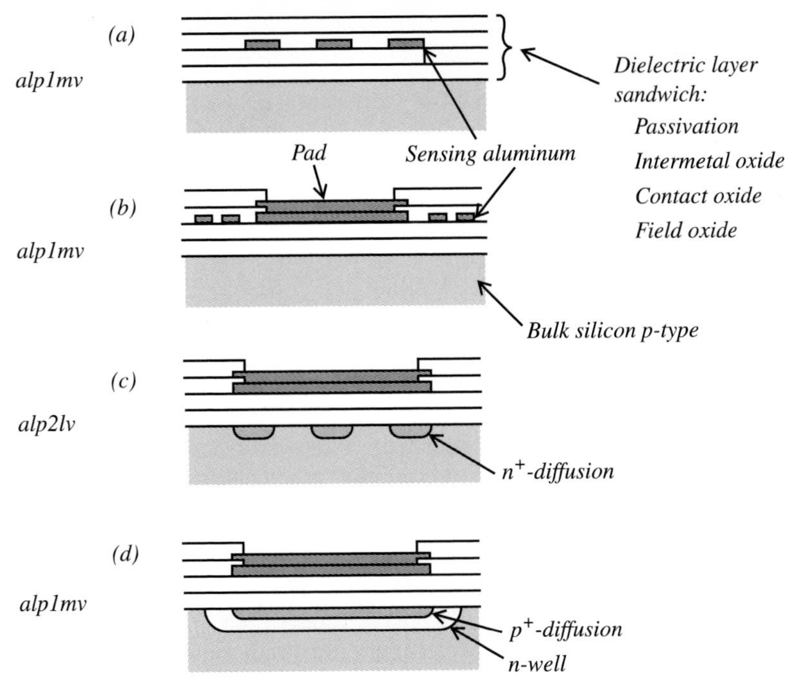

Figure 2.2.10. Schematic cross sections of test structures (**a**)–(**d**) produced with processes *alp1mv* and *alp2lv*. Layers not to scale.

2.2.3.3.2 Integrated Temperature Monitors

There are several possibilities to realize temperature microsensors using VLSI technology. Among them are *resistive temperature detectors* (RTDs), thermistors, *p–n* diodes, and integrated circuits with an output signal proportional to absolute temperature (PTATs). The devices are characterized by a temperature coefficient (TC) which is the relative signal change with temperature. The TC itself depends on temperature. For a resistor, the TC is called TC of resistance (TCR). It is denoted α, and is defined by

$$\alpha(T) \equiv \frac{\frac{\mathrm{d}R}{\mathrm{d}T}}{R(T)} \tag{2.2.1}$$

where T is the temperature and $R(T)$ is the resistance as a function of T.

The measured resistance R of an integrated aluminum RTD and its deviation from a linear fit ΔR_{lin} are shown in Figure 2.2.11 a as a function of temperature. The linear fit matches the measurement to within less than 0.1%. The TCR value is shown as a function of temperature in Figure 2.2.11 b. For a given reference temperature T_{ref}, two definitions are made, $R_{\mathrm{ref}} \equiv R(T_{\mathrm{ref}})$, and $\alpha_{\mathrm{ref}} \equiv \alpha(T_{\mathrm{ref}})$. Thus, the RTD resistance at a temperature T is well approximated by

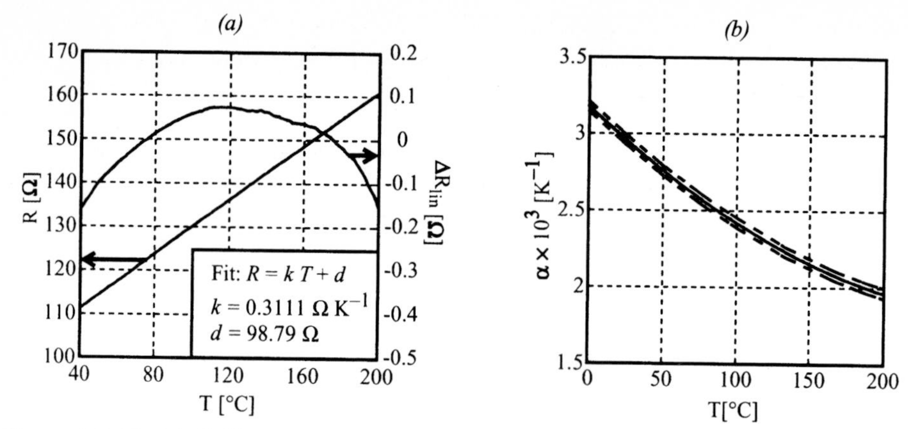

Figure 2.2.11. (**a**) Calibration of an aluminium microsensor: resistance and its deviation from linear fit versus temperature. (**b**) Mean TCR of aluminum microsensor (solid) and mean TCR ± standard deviation (dashed), for six samples.

$$R(T) = R_{\text{ref}}(1 + a_{\text{ref}}\Delta T) \,, \tag{2.2.2}$$

where $\Delta T = T - T_{\text{ref}}$. To increase the temperature prediction accuracy of a single RTD, it is calibrated at a well known calibration temperature T_c by measuring its resistance R_c. The required temperature subsequently is obtained from the resistance R of the RTD using

$$T = T_c + \frac{R - R_c}{R_c} \cdot a(T_c)^{-1} \,. \tag{2.2.3}$$

If only a_{ref} is known, $a(T_c)$ can be calculated using

$$a(T_c) = (a_{\text{ref}}^{-1} + T_c - T_{\text{ref}})^{-1} \tag{2.2.4}$$

Commercial *thermistors* are semiconductor resistors produced using dedicated technologies which are incompatible with CMOS processes. Such devices can have negative TCRs as large as $-60\times10^{-3}\,\text{K}^{-1}$ at 25 °C. Thermistors can be used up to 150 °C, but show significant drift if operated above 125 °C. In comparison, the diffusions and the *n*-well of the CMOS processes described in Section 2.2.3.3.1 have TCRs of about 2×10^{-3} and $5\times10^{-3}\,\text{K}^{-1}$ at 20 °C, respectively.

A *p–n diode* is biased using a constant forward current. The voltage drop across the diode is a measure of temperature. Typical temperature coefficients of the voltage drop at 20 °C of CMOS processes related to this chapter vary from -2.3×10^{-3} to $-5.1\times10^{-3}\,\text{K}^{-1}$ as the forward current varies from 1 to 100 μA, respectively.

A PTAT exploits the temperature dependence of the semiconductor bandgap [23]. CMOS PTATs are limited to temperatures below 150 °C and require a substantial chip surface area.

The main advantage of aluminum RTDs here is their stability up to 400 °C for a long enough time necessary for the measurements. This is in contrast with sensors relying on diffusions. In addition, CMOS processes allow the integration of RTDs with sufficiently thin aluminum lines. Thereby, resistances of several hundred ohms are feasible on an area smaller than a normal bonding pad. Further advantages of aluminum RTDs are that no substrate contact is necessary, they are less sensitive to stress than diffused resistors and *p–n* diodes, and their sensitivity to temperature is higher and more linear than that of diffused CMOS resistors.

2.2.3.3.3 Silicon Piezoresistors

During wire bonding, considerable mechanical stresses are developed at the bonding pad. This is necessary for the deformation and welding of the wire. It is believed that the variation of such stresses is related to process quality. Therefore, ways to integrate stress-sensitive devices as monitors are needed.

Stress sensors often are based on the variation of resistance of a conductor or semiconductor when subjected to a mechanical stress. This resistance change origins from two mechanisms. The first is the geometric deformation. The second is the change in resistivity and is called the *piezoresistive effect*. For semiconductors, the piezoresistive effect predominates. It is due to the dependence of the energy band structure on stress. Doped silicon can be used as piezoresistive material [85]. In particular, the source/drain diffusions of CMOS processes are suited as piezoresistive in situ stress sensors [35]. Details of the piezoresistance of silicon are discussed in the literature [25, 36]. An extensive treatment of the physics of piezoresistance can be found in [86]. A brief summary of the physical principles, mathematical description, and measured material properties is reported in [87]. In the following, the expressions needed here are presented.

Subject to a mechanical stress state,

$$\sigma = \begin{bmatrix} \sigma_{xx} & \sigma_{xy} & \sigma_{xz} \\ \sigma_{xy} & \sigma_{yy} & \sigma_{yz} \\ \sigma_{xz} & \sigma_{yz} & \sigma_{zz} \end{bmatrix},$$

and a temperature change, ΔT, piezoresistors on (100) wafers undergo a relative resistance change [35]

$$\frac{\Delta R}{R} = \sigma_{xx}\frac{\pi_{11}+\pi_{12}+\pi_{44}\cos 2\phi}{2} + \sigma_{yy}\frac{\pi_{11}+\pi_{12}-\pi_{44}\cos 2\phi}{2} + \sigma_{zz}\pi_{12} + \sigma_{xy}(\pi_{11}-\pi_{12})\sin 2\phi + \alpha_1\Delta T + \alpha_2\Delta T^2 + \ldots \tag{2.2.5}$$

where π_{11}, π_{12}, and π_{44} are the piezoresistive coefficients, ϕ is the angle between the current flow in the resistor and the crystal direction [110], as shown in Figure 2.2.12, and α_1 and α_2 are the TCRs of the first and second order, respectively. Piezoresistive coefficients for a dopant density typical of source/drain dif-

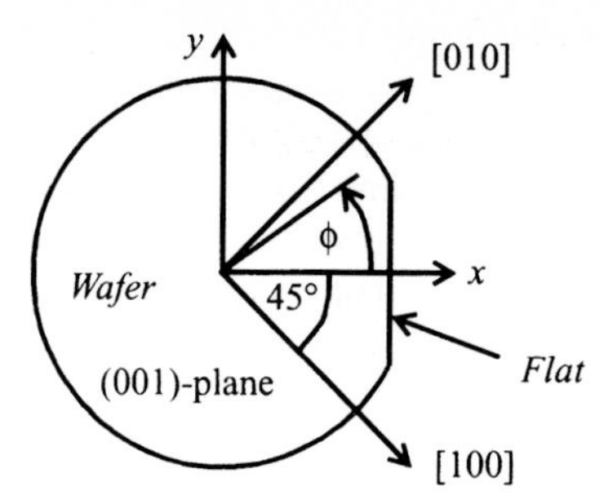

Figure 2.2.12. (100) silicon wafer definitions.

Table 2.2.1. Piezoresistive coefficients for CMOS n^+ and p^+ diffusions [36]

Type	π_{11} (10^{-10} Pa^{-1})	π_{12} (10^{-10} Pa^{-1})	π_{44} (10^{-10} Pa^{-1})
n	–3.8	1.9	–1.9
p	0.57	–0.23	7.1

fusions and at room temperature are given in Table 2.2.1. For resistors directed towards [110], [010], [$\bar{1}$10], and [$\bar{1}\bar{1}$0], and for ΔT=0, Equation (2.2.5) respectively simplifies to

$$\left.\frac{\Delta R}{R}\right|_{\phi=0} = \sigma_{xx}\frac{\pi_{11}+\pi_{12}+\pi_{44}}{2} + \sigma_{yy}\frac{\pi_{11}+\pi_{12}-\pi_{44}}{2} + \sigma_{zz}\pi_{12} \tag{2.2.6}$$

$$\left.\frac{\Delta R}{R}\right|_{\phi=\frac{\pi}{4}} = \sigma_{xx}\frac{\pi_{11}+\pi_{12}}{2} + \sigma_{yy}\frac{\pi_{11}+\pi_{12}}{2} + \sigma_{zz}\pi_{12} + \sigma_{xy}(\pi_{11}-\pi_{12}) \tag{2.2.7}$$

$$\left.\frac{\Delta R}{R}\right|_{\phi=\frac{\pi}{2}} = \sigma_{xx}\frac{\pi_{11}+\pi_{12}-\pi_{44}}{2} + \sigma_{yy}\frac{\pi_{11}+\pi_{12}+\pi_{44}}{2} + \sigma_{zz}\pi_{12} \tag{2.2.8}$$

$$\left.\frac{\Delta R}{R}\right|_{\phi=\frac{3\pi}{4}} = \sigma_{xx}\frac{\pi_{11}+\pi_{12}}{2} + \sigma_{yy}\frac{\pi_{11}+\pi_{12}}{2} + \sigma_{zz}\pi_{12} - \sigma_{xy}(\pi_{11}-\pi_{12})\,. \tag{2.2.9}$$

2.2.3.3.4 Design Considerations

A source of error in resistance measurements using two wires is the lead wire resistance which adds its value to the result. A well known method to overcome the lead wire error is the *four-wire measurement* technique, where a controlled measurement current I is applied via the lead wires. Two additional probe wires are used to measure the voltage drop U across the resistor. There is only a negligible voltage drop along the probe wires if the probe current is sufficiently low compared to the measurement current. The required resistance is given by $R = U/I$. A relative resistance change $x \equiv \Delta R/R$ is obtained via the observed voltage change ΔU, and is $x = \Delta U/(RI)$.

The *Wheatstone bridge* is advantageous when measuring relative resistance changes x. The offset is much lower than for a four-wire measurement. The Wheatstone bridge consists of four resistors with resistances connected as shown in Figure 2.2.13. If one resistance is changed from its original value R to a new value $R + \Delta R$ due to a temperature or stress change, then the bridge is unbalanced and a bridge voltage $U_m \neq 0$ is measured. For bridges with equal resistances, the resulting expressions for U_m for different arrangements and resistance changes and for two biasing conditions are listed in Table 2.2.2 [88]. Note that for the quarter bridge arrangements and for the half bridge type A arrangement with

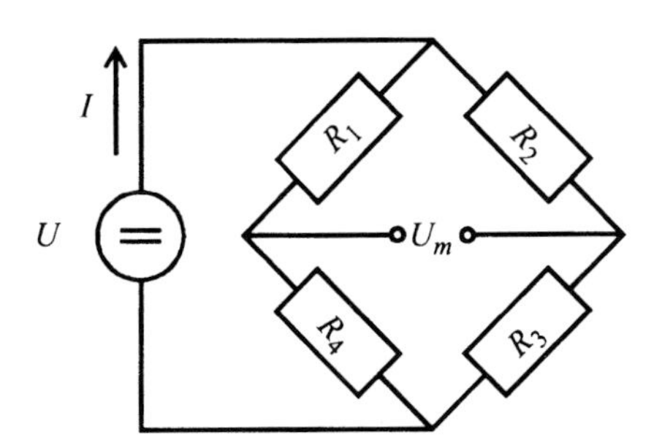

Figure 2.2.13. Wheatstone bridge configuration.

Table 2.2.2. U_m for different Wheatstone bridge arrangements ($x \equiv \Delta R/R$)

Arrangement	R_1	R_2	R_3	R_4	U_m (const. voltage)	U_m (const. current)
Quarter bridge	R	R	$R(1+x)$	R	$U\frac{x}{2(2+x)}$	$IR\frac{x}{4+x}$
Half bridge type A	$R(1+x)$	R	$R(1+x)$	R	$U\frac{x}{2+x}$	$IR\frac{x}{2}$
Half bridge type B	$R(1+x)$	R	R	$R(1-x)$	$U\frac{x}{2}$	$IR\frac{x}{2}$
Full bridge	$R(1+x)$	$R(1-x)$	$R(1+x)$	$R(1-x)$	Ux	IRx

constant bridge voltage, the relation between U_m and x is not linear. However, the smaller the values of x, the smaller is the deviation of U_m from the first-order linear approximation. As an example, this deviation is below 0.5% if $x=0.01$. When operating a bridge with a known constant current, the resistance of the bridge is needed for the determination of $x=U_m/IR$. Operating the bridge with a known constant voltage U allows the direct measurement of x from U_m, eg, $x=U_m/U$ for a full bridge, without knowing the exact value of R. Therefore, a constant bridge voltage is advantageous when comparing results from a set of different samples, because their resistance variation needs not to be measured.

2.2.4 Soft Solder Die Bonding

Microsensor-based characterization of a soft solder die bonding process is reported in this section. After a brief presentation of the equipment used for these investigations, Section 2.2.4.2 reports details of the microsensors, of the setup used for the high-temperature calibration of the microsensors, and of the chip connector enabling electrical contact to the microsensors during soldering. Section 2.2.4.3 reports on the measurement and evaluation procedure, and the variation of machine settings for additional investigations. Microsensor results and SAM pictures are shown in Section 2.2.4.4. Section 2.2.4.5 develops the results of an FE model of the process, and compares different cases with experimental data. Conclusions are drawn in Section 2.2.4.6.

2.2.4.1 Introduction

The experiments were performed on a modified soft solder die bonding system, as shown in Figure 2.2.14. The machine is a high-volume pick-and-place bonder for chips on leadframe (LF). The measurement system comprises a test chip, standard equipment for electrical measurements, and a custom-made pick-and-place tool connecting the chip to the measurement equipment during bonding. Figure 2.2.15 shows the connection tool mounted on the bondhead of the machine. A test chip bonded to an LF is shown in Figure 2.2.16. A large fraction of the results presented here were presented at the 32nd International Symposium on Microelectronics 1999 (IMAPS99) [89].

2.2.4.2 Test Chip and Chip Connector

The test chips were fabricated using the standard single poly, double metal 1 μm medium voltage ASIC CMOS process of EM Microelectronic-Marin SA,

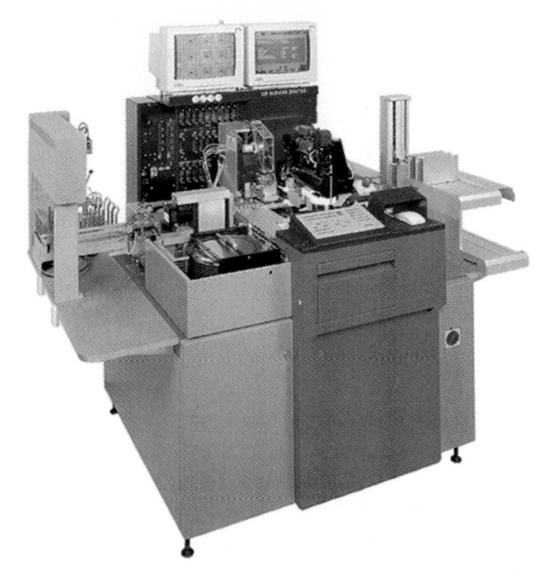

Figure 2.2.14. Soft solder die bonder (Courtesy of ESEC SA, Cham, Switzerland).

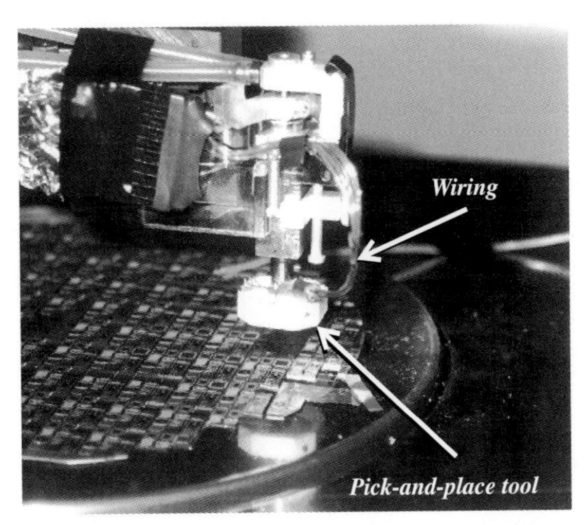

Figure 2.2.15. Pick-and-place connection tool mounted on bondhead in pick position above wafer on adhesive foil.

Switzerland. Layouts of the chip and microsensor are shown in Figure 2.2.17 a and b, respectively. The side length of the square chip is 4 mm. Nine aluminum-based RTDs are integrated in the chip center, at the center of the four sides, and in the four corners, at respective distances of 1.325 and 1.874 mm from the center. The same layers are used as shown in Figure 2.2.10 a. The RTDs are 2 μm

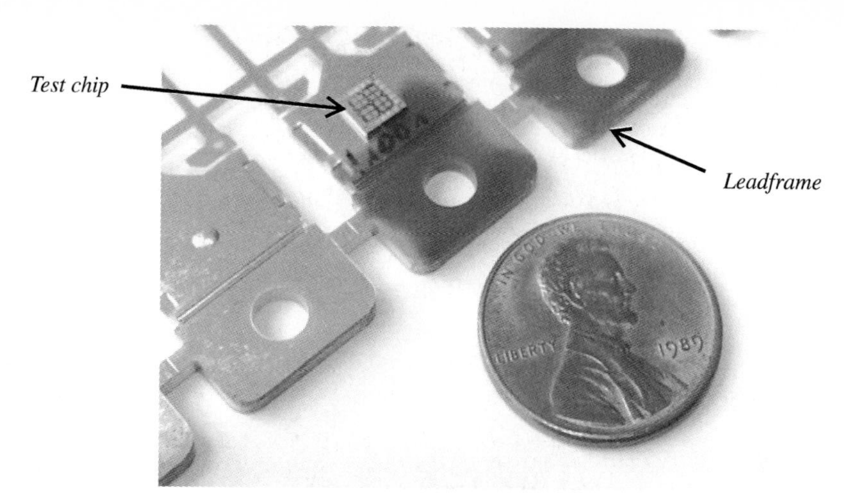

Figure 2.2.16. Text chip bonded to leadframe.

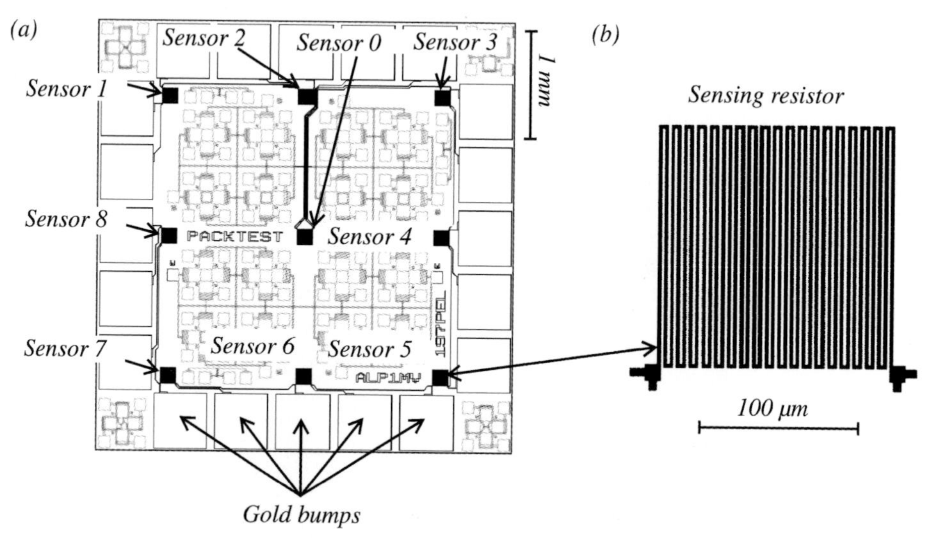

Figure 2.2.17. (**a**) Layout of test chip comprising nine aluminum microsensors. (**b**) Microsensor layout.

wide and 5.7 mm long meandering lines of the lower CMOS metal layer. Each line covers a surface of 150 by 150 μm. The resistance of the meanders at ambient temperature during pick up is 227 ± 49 Ω. The sensors are connected in series. A sensing current is supplied to the entire chain via two pads. Eighteen further pads allow the voltage drop over each sensor to be probed in a four-wire configuration. The pads are squares with 500 μm side length and a pitch of 600 μm and are arranged along the perimeter of the test chip; 20 μm high gold

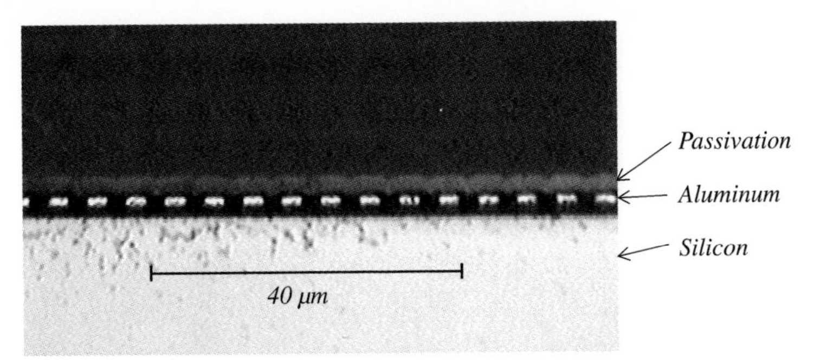

Figure 2.2.18. SEM of cross section of aluminium temperature monitor.

bumps cover the pads. The silicon substrate is 700 μm thick and the thicknesses of the dielectric layer sandwich below the temperature sensors, of the aluminum, and of the capping passivation sandwich are 1.9, 0.73, and 1.5 μm, respectively. A cross section of the sensor is shown in Figure 2.2.18. To allow soft solder die attachment, a Ti–Ni–Au layer stack with respective thicknesses of 0.1, 2.5, and 0.1 μm is sputtered on the bottom of the wafer.

The TCR of the sensors was calibrated on the dedicated setup shown in Figure 2.2.19. This setup is designed to withstand temperatures up to 400 °C. It consists of a metal base with two sites for ceramic substrates. The left and right substrates contain a commercial Pt100 RTD and a test chip, respectively. The sensor is wire bonded to the substrate terminals, which are contacted mechanically by four wires clamped between two ceramic parts. The setup is placed on a hot-plate. In view of the symmetry of the setup, the temperatures on the left and right substrates are assumed to be identical. The resistance of the sensors is measured while cooling the chips from 400 to 80 °C over 94 min. The highest cooling rate was 0.16 K/s. The

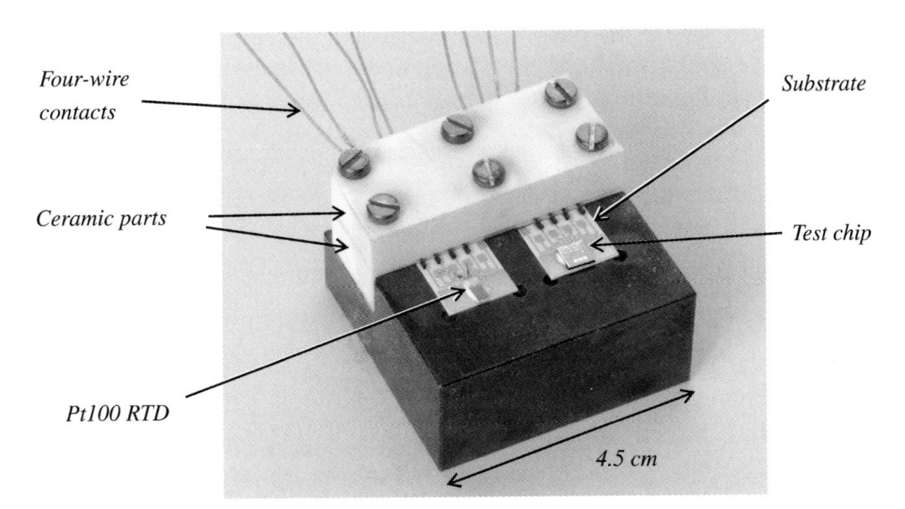

Figure 2.2.19. High-temperature characterization setup.

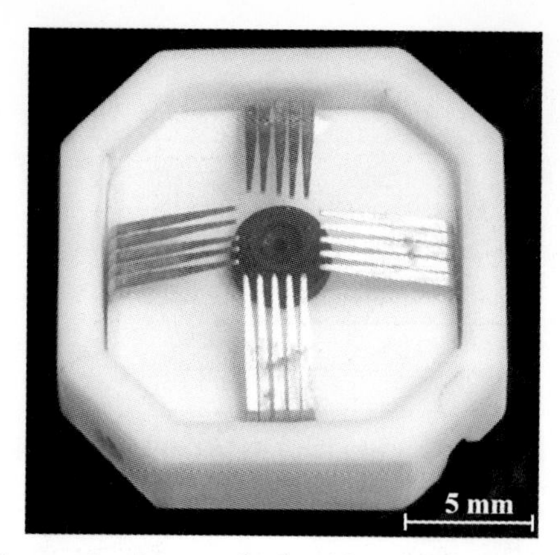

Figure 2.2.20. Connection tool with standard rubber tip in the center and 20 gold-plated copper contact fingers.

accuracy of the Pt100 temperature measurement was better than 1 K. The TCR at 30 °C of the sensors (eight samples) was $\alpha_{30\,°C} = (3.03 \pm 0.02) \times 10^{-3}\ K^{-1}$. The maximum deviation from a linear fit over the entire temperature range was 0.3% of the signal at 400 °C, corresponding to an error of 2.7 K. The overall temperature accuracy of the microsensors at 400 °C was ±7 K.

The custom-made pick-and-place connection tool consists of thermally resistant materials and includes two original components of the bondhead: a steel adapter to the bondhead and a rubber tip for pick-up of the chip. In addition to these standard components, a total of 20 finger contacts were cut out of a 0.1 mm thick copper foil and coated with gold. Two MACOR™ ceramic parts form a structural frame suitable to assemble the rubber, steel, and copper parts. The thickness and geometry of the fingers had to be optimized to obtain satisfactory electrical contact to the pads without detaching the chips from the rubber tip by an excessive spring force. A photograph and a schematic cross section of the tool are shown in Figure 2.2.20 and Figure 2.2.21 a, respectively.

2.2.4.3 Experimental

The measurements were performed using a digital current source and an eight-channel data acquisition card, allowing eight sensors to be read out simultaneously. Sensor 1 was usually not connected. Each channel was differentially connected to the pick-and-place tool. The electrical circuit scheme of the system is shown in Figure 2.2.22. A d.c. current of 1 mA was applied to the sensors, while the voltage signals were scanned at a rate of 1000 samples/s during 1.3 s from pick-up to chip release after bonding.

It was necessary to align the test chips to a predefined position and to determine the heights for the pick-up and bonding. The operator manually synchron-

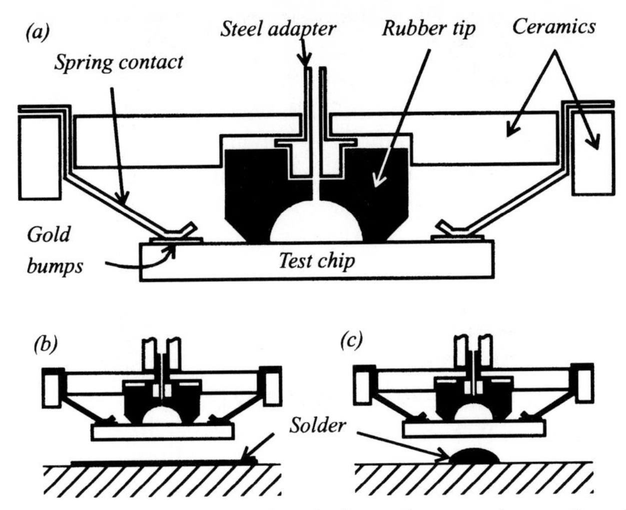

Figure 2.2.21. (**a**) Schematic cross sectional view of connection tool and test chip; die bonding with patterned solder (**b**) and solder drop (**c**).

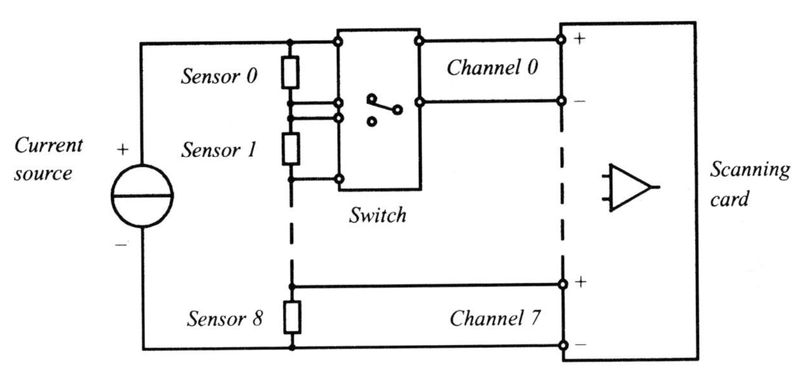

Figure 2.2.22. Electrical circuit scheme of the measurement setup.

ized the start of the electrical measurement and the bonder. An absolute time reference was provided by the interruption of the electrical contact upon chip release. This allowed comparison of different measurements on a common time scale.

The investigated solders were PbSn10 and SnSb8.5 with melting temperatures ranges of 280–300 and 241–246 °C, respectively. Thermal diffusivities were 0.32 and 0.36 cm^2/s, respectively. Prior to bonding, a solder drop was placed on the leadframe and patterned into a thin rectangular layer, 8 mm long and 6 mm wide, with an average height of 70 μm. The test chip was placed in the center of this solder pattern.

During a measurement, the test chip was picked from the adhesive foil and moved to the LF to be placed on the molten solder. During pick-up, the sensors were at room temperature, T_{RT}. This reference temperature, the corresponding reference resistances $R_{j,RT}$ of each sensor (j=0–8), and Equation (2.2.3) were used to determine the temperatures from the subsequently measured sensor resistances. After initial solder contact, the chip was pressed on to the LF for 140 ms. The output of a typical measurement is shown schematically in Figure 2.2.23. The temperature of a sensor was plotted as a function of time. Starting from an initial temperature T_i, the sensor reached a plateau T_p at the steady state. The times needed to reach $T_{50}=(T_p - T_i)\times0.5+T_i$ and $T_{90}=(T_p-T_i)\times0.9+T_i$ were defined as $\Delta t_{50}=t_{50}-t_i$ and $\Delta t_{90}=t_{90}-t_i$, respectively.

Starting from a set of reference parameters given in Table 2.2.3, the sensitivity of the measured temperature variations to changes in a number of process parameters was investigated. Test chips without rear metallization (bare silicon) were used to investigate the interaction of the solders with the rear surface of the chip. Both bare and nickel-coated copper LFs were used. In one experiment the LF temperature was reduced from 336 to 315 °C. In addition to the rectangle-shaped solder patterns, nonpatterned solder drops were also used, as shown schematically in Figure 2.2.21 c. Their height was roughly 0.5 mm. In this case, several

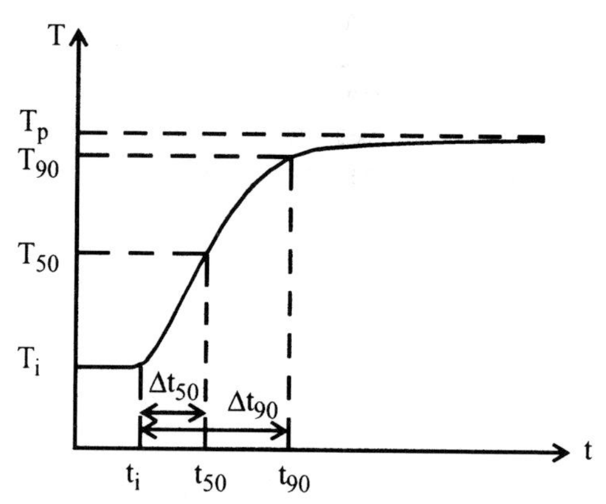

Figure 2.2.23. Schematic variation of microsensor temperature. Times t_i, Δt_{50}, and Δt_{90} denote start of rise, time to reach T_{50}, and time to reach T_{90}, respectively.

Table 2.2.3. Reference process parameter set

Parameter	Value
LF temperature	336 °C
LF surface material	Copper
Solder	PbSn10
Solder volume	1.6 mm^3
Impact time	0.2 s
Bondhead force	0.44 N
Overtravel	–0.15 mm

process parameters were varied individually. Drop volume, dynamic impact parameter (impact time), bondhead force, and overtravel were varied. The resulting solder layers under the bonded chips were investigated off-line using SAM.

2.2.4.4 Results and Discussion

Figure 2.2.24 a shows a set of temperature variations measured during chip placement on hot PbSn10 solder. Process parameters had the reference values listed in Table 2.2.1. After the solder had been patterned, the chip was placed on the solder. The temperatures of the microsensors rapidly rose with slopes between 4.5×10^3 and 11.6×10^3 K/s. At the moment of bondhead lift-off, the chip surface temperature was between 305 °C at the perimeter and 320 °C in the center. The mean rise times of 24 sensors on three test chips were $\Delta t_{50} = 23.0 \pm 4.8$ ms and $\Delta t_{90} = 49.3 \pm 3.6$ ms. After T_{90} was reached, the temperature of the microsensor at the center became 5–15 K higher than that of the sensors at the sides and in the corners. This is due to peripheral chip cooling by the contact fingers. Figure 2.2.24 b shows a SAM micrograph of the solder joint. The void density is low. Such a homogeneous interface is a sign of a successful bond. Mean rise times on chips soldered with SnSb8.5 are comparable, with $\Delta t_{50} = 25.1 \pm 7.1$ ms and $\Delta t_{90} = 55.1 \pm 11.8$ ms.

A measurement performed on a test chip without rear metallization is shown in Figure 2.2.25. Sensor 0 was not connected. The mean rise times of the measured variations were $\Delta t_{50} = 35.1 \pm 1.8$ ms and $\Delta t_{90} = 132.6 \pm 7.0$ ms. These values are much larger than those for rear metallized chips because the absence of wetting results in a higher thermal contact resistance at the interface.

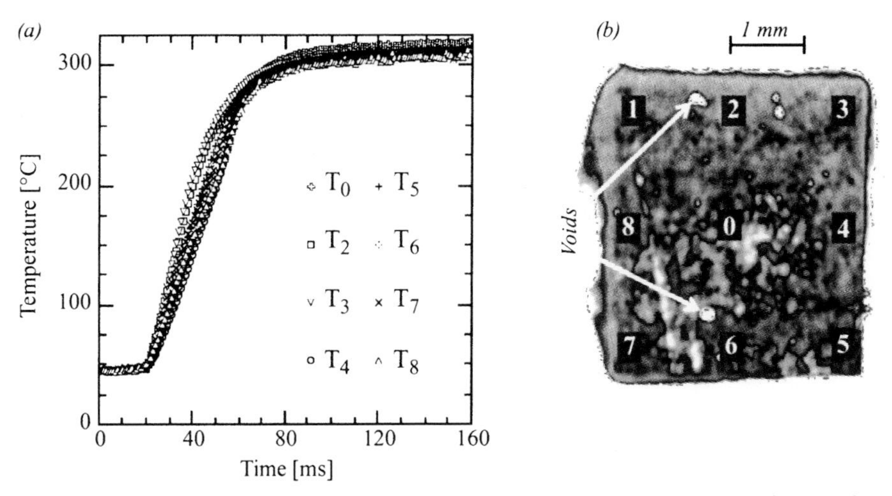

Figure 2.2.24. (**a**) Temperature variation of microsensors during soldering of test chip on patterned solder; (**b**) SAM micrograph of resulting solder layer.

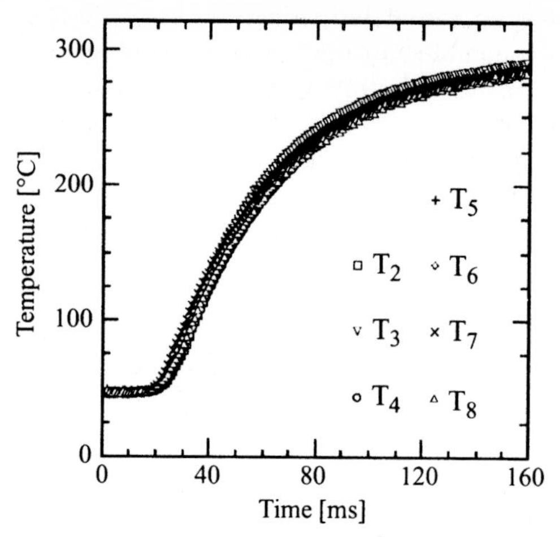

Figure 2.2.25. Temperature variation of selected microsensors on test chips without rear metallization (patterned solder).

Figure 2.2.26 shows a measurement performed on a sample soldered with unpatterned PbSn10. Parameters were those listed in Table 2.2.3. With the solder drop centered below the chip, the first contact with the molten solder is in the center. Solder and heat then spread radially. Since the drop is more than 400 μm higher than a solder pattern, initial contact occurs about 10 ms earlier. The temperature of the center sensor T_0 rises first in Figure 2.2.26 at a time of 10 ms. Next come the side sensors (T_2, T_4, T_6, and T_8). The corner sensors (T_3, T_5, and T_7) are furthest from the point of first contact and consequently show the slowest

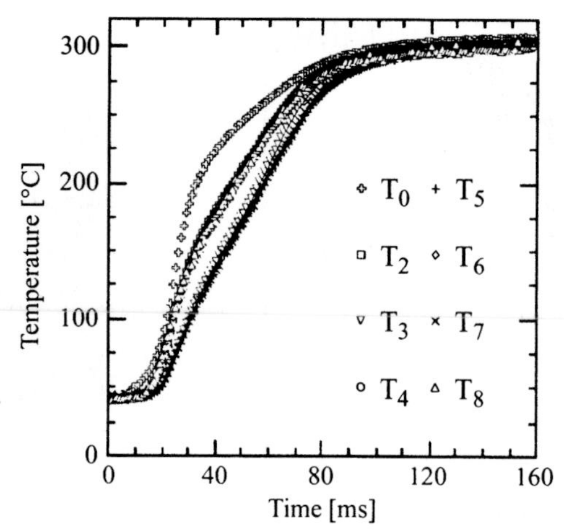

Figure 2.2.26. Temperature variation of microsensors during soldering of test chip on unpatterned solder (drop).

response. The corresponding delays are clearly resolved by the measurements. The delay between center sensor and corner sensors in reaching T_{50} was typically 25–30 ms. Values for Δt_{50} from Figure 2.2.26 are 26, 35±1, and 51±1 ms for microsensors located in the center, on the sides, and at the corners, respectively. For comparison, peripheral sensors usually reach T_{50} first when using patterned solder, and the delay between first and last sensor is shorter than 10 ms.

Different solder drop quantities result in large temperature variations. With 3.2 mm^3 of solder, the rise time was $\Delta t_{50}=25$ ms, whereas with 0.3 mm^3 it was $\Delta t_{50}=60$ ms. The chip temperature rises faster if more solder is used. Varying the dynamic impact parameter (impact time) had no effect on the temperature profile. Measurement variations were also negligible when the bond height and overtravel were varied, except when the test chip was held too high above the LF. In this case, the rise times were longer than $\Delta t_{50}=65$ ms and $\Delta t_{90}=145$ ms.

The presence of large voids in the solder layer is a characteristic of low-quality bonds. Voids reduce the shear strength and cause a locally reduced heat transfer not only during the device operation but also during the bonding. This is evident from microsensor temperature variations as those shown in Figure 2.2.27a. Here, an SnSb8.5 alloy was used. The LF temperature was 336 °C. Sensors 0 and 7 were not connected. When T_4 and T_5 reach $T_{50}=180$ °C, ie, roughly 40 ms after touch down on the LF, T_8 is already more than 50 K higher. This indicates a higher thermal resistance in the region of the chip/solder interface situated below sensors 4 and 5. Confirming this statement, the SAM picture in Figure 2.2.27b reveals a large nonwetted area below sensors 4 and 5. The temperature difference T_8–T_4 is 77 K, 34 ms after initial contact. Consequently, large voids can cause temperature gradients along the chip surface of 30 K/mm measured during soldering.

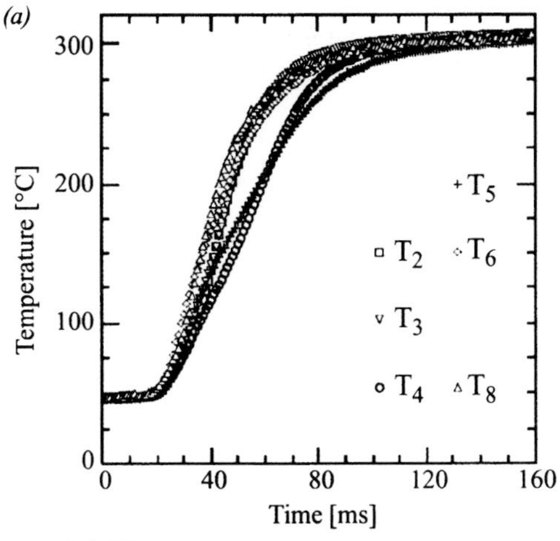

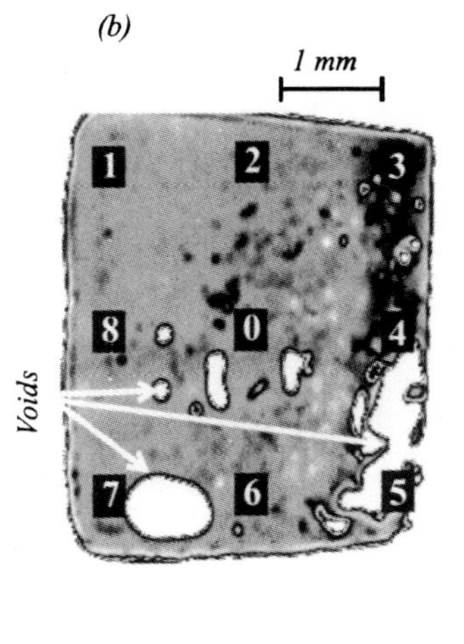

Figure 2.2.27. Low-quality bond: (**a**) temperature variations of selected microsensors during soldering (patterned solder); (**b**) SAM micrograph of resulting solder layer.

2.2.4.5 Transient Thermal FE Model

A transient thermal FE model to simulate the temperature variation allows a comparison of the measurements with the numerical results and finally a better understanding of the wetting behavior, the solder temperature distribution, and the peripheral chip cooling by the contact fingers. The implemented model geometry is shown in Figure 2.2.28 a and b and includes rubber tip, steel, ceramics, contact finger, test chip, solder layer, and LF. It is limited to one eighth of the original structure in view of symmetry, to minimize the computation effort. Influences from the CMOS thin films on the chip top and metallization layers on the chip bottom were assumed to be negligible. Therefore, these layers were not taken into account in the model.

The contribution of heat convection and radiation on the chip temperature during the first 150 ms after initial contact is estimated to be below 2 K. This is less than 1% of the measured temperature change. Therefore, these two effects were neglected in the simulation. The model does not include process effects such as solder solidification [90] and progressive wetting at the solder/chip interface. As simplifying assumptions, the solder is assumed always to be liquid and to wet the chip bottom perfectly immediately after touch down.

Initial temperatures are 43 °C for chip, rubber, steel, fingers, and ceramic, and 336 °C for leadframe and solder. As boundary conditions, the side wall of the LF was kept at a constant temperature of 336 °C, and the top plane of the steel adapter was kept at 43 °C. Thermal contact resistance was neglected at all interfaces. Material properties are listed in Table 2.2.4. Values for silicon, copper, gold, and stainless steel were taken from [91]. Values for solder were taken from [90, 92]. For silicon, temperature-dependent properties were implemented. The

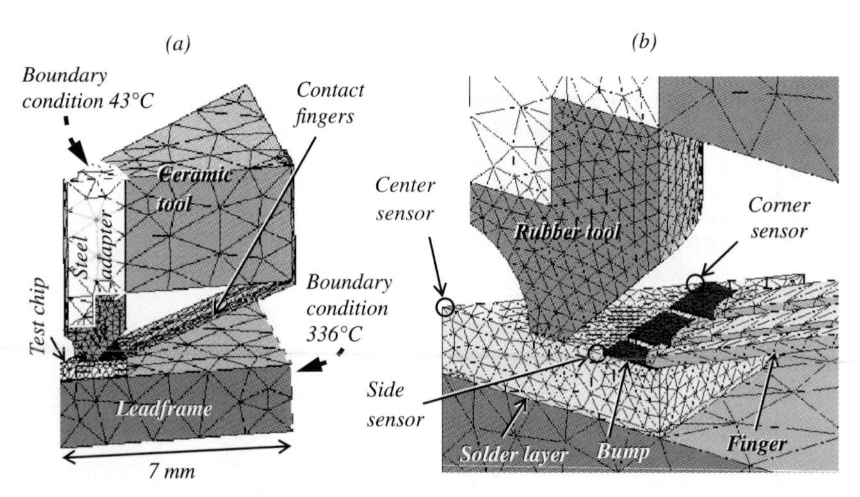

Figure 2.2.28. FE model for one eighth of die bonding geometry with in situ chip connection. (**a**) Total structure (coarse mesh). Boundary conditions are indicated by dashed lines. (**b**) Close-up of chip, rubber tip, fingers, and gold bumps (refined mesh). Locations of microsensors are indicated by circles.

Table 2.2.4. Material parameters used in simulation

Material	Density ($\times 10^3$ kg/m^3)	Thermal conductivity (W/m/K)	Specific heat (J/kg/K)
Silicon	2.33	168.9 − 0.8 T	658.2 + 2.2 T
Copper at 600 K	8.93	398	383
Gold at ambient	19.29	319	129
Stainless steel at ambient	7.50	15	420
PbSn10 liquid [90, 92]	10.10	20	62
Rubber	0.92	0.15	2000
Ceramic MACOR™	2.52	1.6	600

simulation was made using ANSYS®5.4. For the refined mesh, a total of roughly 10000 SOLID70 tetrahedron-shaped elements were used. The transient temperature distribution was calculated using minimum and maximum time steps of 0.1 and 3 ms, respectively.

Simulated temperature distributions at times t=0.1, 1, 5, 10, 20, and 100 ms are shown in Figure 2.2.29. Dark and light colors denote cold and hot areas, respectively. In the course of the process, chip and contact fingers are heated to largely the same temperature as that of the LF, whereas the rubber tool, owing to its low thermal conductivity, largely remains at its initial temperature. At times t=5, 10, and 20 ms, a significant temperature gradient occurs between the middle of the chip top surface and the edge, revealing chip cooling due to the contact fingers.

For comparison, a measured temperature variation from Figure 2.2.24 is shown together with two simulated variations in Figure 2.2.30. The simulated

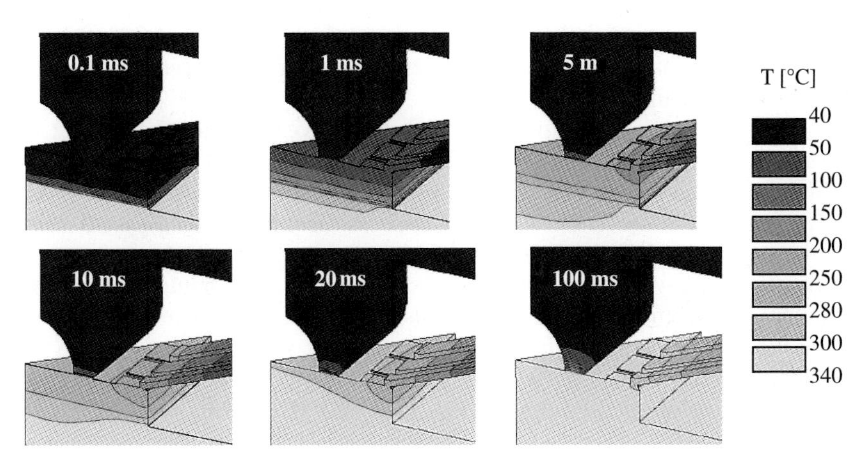

Figure 2.2.29. Temperature distribution at various times after initial contact. One eighth of the real structure is shown. contact fingers are touching the bumps at the right edge of the chip.

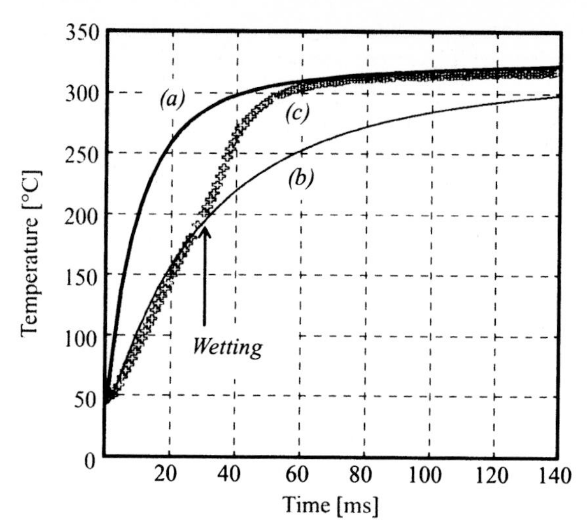

Figure 2.2.30. Temperature variations of center sensor: simulated for (**a**) ideal and (**b**) reduced thermal contact at the chip-solder interface, and (**c**) measured, from Figure 2.2.24.

variations correspond to two different situations. Curve (a) is obtained on the assumption of ideal thermal contact between chip and solder. This differs from the experiment where some time elapses before an intimate contact is obtained by wetting. Curve (b) was obtained by implementing reduced thermal contact between chip and LF. This was achieved by changing the density, thermal conductivity, and specific heat of the solder layer to $\rho = 1.2 \times 10^3$ kg/m^3, $\kappa = 2.5$ W/mK, and $c = 850$ J/kgK, respectively, which corresponds to a mixture of air and solder. During the first 25 ms, the measured temperature variation follows the reduced thermal contact simulation. After roughly 30 ms, the measured temperature variation rises faster and approaches the ideal thermal contact simulation result. This behavior suggests that for the reference set of process parameters, wetting occurs around 30 ms after initial contact.

The contact fingers of the modified pick-and-place tool have an impact on the temperature variation of chip and solder, as compared with the standard die attach situation. This cooling effect was quantified using a modified FE geometry. From the original model the ceramic tool and contact fingers were removed while the other components were kept for the new model. The microsensor and solder temperature variations resulting from a simulation using the first model were subtracted from the variations obtained with the second model. The results in Figure 2.2.31 show that the additional cooling never exceeds –10 K at the chip/solder interface. For the side and corner microsensors, ie, close to the contact fingers, the additional cooling is less than –22 and –25 K, respectively.

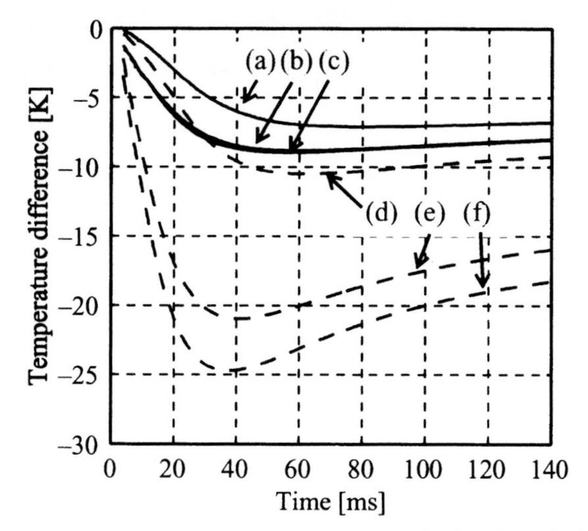

Figure 2.2.31. Simulated additional cooling due to modified pick-and-place tool. Cooling of chip-solder interface (solid lines): (**a**) center, (**b**) side, and (**c**) corner below microsensors. Cooling of chip top surface (dashed lines): (**d**) center, (**e**) side, and (**f**) corner microsensor.

2.2.4.6 Summary

The system based on test chips and a chip connector allows one to measure the in situ temperature during soft solder die bonding. A pick-and-place tool modification permits in situ electrical contact to the test chip. Real-time temperature variations were reported for several bonding conditions. Wetting and spreading of the solder under the chip were resolved. The microsensors were typically heated from 41 to 310 °C, and the time needed to reach 283 °C (90% of the total temperature rise) was typically 50 ms. From the measurements taken for various sets of machine parameters, the solder volume was found to have a large influence on the chip temperature.

Results from simplified transient thermal FE simulations agree well with the temperature variations of a test chip without rear metallization, whereas the agreement of the simplified model is limited for chips with rear metallization. This deviation of experimental from simulated data is a measure of the progress of wetting, and the time needed to start wetting is found to be roughly 30 ms. The FE model also showed that the additional cooling of the chip/solder interface by the contact fingers does not exceed –10 K. Therefore, the die bonding process is only slightly perturbed by the presence of the chip connector.

2.2.5 Thermosonic Ball Bonding

Thermosonic ball bonding is a complex operation compared with other packaging processes because of the variety of bonding parameters and their interactions. Process research, in this field, is still ongoing and heading towards a thorough understanding of the relations between bond quality and interacting process parameters. This section reports on the application of microsensors to acquire in situ experimental results contributing to ball bonding process research. After introductory remarks on the experimental setup including the custom-made contacting device in Section 2.2.5.1, an aluminum temperature sensor is reported in Section 2.2.5.2. Section 2.2.5.3 deals with piezoresistors integrated below bonding pads designed to measure ultrasonic tangential force. Experimental results and interpretations are described in Section 2.2.5.4. Early sensor designs for measuring the out-of-plane force were sensitive to temperature changes as reported in [94]. An advanced force sensor type, however, is able to measure simultaneously forces in the x-, y-, and z-directions while largely reducing temperature influences, as reported in Section 2.2.5.5, followed by a summary in Section 2.2.5.6.

2.2.5.1 Experimental

Test chips were produced using commercial CMOS technology. They are attached by a die bonder to custom-made BGA-type substrate strips [93, 94] using a commercial silver-filled epoxy. These BGA-type substrates comprise six gold-plated die pads with adjacent wire bond terminals. A photograph of a bonded test chip on a substrate is shown in Figure 2.2.32 a. After die attachment, the chips and substrates are usually cleaned using an argon–hydrogen plasma to ensure repeatable surface conditions for wire bonding.

The setup for electrical contact to the microsensors is shown in Figure 2.2.32 b and schematically in Figure 2.2.33 a and b. Elastic copper fingers were soldered to a 15×10 mm printed circuit board (PCB). This PCB is screwed to the clamping plate of the wire bonder. Wires were soldered to the PCB and connected to the measurement equipment. During automatic loading of the substrate, the clamping plate is in an upper position, as shown in Figure 2.2.33 a. As soon as the test chip has reached the bonding area, the plate moves to the lower position in order to clamp the substrate to the heater stage, as shown in Figure 2.2.33 b. Simultaneously, electrical contact to the substrate terminals via the fingers is established. The contacting technique is suited for consecutive measurements with several test chips. The microsensor signals are amplified, filtered, logged on an oscilloscope, and subsequently transferred to a PC for data evaluation, as shown schematically in Figure 2.2.34. Geometry, shear force, and shear strength of bonded balls were determined as described in [94]. The bonding force is measured on-line by the piezoelectric sensor system of the wire bonder [95]. For all experiments, 25 μm diameter AW14 wire and an automatic ball bonder are used.

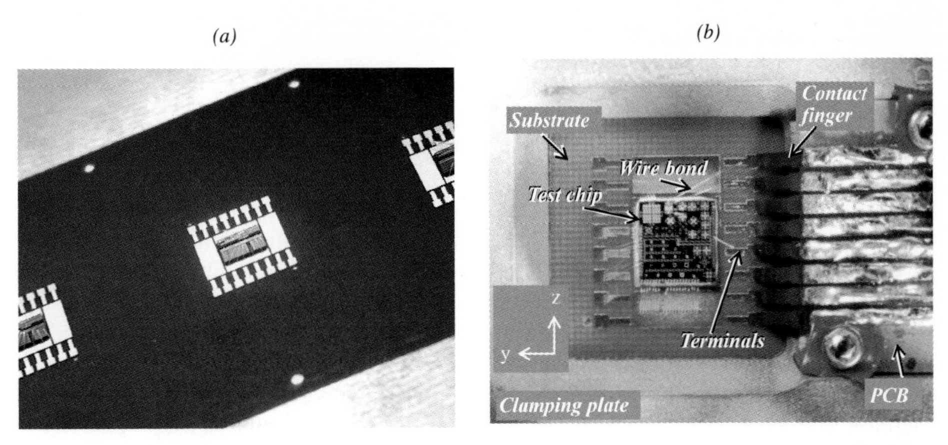

Figure 2.2.32. (**a**) Test chip on substrate. (**b**) Microsensor is connected to terminals using wire bonds. Contact fingers on PCB, mounted on clamping plate, contacting to substrate terminals.

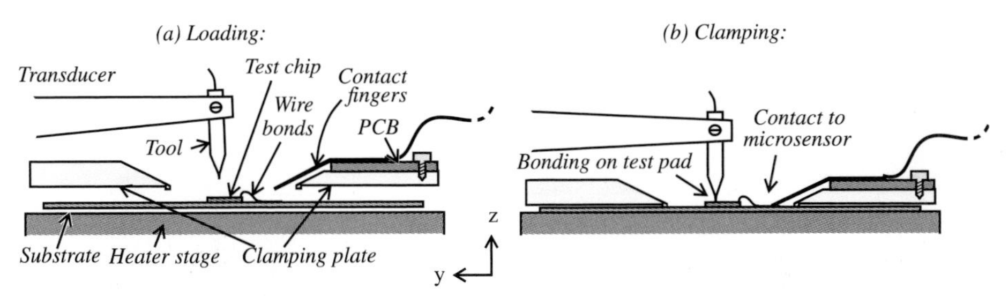

Figure 2.2.33. Contact to test chip on wire bonder. Clamping plate and transducer in (**a**) upper and (**b**) lower position.

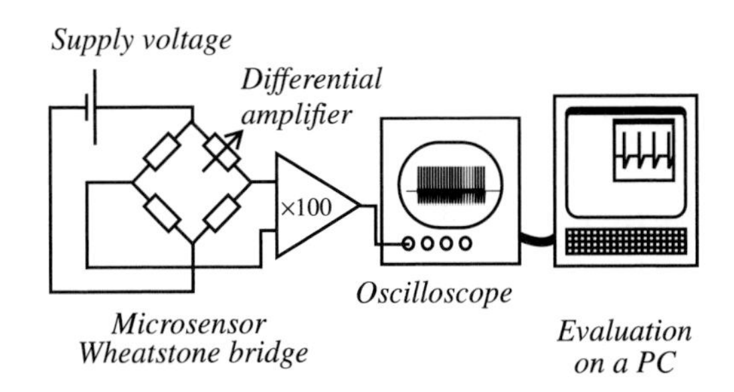

Figure 2.2.34. Schematic diagram of measurement setup.

2.2.5.2 Temperature Microsensor

This sensor is an RTD consisting of an aluminum line of circular shape integrated around a circular bonding pad. An scanning electron microscope (SEM) picture, schematic layout, and cross section of the sensor are shown in Figures 2.2.35 a, b, and Figure 2.2.10 b, respectively. On the test chip, eight microsensors are grouped into two Wheatstone bridges, as schematically shown in the layout in Figure 2.2.36. The temperature change caused by the bonding is expected to be very localized so that the bridge output only indicates the response of the resistor surrounding the ball being bonded. The test bonding pad material is $AlSi_{1.0}Ti_{0.15}$. It has a circular shape and a nominal thickness of 1 μm. The diameter of the pad opening is 75 μm. It covers the sensing aluminum line in order to increase the heat flux between bond contact zone and microsensor. The oxide layer with a thickness of 0.58 μm electrically isolates the pad from the microsensor aluminum. The inner and outer radii of the area covered by the sensing alu-

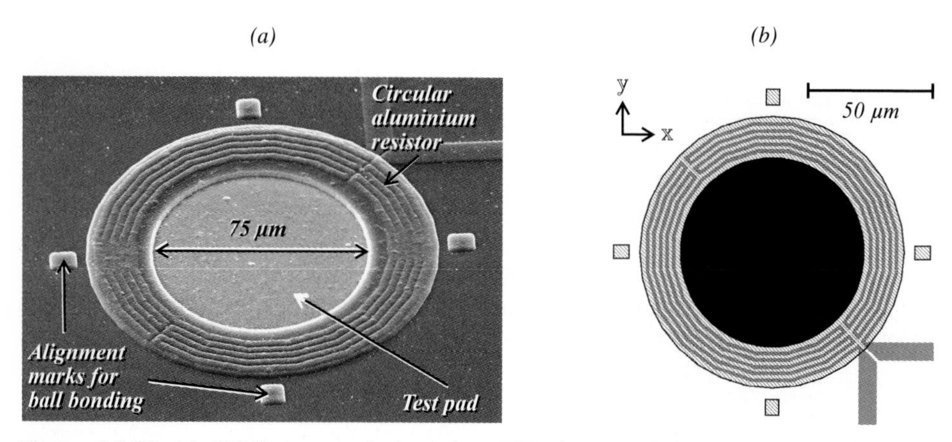

Figure 2.2.35. (**a**) SEM picture of aluminium RTD integrated around a test pad and (**b**) schematic layout of integrated aluminium RTD.

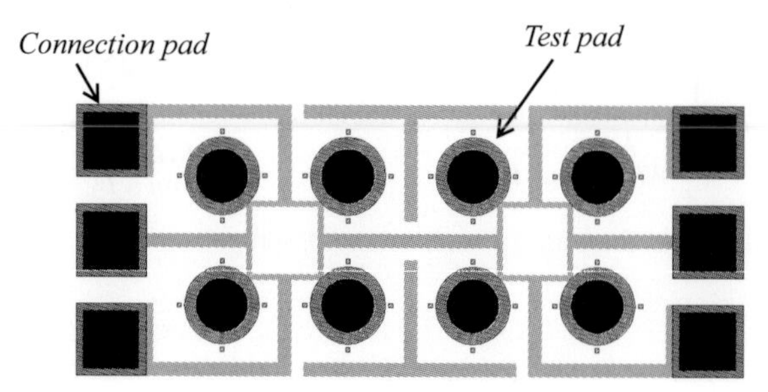

Figure 2.2.36. Schematic layout of eight microsensors with connection pads grouped in two Wheatstone bridges. Test pad pichted is 200 μm.

minum line are 88 and 115 μm, respectively. The line width and spacing are both 1.5 μm.

Resistance and the TCR were determined for six samples for temperatures ranging from 40 to 200 °C. The mean resistance and its standard deviation at 40 °C are 113.5 and 5.5 Ω, respectively. The mean TCR value with standard deviation was measured at 34 and 106 °C as 2900±60 and 2400±60 ppm/K, respectively.

The bonding force induces a strain to the aluminum meander of the sensor. A corresponding strain signal is thus superimposed on the temperature responses of the device. For the circular resistor located around the test pad, numerical simulations show that this strain increases the microsensor resistance up to about 0.12% for a bonding force of 300 mN. As the bonding force possibly changes during ultrasound dissipation, a time-dependent bonding force-induced resistance change may be superposed on the temperature signals. Such a change has only a slight effect.

2.2.5.3 Tangential Force Microsensor

During bonding, the bonding force induces a transient stress distribution in the chip. In addition, the ultrasonic vibration generates an interfacial (out-of-plane) shear stress field at the bonding pad. It is similar to the stress field created by a circular sliding contact, which has been calculated previously [96]. A result of this analysis is that nonzero normal stress fields have a similar order of magnitude as the out-of-plane shear stress field. This is also valid for the bonding pad during ultrasound dissipation. As an example, the normal stress component in the sliding direction, σ_{yy}, is shown schematically in Figure 2.2.37 a. It is compressive in front of the contact zone and tensile behind. This motivated the development of an O-shaped p^+-diffused piezoresistor, contacted in a Wheatstone bridge configuration [77]. A micrograph of the device is shown in Figure 2.2.38. The lateral contacts subdivide the vertical branches of the structure into the resistor pairs R_1+R_4 and R_2+R_3, as shown schematically in Figure 2.2.37 b. Bonding is performed on the zone between R_1 and R_4. During bonding, these two resistors experience opposite stress changes and therefore opposite resistance changes. The ultrasonically induced stress fields change their signs periodically. The bridge is thus periodically deflected, and an oscillating signal is recorded. The bridge acts like a half bridge, the sensitivity of which is given by

$$\frac{U_\mathrm{m}}{U_0} = \frac{\Delta R}{2R} , \tag{2.2.10}$$

where $R=R_i$ and $i=1$–4. Signals due to bonding force and temperature changes are expected to be the same on resistors R_1 and R_4 and therefore are cancelled by the bridge. The stresses at resistors R_2 and R_3 are much smaller, so the changes in these resistances during bonding are considered to be negligible.

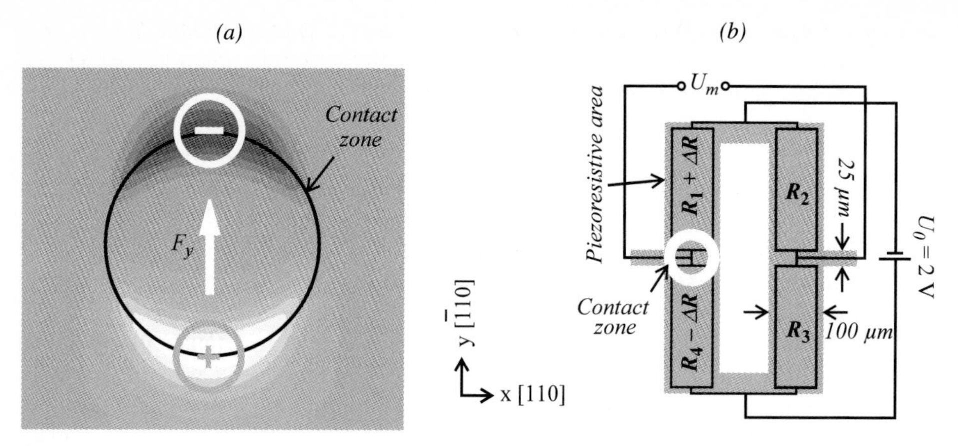

Figure 2.2.37. (**a**) Schematic stress σ_{yy} in chip surface plane, produced by ultrasonic shear force F_y in the y direction. Light and dark colour indicate tensile (+) and compressive (–) stress, respectively. (**b**) Illustration of operating principle of shear force sensor. R_1, R_2, R_3, and R_4 denote the resistors of the Wheatstone bridge. Piezoresistive area is shaded gray.

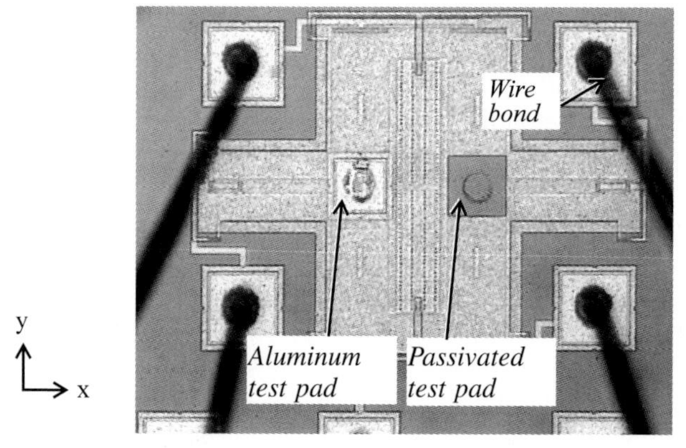

Figure 2.2.38. Micrograph of ultrasound tangential force sensor. Contact pads connected with wire bonds. Bond on aluminium test pad was qualified after measurement using a shear test.

2.2.5.4 Results

2.2.5.4.1 Temperature Change

For the measurements, an ultrasound frequency of $f_{US} = 100$ kHz was used. The transducer amplitude, measured at the freely vibrating horn tip, was $A_{HT} = 0.62$ μm. Chip temperatures and bonding forces were $T_{Chip} = 106$ and 34 °C and $F_N = 250$ and 300 mN, respectively. An initial ball diameter of 55 μm and a capillary suitable for an 100 μm pitch process were used.

An example of a temperature microsensor signal is shown in Figure 2.2.39. The resistance change is proportional to the temperature change during ball bonding. The ball bonding process on test pads is modified by insertion of two additional time delays. This allows the separation of the thermal signal contributions due to cooling and ultrasonic heating. The first time delay is programmed between impact and start of ultrasound. The second is programmed between end of ultrasound and the lift-off of the capillary. Both time delays are 40 ms. The temperature microsensor signal shows roughly three stages of the bonding process, indicated by letters A, B, and C. The first, second, and third stages occur before, during, and after ultrasound dissipation, respectively. Stage A begins with the touch down of the ball on the pad. At this time, four temperature effects can be considered: (1) the rest heat from electrical flame-off heating the pad, (2) the initial ball deformation heating the pad, (3) the capillary cooling the pad, and (4) the heat from the heater stage reservoir reducing the capillary cooling. The experiment shows that the resistance and, thus, the temperature decrease abruptly. The cooling effect absorbs the heat produced by the ball deformation and the rest heat from electrical flame-off if bonding continuously. A minimum is observed during the first 5 ms after touch down. This is because the cooling rate by the capillary and wire is faster than the chip heating rate by the supporting heater stage. This interpretation is supported by the results of a transient thermal simulation of a cold, deformed ball pressed by the capillary to the hot chip [66]. Stage B is characterized by the dissipation of ultrasound. Responses during this stage for bonds realized with different bonding parameters are shown in Figure 2.2.40 a–c. The cooling and time offsets on the figure axes have been removed for easier comparison and evaluation. The measurements exhibit three distinct phases of the welding process, denoted α, β, and γ. They are discussed using Figure 2.2.40.b. Phase lasts for about 4 ms and is characterized by a monotonic increase in temperature. Phase β is characterized by a reduced heating rate. It lasts for about 2 ms. Less heat is delivered to the chip because the amplitude of rela-

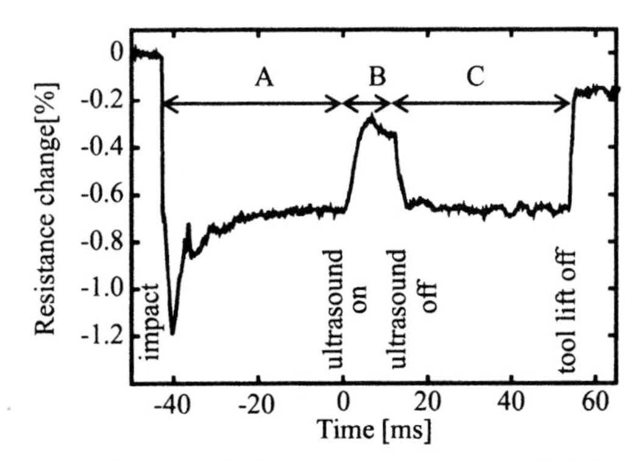

Figure 2.2.39. Resistance change of microsensor during test ball bond operation. Letters A and C denote periods without ultrasonic energy. Letter B denotes period when ultrasound is dissipated. $F_N = 250$ mN, $T_{Chip} = 106\,°C$.

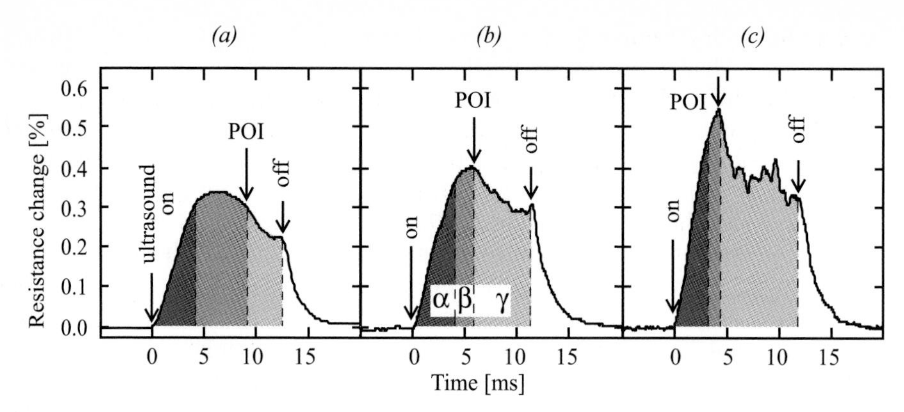

Figure 2.2.40. Microsensor temperature signals during ultrasound dissipation. Dark-, medium-, and light-shaded areas indicate phases α, β, and γ, respectively. POI is point of inflection. A_{HT} is (**a**) 1.20, (**b**) 1.24, and (**c**) 1.45 µm. Other parameters: $F_N = 300$ mN, $t_{US} = 12$ ms, $T_{Chip} = 34\,°C$.

tive motion decreases due to higher friction forces. Phase γ begins with a point of inflection (POI), characterized by the transition from convex to concave. This POI indicates that cooling has become the predominant thermal effect. In particular, it indicates that the interfacial motion and the accompanying heating have ended. Ultrasonic ball deformation increases rapidly right after the POI is reached [76, 94]. For sufficiently high ultrasound levels, the POI occurs at the maximum, as in Figure 2.2.40 b and c. The times to reach the POI were measured for transducer amplitudes A_{HT} of 1.20, 1.24, and 1.45 µm. The resulting average times with standard deviations are 7.1 ± 1.1, 5.9 ± 0.7, and 4.3 ± 1.1 ms, respectively. It is concluded that the POI indicates a *lower limit of ultrasound time* needed for optimal bonding. During phase γ, the temperature decreases to a constant level which is reached for ultrasound times longer than 12 ms. The heat produced during this phase possibly is due to an internal or external friction effect produced by the ultrasound inside or next to the ball. After the end of ultrasound dissipation, the temperature falls to a lower value, defined by the equilibrium of pad cooling and chip heating. At the end of stage C, upon capillary lift-off, the temperature rises again to a constant level which is lower than the initial value. This is because the bonded wire acts as a heat sink until the second bond is made. The measured rapid increase of temperature due to the instantaneous lift-off of the capillary indicates a sub-millisecond response time of the microsensor.

A quality parameter can be derived from the temperature signal variation [76, 94]. This parameter correlates with shear strength and thus can be used to determine the bonding force process window without destructive shear testing.

2.2.5.4.2 Tangential Force

An ultrasound frequency $f_{US}=128$ kHz was used, the transducer amplitude was $A_{HT}=0.62$ μm, the substrate temperature and bonding force were $T_{sub}=60\,°C$ and $F_N=300$ mN, respectively, and an initial ball diameter of 50 μm and a capillary suitable for an 80 μm pitch process were used.

Figure 2.2.41 shows the signal obtained by the microsensor during ball bonding. The signal has been high pass filtered for clarity. The signal consists of about 1660 ultrasonic oscillations. Therefore, it appears as a filled area. Zooming in, the wave form of the signal is found to vary with time as shown in Figure 2.2.42. To investigate this effect further, the signal was digitally filtered [97] at its fundamental and harmonic frequencies. Resulting amplitudes are shown in Figure 2.2.43. The three largest components are the fundamental (A_1), third (A_3), and fifth (A_5) harmonics at frequencies of 128, 384, and 640 kHz, respectively. Based on the time variation of the harmonics, the following interpretation distinguishing four different time phases during ultrasonic dissipation is proposed.

Phase 1 – As soon as the ultrasound is switched on, the microsensor signal emerges and increases in parallel with the vibrational amplitude of the horn tip. For the transducer rise time used, there is no measurable time lag of the microsensor signal to the horn amplitude. In this phase, the ultrasound has not yet reached a large enough amplitude to overcome the initial stiction (static friction). The ball adheres to the pad possibly also due to mechanical interlocking of the two contact surfaces. Such a connection may arise during the initial deformation during touchdown of the ball on to the pad (impact). The wave form is sinusoidal. As a consequence, the harmonics are small. As there is no sliding to produce friction heat, this phase is not visible in the temperature signals shown earlier.

Phase 2 – This phase starts after about 0.5 ms. It is characterized by a rapid increase in A_3. At the same time, the fundamental amplitude A_1 reaches a local maximum, then decreases abruptly, and reaches a local minimum. The sinusoidal wave form becomes cropped. This phase lasts for less than 1 ms. It is believed that interfacial friction starts in this period, and the aluminum oxide layer is removed. The *lower limit of the transducer current* required for interfacial friction occurs at the time of the local maximum. This time also is a *lower limit for the ultrasound time* needed for friction bonding.

Phase 3 – The metallurgical bond is forming during this phase. Owing to the increasing bond strength, friction force also increases. The pad experiences increasing stress, and the microsensor signal A_1 consistently increases. The bond under formation damps the relative motion. The damping is expressed by a decrease in A_3 and A_5. This phase lasts until A_3 and A_5 are constant. Phases 2 and 3 here correspond to the first two phases, α and β, of the thermal signal.

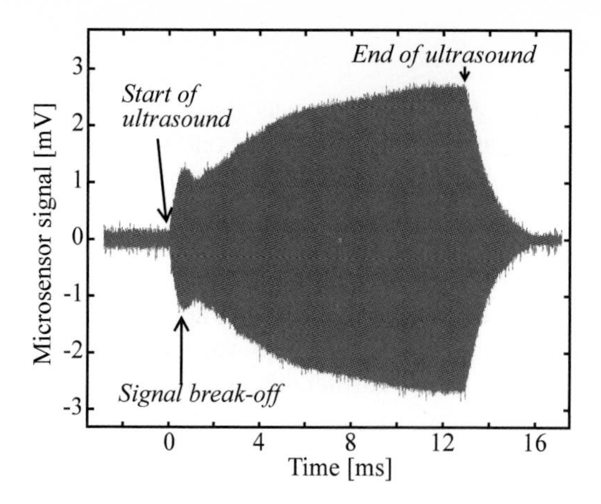

Figure 2.2.41. Ultrasound sensor signal below bonding pad measured during ball bonding.

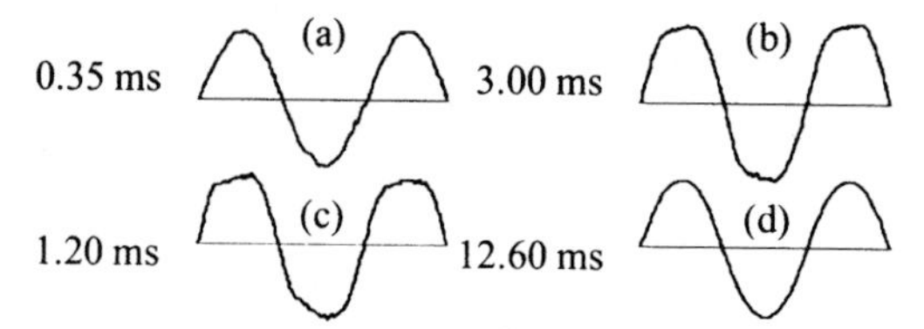

Figure 2.2.42. Wave forms at various times after start of ultrasound.

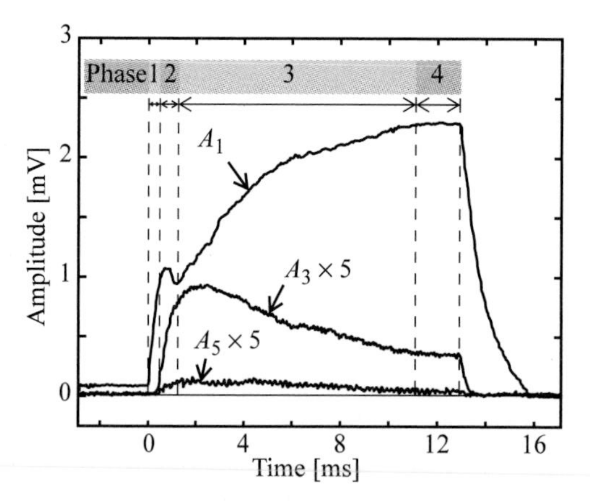

Figure 2.2.43. Harmonics of ultrasound sensor stress signal during ball bond formation.

Phase 4 – The harmonics undergo little change. The wave form returns to a sinusoidal shape.

During the sliding phases, the stress signal has a cropped sinusoidal wave form, as shown in Figure 2.2.42 b and c. Such a wave form is typical for the deflection of a

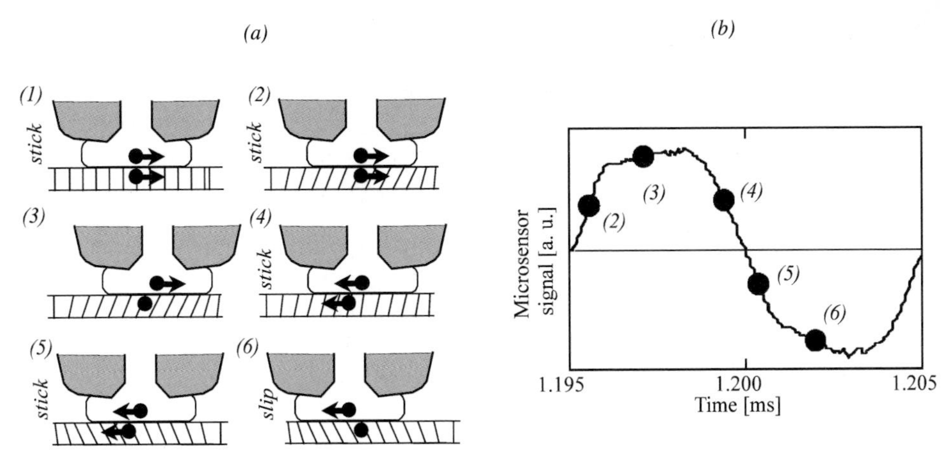

Figure 2.2.44. (**a**) Schematic illustration of capillary tip, ball, and pad during stages of oscillation period: ball and silicon (1) at origin, (2) moving simultaneously, (3) sliding on each other, (4) moving back to origin, (5) having passed origin, and (6) sliding. Arrows indicate velocity. Inclined lines indicate elastic deformation. (**b**) Microsensor signal during one ultrasound cycle: wave form after 1.2 ms, exhibiting stick-slip behavior. Numbers in parentheses refer to part (**a**).

harmonically forced friction oscillator [98]. It arises due to a stick-slip motion. To illustrate such a behavior for ball and pad, several stages are shown schematically in Figure 2.2.44 a [78]. A measured stick-slip wave form is reproduced in more detail in Figure 2.2.44 b with numbers referring to the various stages in Figure 2.2.44 a.

A harmonic finite element analysis is reported in [94] to calibrate the experimental shear force signal. According to this reference, a typical maximum ultrasonic shear force during bonding on a contact zone with a diameter of 50 μm is 0.12 N.

2.2.5.5 Optimized *xyz*-Force Microsensor

The first integrated piezoresistive microsensor capable of measuring in situ simultaneously the forces in all three axes during ultrasonic ball wire bonding was reported in [99]. Its bandwidth reaches from low-frequency forces as the bonding force up to the ultrasonic forces generated by the transducer system. The microsensor is optimized to the requirements of wire bonding to achieve high measurement repeatability and insensitivity to noise and temperature.

Besides the principal feature – the simultaneous three axes measurement capability – several other requirements have been taken into account for the design of this advanced wire bonding microsensor. A high signal-to-noise ratio assures compatibility with any industrial surroundings, in which the electrical noise level can be high. Furthermore, the sensor supports measurements on a wide bonding parameter range (temperature, ultrasound power, bonding force). By using extreme bonding parameters, the influence of certain physical effects on the sensor signal is more pronounced and signal interpretation is simplified. Installed on a commercial wire bonder, the measurement equipment allows continuous bonding without dead-times in order to measure the dynamic behavior of the bonder. In order to use the sensor as calibration or analysis tool, short setup times with no need for adjustment are possible. A calibration application demands high repeatability and robustness on the measurement method.

On-chip integrated microsensors can only be used once for bond process inspection because of the resulting bond between the wire and aluminum pad. Consequently, the sensor design is based on a standard IC technology for cost reduction. In addition, the test pad design is not restricted because the diffusion and connection lines of the sensor are integrated outside the pad area.

2.2.5.5.1 Sensor Principles

For local stress fields with symmetry, piezoresistors can be arranged to be selective to stress fields generated by uniaxial forces applied to the contact zone. High resistance in regions of strong stress fields (eg, under the contact zone) results in high sensor response but it also entails large stress field gradients which diminish the repeatability. Thus, placing the sensing areas outside the contact zone assures more reliable signals.

The coordinate system is defined by the x-, y-, and z-axes along [110], $[\bar{1}10]$, and [001], respectively. The wire bonder ultrasound oscillation is along the y-axis. Only the stress fields σ_{xx}, σ_{yy}, σ_{zz}, and σ_{xy} can be detected by piezoresistors on a (100) wafer [35]. The integrated sensor consists of x-, y-, and z-force sensor elements. A shear stress σ_{yz} applied on a circular contact area also induces σ_{yy} and σ_{xx} stress fields that are maximum at the border of the contact area (compare Section 2.2.5.3). The decrease of the stress field in direction y away from the contact is smaller for σ_{yy} than for σ_{xx}. As the piezoresistive coefficient of p^+-doped silicon for in-plane main stresses is large, it is favorable to choose p^+-doped line-shaped resistors parallel to the y-axis as sensing elements. Figure 2.2.45 schematically shows the x- and y-force sensing resistors. The arrangement of the x-force resistors is the same as that of the y-force resistors except for a rotation by 90°.

The sensor design is based on the 0.8 μm CMOS process CXQ of Austria Mikro Systeme International AG (Unterpremstaetten, Austria), and uses the source/drain implantations as piezoresistors. Four serpentine-shaped resistors R_1 to R_4 are grouped together in a Wheatstone bridge configuration, as shown in Figure 2.213. The wiring between the sensing resistors consists of aluminum lines.

The operation principle is similar to that described in Section 2.2.5.3. Under an applied shear force (eg, ultrasound oscillation of the capillary), R_1 and R_3 change their resistances opposite to the resistances of R_2 and R_4. The stress field caused by the bonding force has point symmetry and therefore does not contribute to the signal. A circular punch (rigid capillary pressed on the silicon surface)

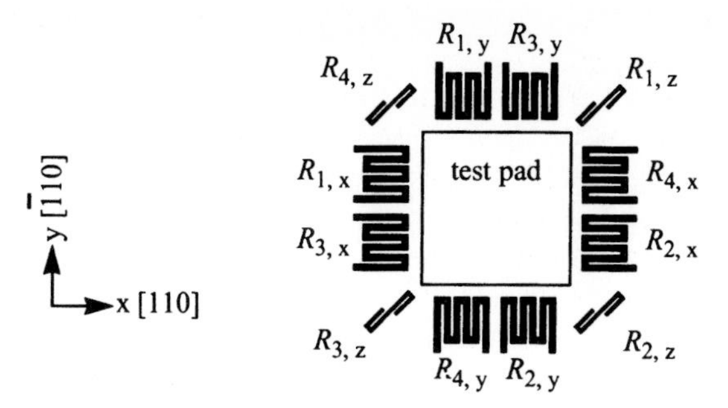

Figure 2.2.45. Design of the diffused piezoresistors.

results in strong stress fields under the contact area. The shear stress σ_{xy} extends outside the contact zone and has negative values on the diagonal axis [100] and positive values on the [010] axis, as shown in Figure 2.2.46. To take advantage of this property, n^{+}-doped piezoresistive resistors are used. The four diagonal serpentine-shaped resistors R_1 to R_4 are placed next to the corners of the test pad. Under an applied normal force, R_1 and R_3 change their resistances opposite to the resistances of R_2 and R_4. Each of the three sensor Wheatstone bridges is a full bridge with a sensitivity of

$$\frac{U_{\mathrm{m}}}{U_0} = \frac{\Delta R}{R}, \tag{2.2.11}$$

where $R = R_i$ and $i = 1$–4. The cut-off frequency is mainly determined by the capacitance of the external wiring, as shown with the microsensor in Figure 2.2.47,

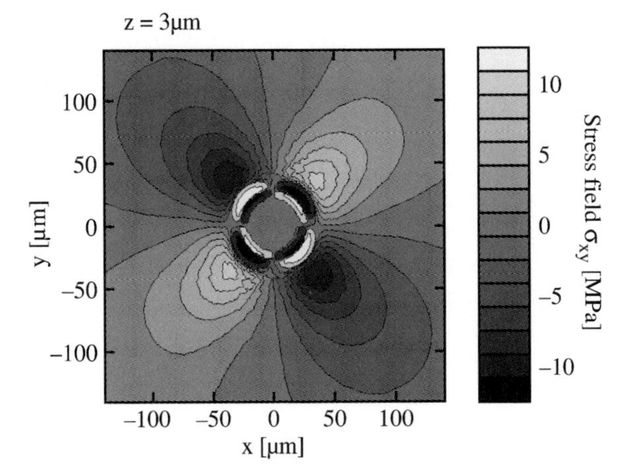

Figure 2.2.46. (a) Schematic of stress field σ_{xy} as caused by a 50 μm diameter circular punch.

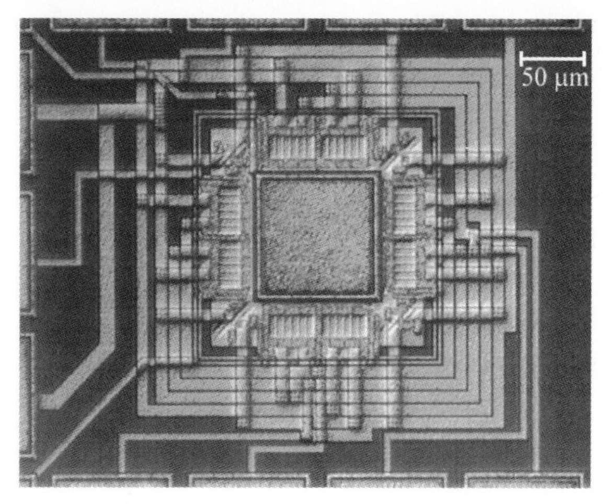

Figure 2.2.47. Micrograph of *xyz*-force sensor with metallized test pad.

and the source resistance on the sense line. Sensor signal buffering near the chip is hindered by the restricted space due to the bondhead movement and by operation temperatures up to 200 °C. Therefore, the cut-off frequency has to be controlled by an appropriate source resistance of the sensor. There are no diffusions and metal lines under the test bond pad. As a result, no limitation exists for the test pad design. The diffusion layers are electrically and optically shielded by an additional metal layer. A ground connection to the test pad keeps the pad metallization on a defined potential and reduces capacitive coupling of distortions.

Figure 2.2.48 a–c shows an example measurement of the sensor during ball bonding on the aluminum test pad. The three force signals are given in normalized units of mV/V, ie, the respective bridge deflection voltages (mV) are divided by the supply voltage (V). The individual oscillations of the ultrasound stress signal are not resolved in time in the figure. The stress signal envelope deviates significantly from that of the transducer current during bonding of a gold wire on an aluminum pad, as shown in Figure 2.2.49. Owing to physical processes at the bond interface and in the gold ball, higher harmonics appear during particular periods of bonding as described in Section 2.2.5.4.2. The fundamental and the higher harmonics constitute a signature of the ball bond. This resulting signature is repeatable for fixed machine parameters, whereas a change of machine parameters significantly alters the signature [77, 94]. The strong third harmonic at the beginning of bonding in Figure 2.2.49 is due to friction between gold ball and aluminum pad.

2.2.5.5.2 Sensor Characterization

To characterize the microsensor, measurements are carried out without wire on passivated pads to exclude any bonding and deformation effects. The capillary is

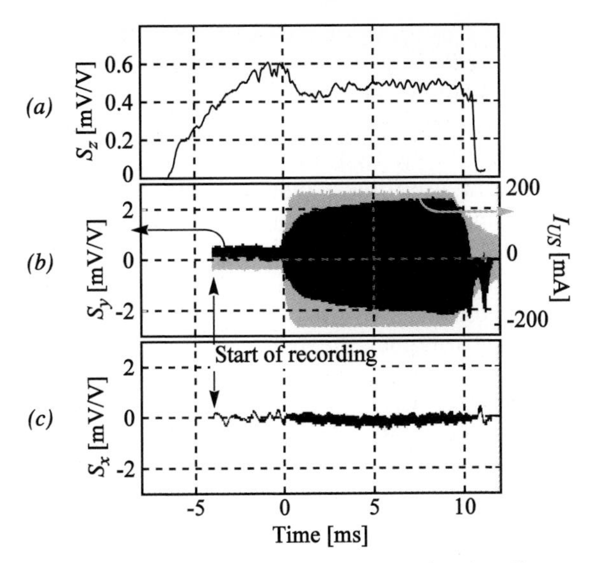

Figure 2.2.48. Example microsensor signals obtained during bonding an Au ball to an Al pad. (**a**) S_x, (**b**) S_y, (**c**) S_z, and (**b**) I_{US} are *x*-, *y*-, *z*-force signals and ultrasonic transducer current (gray).

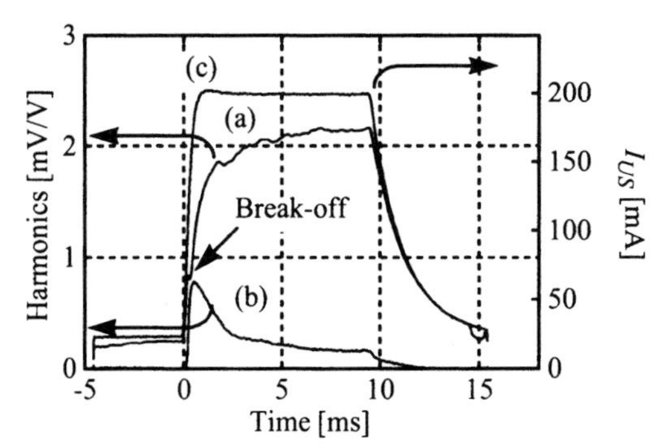

Figure 2.2.49. Envelopes of (**a**) fundamental of *y*-force sensor signal, (**b**) 3rd harmonic scaled by factor 5, and (**c**) fundamental of ultrasonic transducer current I_{US}.

pressed directly on the passivation of the chip ('capillary bonds') using a high normal force and a low enough ultrasonic amplitude to avoid sliding. This induces only very slight wear of the passivation that does not reduce the sensor performance even after more than 100000 tests on the same passivated pad.

The electrical resistances of the *x*- and *y*-force resistors are 1.63 kΩ and that of the *z*-force resistors is 1.05 kΩ. The values of the resistors are chosen high enough to exclude the influence of the measurement lines. The measured bonding force sensitivity (*z*-axis) is 2.02±0.05 mV/V/N, and the ultrasound sensitivity with respect to the driving transducer current is 15 mV/V/A. The sensor signal amplitude is highly linear to the amplitude of the driving current of the transducer, as shown in Figure 2.2.50. The ultrasound sensitivity was found to be independent of the applied bonding force which was varied in the range from 600 to 1200 mN. As the piezoelectric transducer system has a well defined resonance frequency, digital filtering in the frequency range of the fundamental and the harmonics is carried out. The used linear quadrature filter offers high-frequency selectivity together with the ability to examine transient processes. The signal-to-noise ratio of the x- and y-force sensor elements is better than 50 dB for typical bonding parameters. Harmonic distortions are below –44 dB for the second harmonic, –47 dB for the third harmonic, and –49 dB for the fifth harmonic. The distortions of the other harmonics are below the measurement limit. The noise equivalent bonding force of the z-force sensor element is 2 mN.

During capillary bonding, the rigid ceramic of the capillary tip is pressed directly on the passivation of the die. The extension of the contact zone is not exactly known, and the contact force is generally not equally distributed over the whole contact zone. The main supporting point of the capillary may therefore be off-centered even though the capillary is centered. To specify the influence of the contact position on the sensor signal, measurements are performed with a 25 µm tip diameter test capillary placed at various off-center positions at distances up to 36 µm. Figure 2.2.51 shows the result of such a placement sensitivity investigation for the x- and y-force sensor element. The placement response of the sensitivity curve is equivalent to a saddle surface. This property of the placement sensitivity holds also for standard capillaries. The surface shape can be used for extraction of a precise value (saddle point) useful for ultrasonic system calibration, even if the supporting point of the capillary is not accurately known. This calibration procedure is worked out in detail in [100]. An off-center displacement of 10 µm results in an error of 7% in case of the x- and y-force sensor elements, whereas the sensitivity change of the z-force sensor element is 5% for a 20 µm displacement. As long as the axes of symmetry of the contact forces coincide with the axes of symmetry of the sensor, there is no cross-talk. For off-center placements of 10 µm, the maximum cross-talk between the x- and y-force sensor

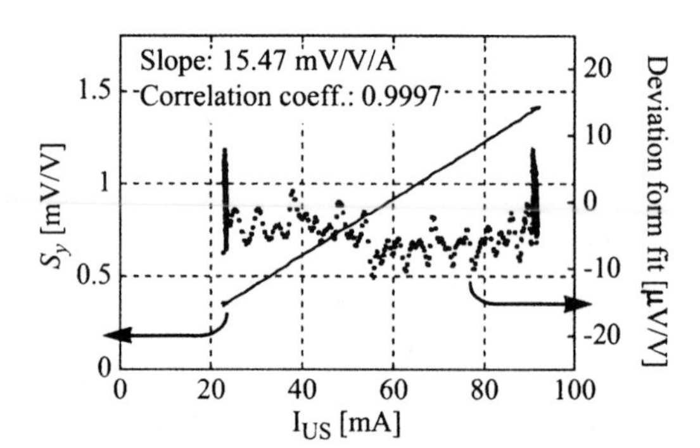

Figure 2.2.50. Linearity of microsensor signal versus transducer current for an 80 µm pitch capillary.

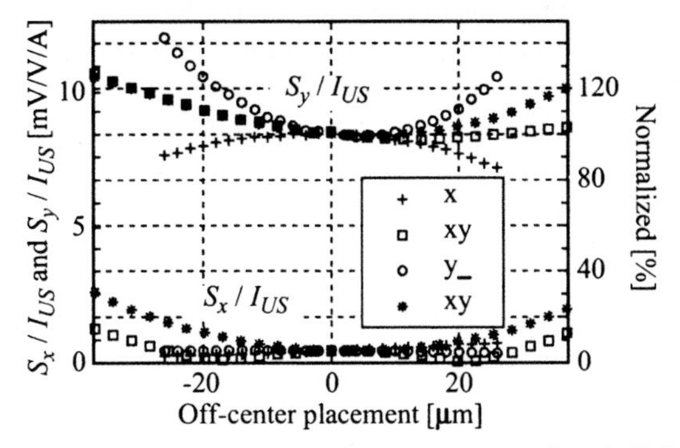

Figure 2.2.51. Placement sensitivity: *x*- and *y*-force sensor signal divided by transducer current for various placements of the test capillary tip relative to the microsensor center.

element is 1%. All sensing resistors of the Wheatstone bridge are placed at the same distance to the test bond pad. As a consequence, all temperature dependences except those of the piezoresistive coefficients are canceled out. The TC for the n^+ sensor sensitivity is found to be 0.05%/K in the range 30–200 °C.

To characterize the sensor repeatability, various measurement series were performed. To exclude any inaccuracy due to capillary misplacement, the sensitivity at the saddle point is numerically extracted out of a set of measurements. The resulting average sensitivity and standard deviation of a single sensor are 15.40 and 0.053 mV/V/A (0.4%), respectively, for a standard 80 µm pitch capillary. The sensor to sensor standard deviation is 0.15 mV/V/A (0.98%) for a set of nine sensors.

2.2.5.6 Summary

Temperature and stress microsensors are suitable for in situ wire bond monitoring. The sensors reported in this section include a temperature sensor and various force sensors. They all are designed to monitor the ball bonding process on aluminum pads on a test chip. An on-chip sensor suitable for the characterization of second (stich) bonds on aluminum pads is reported in [101]. Its design is similar to that of the *xyz*-force sensor. Compared with the distinct force signals, the temperature sensor yields a signal that is a superposition of several temperature effects, each being a measure for a certain heat flow between bonding tool and chip. Evaluating a bond quality parameter from the temperature signal [76] appears to be a less complex task compared with attempts with the force signals [102]. A modified version of the temperature sensor has a passivated test area instead of the standard aluminum pad layers. This modified *xyz*-force sensor can be used for wire bonder ultrasonic system calibration [100].

The full Wheatstone bridge configurations of the *xyz*-force sensor cancel out all temperature dependences except that of the piezoresistive coefficient, which is small for highly doped silicon as used for the sensors. Compared with the temperature sensor, the *xyz*-force sensor yields more detailed signals of the bond process. The signal-to-noise ratio of its filtered harmonics is higher than that of the temperature sensor, especially at elevated temperatures. In addition, friction effects can be investigated using the *xyz*-force sensor.

In consequence of the good measurement repeatability and stability, the *xyz*-force sensor offers the opportunity to measure the ultrasound force reduction due to elastic and plastic ball deformation. This can be used to examine the ultrasound softening effect.

The three-axes piezoresistive microsensor (*xyz*-force sensor) is optimized for wire bonding applications ranging from calibration of bonding force and ultrasound amplitude to in situ examination of the bond process. The sensor's ability for locally measuring forces in all three axes may furthermore be applied for other electronic packaging processes.

2.2.6 Conclusions and Outlook

Test chips fabricated using commercial CMOS processes were successfully applied for real-time monitoring of standard bonding processes in microelectronic packaging. The test chips contain integrated microsensors relying on well proven sensor principles such as resistive temperature detection and piezoresistive stress sensing.

The microsensors made it possible to record in situ variations of temperature and stress. Selectivity of the sensors was effectively achieved by an appropriate choice of design and material. Integrated aluminum resistors performed well as temperature detectors, whereas n^+- and p^+-diffusions were suited for stress measurements.

The challenge of establishing electrical contact to the microsensors during the bonding processes is solved by the development of special connector tools. Their central component is an arrangement of elastic finger contacts. Such tools work satisfactorily during thermosonic ball bonding for process temperatures up to 220 °C. Similarly, they were successfully applied during pick-and-place of soft solder die bonding, where the maximum process temperature was 340 °C. The tools minimally influenced the processes being monitored.

The *soft solder die bonding* process was investigated by measuring the in situ temperature during the placement of the test chip on the leadframe. Results from simplified transient thermal FE simulations agree well with the temperatures measured on a test chip without rear metallization. The agreement with results from rear-metallized chips undergoing wetting is limited. In this case, the deviation of experimental results from simulated data provides a method of determining the beginning of wetting after the solder-chip contact.

The real-time in situ monitors for the *thermosonic ball bonding* process produce a surprisingly vast range of results, revealing thermal, mechanical, and tribological effects of the welding process in great detail. When compared with the integrated sensors in this chapter, no other technique is able to obtain a comparable spatial resolution and vicinity to the process while permitting real-time data collection without disturbing the process. The monitors include temperature, bonding force, and ultrasound shear (tangential) force sensors.

This microsensor technology can help to reduce time-consuming off-line inspection and to optimize the setup times of wire bond processes.

Multiplexers can be co-integrated on test chips comprising a large number of test pads with sensors. This allows statistical evaluations based on a larger number of measurements [103].

Integrated microsensors can be used to record ball bond *fingerprints*, ie, sets of characteristic curves including temperature, bonding force, and ultrasonic stress harmonics. Such fingerprints serve as highly accurate process indicators and can be used to characterize process materials, eg, of wire, bonding pad, die attachment, and substrate. They also quantify the robustness of processes.

In view of process window determination, further investigations concentrate on correlating conventional shear strength data to the ultrasound stress sensor signal or its derived quantities.

Beyond wire bonding, the ultrasonic stress sensors can be used for studying tribology and welding of other material combinations subjected to harmonically driven interfacial stick-slip oscillations.

As the microsensors yield access to the welding mechanisms during ball bonding, they are currently successfully used in wire bonding process development. They are also promising candidates for the characterization of other microelectronic packaging processes.

2.2.7 Acknowledgments

The authors are grateful for contributions from Professor Dr. Paul, O., L. Plattner, Dr. O. Brand, and Dr. D. Bolliger. This work was funded by the Swiss Federal Priority Program MINAST (Micro & Nano System Technology) and by the Swiss Federal Commission for Technology and Innovation (CTI).

2.2.8 References

[1] Lee, Y.C., Chen, W.T. (eds.), *Challenges in Electronics Manufacturing*; London: Chapman & Hall, 1998.

[2] Tummala, R.R., Rymaszewski, E.H., Klopfenstein, A.G. (eds.), *Microelectronics Packaging Handbook, Part I*, 2nd edn., New York: Chapman & Hall, 1997, pp. I-37–I-38.

[3] Brown, W.D. (ed.), *Advanced Electronic Packaging*, New York: IEEE Press, 1999.
[4] Mahajan, R.L., in: *Challenges in Electronics Manufacturing*, Lee, Y.C., Chen, W.T. (eds.); London: Chapman & Hall, 1998, pp. 185–220.
[5] Collins, G.J., Sullivan, J., Loiterman, R.S., in: *Semicond. Fabtech,* 7th edn., London: ICG Publishing, 1998, pp. 261–276.
[6] Bose, A.F., *PhD Thesis*, No. 11224, ETH Zurich, Zurich, 1995.
[7] Baker, M.D., Williams, F.R., May, G.S., *IEEE Trans. Semicond. Manuf.*, **2** (1998) 254–265.
[8] Pecht, M., Evans, J., Evans, J. (eds.), *Quality Conformance and Qualification of Microelectronic Packages and Interconnects*; New York: Wiley, 1994.
[9] Pecht, M.G., Nguyen, L.T., Hakim, E.B. (eds.), *Plastic-Encapsulated Microelectronics*; New York: Wiley, 1995.
[10] Carrass, A., Jaecklin, V.P., in: *Proc. 2nd Eur. Conf. Electronic Packaging Technol., EuPac'96;* 1996, pp. 135–139.
[11] Nguyen, L.T., Gee, S.A., v. d. Bogert, W.F., *ASME J. Electron. Packaging*, **113** (1991) 397–404.
[12] Tsao, P.-H., Voloshin, A.S., *IEEE Trans. Comput. Packaging Manuf. Technol., Part A*, **18** (1995) 201–205.
[13] Kurabayashi, K., Goodson, K.E., *IEEE Trans. Comput. Packaging Manuf. Technol., Part A*, **21** (1998) 506–514.
[14] Nguyen, L.T., *ASME J. Electron. Packaging*, **115** (1993) 346–355.
[15] Dickerson, S.D., Du, W.Y., Volcy, J., in: *Sensors in Electronic Packaging*, Ume, C., Pin-Yeh, C. (eds.), MED-vol. 3/EEP-vol. 14; New York: ASME, 1995, pp. 73–78.
[16] Herman, C.R., Skormin, V. A., Westby, G.R., *ASME J. Electron. Packaging*, **15** (1993) 44–54.
[17] Behler, S., in: *Proc. 1st IEEE/CPMT Electronic Packaging Technol. Conf. EPTC'97*, Singapore; 1997, pp. 141–145.
[18] Lim, F.J., Nguyen, L.T., in: *Proc. Technical Sessions, MEPPE Focus 1991*, Santa Clara, CA; 1991, pp. 246–270.
[19] Sato, M., Yokoi, H., in: *Abstr. PPS 14th Annu. Meeting*, Yokohama, Japan; 1998, pp. 91–92.
[20] Sato, M., Yokoi, H., in: *Abstr. PPS North American Meeting*; 1998, pp. 62–63.
[21] Pecht, M.G., Govind, A., *IEEE Trans. Comput. Packaging Manuf. Technol., Part C*, **20** (1997) 207–212.
[22] Farassat, F., *PhD Thesis*, Technical University, Berlin, 1996.
[23] Gardner, J.W., *Microsensors – Principles and Applications;* Chichester: Wiley, 1994.
[24] Middelhoek, S., Audet, S.A., *Silicon Sensors*; Delft: Delft University of Technology, 1994.
[25] v. Gestel, H.C.J.M., *PhD Thesis*, Delft University of Technology, Dept. Electrical Eng., Delft, 1994.
[26] Miura, H., Kitano, M., Nishimura, A., Kawai, S., *ASME J. Electron. Packaging*, **115** (1993) 9–15.
[27] Zou, Y., Suhling, J.C., Johnson, R.W., Jaeger, R.C., Mian, A.K.M., *IEEE Trans. Electron. Packaging Manuf.*, **22** (1999) 38–52.
[28] Berg, H.M., Paulson, W.M., *Microelectron. Reliab.*, **20** (1980) 247–263.
[29] Emerson, J.A., Peterson, D.W., Sweet, J.N., in: *Proc. 42nd Electronic Comput. Technol. Conf., ECTC'92*, 1992.

[30] Hullinger, A.K., Domer, S.M., Duffalo, J.M., Hollstein, R.L., Niederkorn, A.J., in: *Sensors in Electronic Packaging,* Ume, C., Pin-Yeh, C. (eds.), MED-vol. 3, EEP-vol. 14; New York: ASME, 1995, pp. 43–54.
[31] O'Mathuna, S.C., Moran, P.L., in: *Proc. IMC 1986*; 1986, pp. 378–386.
[32] Miura, H., Nishimura, A., Kawai, S., Nakayama, W., in: *Proc. InterSociety Conf. Therm. Phenomena in the Fabrication and Operation of Electronic Components, I-THERM'88*; 1988, pp. 50–59.
[33] Edwards, D.R., Heinen, K.G., Groothuis, S.K., Martinez, J.E., *IEEE Trans. Comput. Hybr. Manuf. Technol.*, **CHMT-12** (1987) 618–627.
[34] Edwards, D.R., Heinen, G., Bednarz, G.A., Schroen, W.H., *IEEE Trans. Compon. Hybr.*, **6** (1983) 560–567.
[35] Bittle, D.A., Suhling, J.C., Beaty, R.E., Jaeger, R.C., Johnson, R.W., *ASME J. Electron. Packaging*, **113** (1991) 203–215.
[36] Sweet, J.N., in: *Thermal Stress and Strain in Microelectronics Packaging,* Lau, J.H. (ed.); New York: Van Nostrand Reinhold, 1993, pp. 221–271.
[37] Sweet, J.N., Burchett, S.N., Peterson, D.W., Hsia, A.H., Chen, A., in: *Advances in Electronic Packaging*, vol. 2, EEP 19-2; 1997, pp. 1731–1740.
[38] MCNC Bumped Die Test Vehicle (BDTV); *http://api.mcnc.org/bdtv.html* (current May 29, 2002).
[39] The Advanced Thermal and Mechanical Stress Test Chip; *http://www.mesmeric.org* (current June 4, 2002).
[40] Sandia Assembly Test Chips; *http://www.mdl.sandia.gov/MDL/Packaging/ATC.html* or *http://www.sandia.gov/mems/microelectronics/Packaging/ATC.html* (current May 29, 2002).
[41] *JEDEC Standard 33-A. Standard Method for Measuring and Using the Temperature Coefficient of Resistance to Determine the Temperature of a Metallization Line*; Arlington, VA: EIA/JEDEC, 1995.
[42] Rey, P., Woirgard, W., Thébaud, J.-M., Zardini, C., *IEEE Trans. Comput. Packaging Manuf. Technol., Part A*, **21** (1998) 365–372.
[43] Manzione, L.T., *Plastic Packaging of Microelectronic Devices*; New York: Van Nostrand Reinhold, 1990, pp. 28–35.
[44] Klein Wassink, R.J., *Soldering in Electronics*, 2nd edn.; British Isles: Electrochemical Publications, 1989.
[45] Lee, C.C., Wang, C.Y., Matijasevic, G., *J. Electron. Packaging*, **115** (1993) 201–207.
[46] Matijasevic, G.S., Wang, C.Y., Lee, C.C., in: *Thermal Stress and Strain in Microelectronics Packaging,* Lau, J.H. (ed.); New York: Van Nostrand Reinhold, 1993, pp. 194–220.
[47] Anderman, J., Tustaniwskyj, J., Usell, R., *IEEE Trans. Comput. Hybrids Manuf. Technol.*, **CHMT-9** (1986) 410–415.
[48] Nicolics, J., Musiejovsky, L., Schrottmayer, D., *Sens. Actuators A*, **42** (1994) 511–515.
[49] Harman, G.G., *Wire Bonding in Microelectronics*, 2nd edn., New York: McGraw-Hill, 1997.
[50] Tummala, R.R., Rymaszewski, E.H., Klopfenstein, A.G. (eds.), *Microelectronics Packaging Handbook, Part II*, 2nd edn.; New York: Chapman & Hall, 1997, pp. II-186–II-217.
[51] Pecht, M., Lall, P., in: *Thermal Stress and Strain in Microelectronics Packaging,* Lau, J.H. (ed.); New York: Van Nostrand Reinhold, 1993, pp. 729–802.

[52] Green, T.J., Launsby, R.G., in: *Proc. 27th Int. Symp. Microelectron., ISHM'94*; 1994, pp. 60–65.
[53] Chen, G.K.C., in: *Proc. Int. Symp. Microelectronics*; 1972, pp. 5-A-1-1–9.
[54] Harthoorn, J.L., in: *Proc. Ultrasonics Int. Conf.* 1973, pp. 43–51.
[55] Weiner, J.A., Clatterbaugh, G.V., Charles Jr., H.K., Romenesko, B.M., in: *Proc. 33rd Electron. Comput. Conf., ECC'83*; 1983, pp. 208–220.
[56] Shu, W.K., in: *Proc. 45th Electron. Comput. Technol. Conf., ECTC'95*: 1995, pp. 91–101.
[57] Tiederle, V., in: *DVS-Berichte*, Band 201; Düsseldorf: DVS-Verlag, 1999.
[58] Aguila, M.T.,. Felipe, R.C., Velard, A.F., Edpan, J.B., in: *Proc. 1st Electronic Packaging Technol. Conf., EPTC'97;* 1997, pp. 46–51.
[59] Chen, Y.-S., Fatemi, H., *J. Hybrid Microelectron.*, **10** (1987) 1–7.
[60] Sheaffer, M., Levine, L., *Solid State Technol.,* Nov. (1990) 119–123.
[61] Sheaffer, M., Levine, L., *Solid State Technol.,* Jan. (1991) 67–70.
[62] Onda, N., Jaecklin, V.P., Arsalane, S., in: *Proc. Semicon Singapore'98, Assembly and Packaging Seminar*; 1998, pp. 159–168,.
[63] Arsalane, S., Jaecklin, V.P., in: *Proc. Semicon Taiwan '98, Packaging Seminar*; 1998, pp. 107–114.
[64] Onda, N., Domman, A., Zimmermann, H., Luechinger, C., Jaecklin, V., Zanetti, D., Beck, E., Ramm, J., in: *Proc. Semicon Singapore*; 1996, pp. 147–153.
[65] Prasad, S., Walker, J., *Adv. Packaging*, **6**, No. 8 (1998).
[66] Budweiser, W., *PhD Thesis*, Technical Univ. Berlin, 1993, p. 133.
[67] Ikeda, T., Miyazaki, N., Kudo, K., Arita, K., Yakiyama, H., *J. Electron. Packaging*, **121** (1999) pp. 85–91.
[68] Takahashi, Y., Shibamoto, S., Inoue, K., *IEEE Trans. CPMT, Part A,* **19** (1996) 213–223.
[69] Gibson, O.E., Gleeson, W.J., Burkholder, L.D., Benton, B.K., *US Pat. 4998664*, 1991.
[70] Pufall, R., in: *Proc. 43rd IEEE Electronic Comput. Technol. Conf., ECTC'93*; 1993, pp. 159–162.
[71] Or, S.W., Chan, H.L.W., Lo, V.C., Yuen, C.W., *Sens. Actuators A*, **65** (1998) 69–75.
[72] Osterwald, F., presented at 4th Laser Vibrometer Seminar, Waldborn, Germany, 1997.
[73] Lang, K.-D., Osterwald, F., Schilde, B., Reichl, H., in: *Proc. Semicon West '98*; 1998, pp. F1–F9.
[74] Schneuwly, A., Gröning, P., Schlapbach, L., Müller, G., *J. Electron. Mat.*, **27** (1998) 1254–1261.
[75] Mayer, M., Paul, O., Baltes, H., in: *Proc. 2nd Int. Conf. Emerging Microelectron. and Interconn. Technol., EMIT'98*; 1998, pp. 129–133.
[76] Mayer, M., Paul, O., Bolliger, D., Baltes, H., *IEEE Trans. Comput. Packaging Technol.*, **23** (2000) 393–398.
[77] Mayer, M., Schwizer, J., Paul, O., Bolliger, D., Baltes, H., in: *Proc. Intersociety Electron. Pack. Conf. (InterPACK99);* 1999, pp. 973–978.
[78] Schwizer, J., Mayer, M., Bolliger, D., Paul, O., Baltes, H., in: *Proc. 24th IEEE/CPMT Int. Electronic Manufacturing Technology Symposium, IEMT'99*, Austin, Texas, Oct. 18–19; 1999, pp. 108–114.
[79] Hizukuri, M., Wada, Y., Watanabe, N., Asano, T., in: *6th Symp. Microjoining and Assembly Technol. in Electronics 2000*, Yokohama; 2000, pp. 169–174.
[80] Hizukuri, M., Asano, T., *Jpn. J. Appl. Phys.,* **39** (2000) 2478–2482.

[81] Gee, S. A., Nguyen, L. T., Akylas, V. R., in: *Proc. Technical Sessions, MEPPE Focus 1991*, Santa Clara, CA; 1991, pp. 156–170.
[82] Trapp, O. D., Loop, L. J., Blanchard, R. A., *Semiconductor Technology Handbook*, 6th edn.; Portola Valley, CA: Technology Associates, 1992.
[83] Jäggi, D., *PhD Thesis*, No. 11567, ETH Zurich, Zurich, 1996.
[84] v. Arx, M., *PhD Thesis*, No. 12743, ETH Zurich, Zurich, 1998.
[85] Smith, C. S., *Phys. Rev.*, **94** (1854) 42–49.
[86] Nathan, A., Baltes, H., *Microtransducer CAD;* Vienna: Springer, 1999.
[87] Kanda, Y., *Sens. Actuators A*, **28** (1991) 83–91.
[88] Pallás-Areny, R., Webster, J. G., *Sensors and Signal Conditioning*; New York: Wiley, 1991, p. 106.
[89] Plattner, L., Mayer, M., Lüchinger, C., Paul, O., Baltes, H., in: *Proc. 32nd Int. Symp. Microelectronics, IMAPS'99*, Chicago, Oct. 26–28; 1999, pp. 213–220.
[90] Schneider, M. C., Beckermann, C., *Int. J. Heat Mass Transfer*, **38** (1995) 3455–3473.
[91] Touloukian, Y. S., Powell, R. W., Ho, C. Y., Nicolaou, M. C., *Thermophysical Properties of Matter;* New York: IFI/Plenum, 1973.
[92] Lloyd, J. R., Zhang, C., Tan, H. L., Shangguan, D., Achari, A., in: *Proc. 17th IEEE/CPMT Int. Electronics Manuf. Technol. Symp.*, New York; 1995, pp. 252–262.
[93] Mayer, M., Paul, O., Bolliger, D., Baltes, H., in: *Proc. 2nd IEEE Electron. Packaging Technol. Conf., EPTC'98*; 1998, pp. 219–223.
[94] Mayer, M., *PhD Thesis*, No. 13685, ETH Zurich, Zurich, 2000; also Hartung-Gorre, Konstanz, Germany, ISBN 3 89649 620 4, 2000.
[95] Meisser, C., *Der Elektroniker*, No. 12 (1990), 67–74.
[96] Hamilton, G. M., Goodman, L. E., *ASME J. Appl. Mech.*, **33** (1966) 371–376.
[97] Schwizer, J., *Diploma Thesis*, Physical Electronics Laboratory, ETH Zurich, Zurich, 1999.
[98] Liang, J. W., Feeny, B. F., in: *Elasto-Impact and Friction in Dynamic Systems*, DE-vol. 90; New York: ASME, 1996, pp. 85–96.
[99] Schwizer, J., Mayer, M., Brand, O., Baltes, H., in: *Proc. Transducers '01 / Eurosensors XV;* 2001, pp. 1426–1429.
[100] Mayer, M., Schwizer, J., in: *Proc. SEMI Technical Symposium, Advanced Packaging Technologies II;* Singapore: SEMI, 2002, pp. 169–175.
[101] Schwizer, J., Mayer, M., Brand, O., Baltes, H., in: *Proc. Int. Symposium on Microelectronics, IMAPS;* 2001, pp. 338–346.
[102] Mayer, M., Schwizer, J., in: *Proc. Int. Symposium on Microelectronics, IMAPS*, Denver, CO; 2002, pp. 626–631.
[103] Schwizer, J., Füglistaller, Q., Mayer, M., Althaus, M., Brand, O., Baltes, H., in: *Proc. SEMI Technical Symposium, Advanced Packaging Technologies I;* Singapore: SEMI, 2002, pp. 163–167.

List of Symbols and Abbreviations

Symbol	Designation
Δt	rise time
A	amplitude
A_{HT}	amplitude at horn tip
c	specific heat
F_N	bonding force
f_{US}	ultrasound frequency
F_y	ultrasonic tangential force
I	current
R	resistance
T	temperature
T_c	calibration temperature
T_{Chip}	chip temperature
T_i	initial temperature
T_p	plateau temperature
T_{ref}	reference temperature
T_{RT}	room temperature
T_{sub}	substrate temperature
U	supply voltage
U_m	sensitivity of bridge
x	relative resistance change
α	temperature coefficient of resistance
κ	thermal conductivity
ϕ	angle between current flow in a resistor and the [110] crystal direction
π	piezoresistive coefficient
ρ	density
σ	stress

Abbreviation	Explanation
BPSG	borophosphosilicate glass
CMOS	complementary metal oxide semiconductor
COB	chip on board
DIP	dual in-line package
DOE	design of experiment
EFO	electrical flame-off
FE	finite element
IC	integrated circuit
LF	leadframe
PCB	printed circuit board

Abbreviation	Explanation
POI	point of inflection
PTAT	proportional to absolute temperature
QFP	quad-flat-pack
RIE	reactive ion etching
RTD	resistive temperature detector
SAM	scanning acoustic microscopy
SEM	scanning electron microscope
SOG	spin on glass
TC	temperature coefficient
TCR	temperature coefficient of resistance
USG	undoped silicate glass
VLSI	very large-scale integration

2.3 Electronic Tongues and Combinations of Artificial Senses *

F. Winquist, C. Krantz-Rülcker and I. Lundström,
The Swedish Sensor Centre and the Division of Applied Physics,
Department of Physics and Measurement Technology,
Linköping University, Linköping, Sweden

Abstract

The technique of electronic tongues or taste sensors has developed very fast during the last years due to its large potential, and the interest for the concept is steadily increasing. In principle, they function in the same way as the electronic nose, but are used in the aqueous phase. In this article, the technique as such is described, as well as different types of electronic tongues, based on potentiometry or voltametry. Also the combination of different artificial senses is discussed.

Different sensing principles can be used in electronic tongues or taste sensors, such as electrochemical methods like potentiometry or voltametry, optical methods or measurements of mass changes based on e.g. quartz crystals. The first concept of an electronic tongue or taste sensor was based on ion sensitive lipid membranes and developed to response to the basic tastes of the tongue, that is sour, sweet, bitter, salt and "umami". It has been further developed and is commercialized. The detecting part is an eight-channel multi sensor, placed on a robot arm and controlled by a computer. This taste sensing system has mainly dealt with discrimination and estimation of the taste of different drinks.

An electronic tongue, based on potentiometric sensor arrays of two general kinds, conventionally ones such as pH, sodium and potassium selective electrodes, and specially designed ones, has also been described. This system has been used for recognition of different kinds of drinks such as tea, soft drinks, juices and beers and compounds of relevance for pollution monitoring in river water.

The use of voltametry as sensing principle in an electronic tongue has also been developed. This electronic tongue consists of a number of working electrodes made of different materials, a reference electrode

---

* This contribution is also featured in the forthcoming *Handbook of Machine Olfaction* edited by T.C. Pearce, J.W. Gardner, S.S. Schiffman and H.T. Nagle.

and an auxiliary electrode. Different types of pulsed voltametry were applied. Application examples include classification of fruit juices, test for bacterial growth in milk or water quality.

A hybrid electronic tongue has also been developed, based on the combination of the measurement techniques potentiometry, voltametry and conductivity.

By using a combination of different artificial senses, the analytical capability will be considerably increased. Thus, the combination of an electronic tongue and an electronic nose for classification of different fruit juices was investigated. Furthermore, a special "artificial mouth" or "crush chamber" has been designed, in which information corresponding to three senses could be obtained – "auditory" by a microphone, "tactile" by a force sensor and "olfaction" by a gas sensor array, thus collecting information mimicking these three human sense. In this artificial mouth crispy products could be crushed under controlled conditions to make a complete sensory evaluation, all five human senses are involved. A new approach for the assessment of human based quality evaluation has been obtained by the design of an electronic sensor head. In this system, the artificial analogues to all the five human senses are used for quality evaluation of a sample could be collected.

Keywords: Electronic tongue, Taste sensor, Potentiometry, Voltametry, Principal component analysis, artificial mouth, Combination of senses

Contents

2.3.1 Introduction

The field of measurement technology is rapidly changing due to the increased use of multivariate data analysis, which has led to a change in the attitude of how to handle information. Instead of using specific sensors for measuring single parameters, it has in many cases become more desirable to get information of quality parameters, such as sample condition, state of a process, or expected human perception of, for example, food. This is done by using arrays of sensors with partially overlapping selectivities and treating the data obtained with multivariate methods. These systems are often referred to as artificial senses, since they function in a similar way as the human senses. One such system, the electronic noses, has attracted much interest [1–3]. This concept is based on the combination of a gas sensor array with different selectivity patterns with pattern recognition software. A large number of different compounds contribute to a measured smell; the chemical sensor array of the electronic nose then provides an output pattern that represents a combination of all the components. Although the specificity of each sensor may be low, the combination of several specificity classes allows a very large number of odors to be detected.

Similar concepts, but for use in aqueous surroundings have also recently been developed. These systems are related to the sense of taste in a similar way as the electronic nose to olfaction, thus, for these systems the terms 'electronic tongue' or 'taste sensor' have been coined [4–6].

In some applications, there are advantages when measuring in the aqueous phase compared to measurements in the gas phase; gas analysis is an indirect method that gives the final information about the aqueous phase via measurements in the gas phase. Many compounds such as ions or those having a low vapor pressure can only be measured in the aqueous phase, also for many online or inline applications it is only possible to use systems that measure directly in the solution. Furthermore, the development of electronic tongues offers an intriguing possibility to study their combinations with other types of artificial senses.

In principle, the electronic tongue or taste sensor functions in a similar way to the electronic nose, in that the sensor array produces signals that are not necessarily specific for any particular species; rather a signal pattern is generated that can be correlated to certain features or qualities of the sample. Electronic noses and tongues are normally used to give qualitative answers about the sample studied, and only in special cases to predict the concentration of individual species in the sample.

Different sensing principles can also be used in electronic tongues or taste sensors, such as electrochemical methods such as potentiometry or voltametry, optical methods, or measurements of mass changes based on, for example, quartz crystals.

The sense of taste may have two meanings. One aspect denotes the five basic tastes of the tongue; sour, salt, bitter, sweet, and 'umami'. These originate from different, discrete regions on the tongue containing specific receptors called papillae. This aspect of taste is often referred to as the sensation of basic taste. The other aspect of taste is the impression obtained when food enters the mouth. The basic taste is then merged with the information from the olfactory receptor cells, when aroma from the food enters the nasal cavities via the inner passage. This merged sensory experience is referred to as the descriptive taste by sensory panels.

The approach to more specifically mimic the basic taste of the tongue is made by the taste sensor system [4, 7, 8], in which different types of lipid membranes are used to determine qualities of food and liquids in terms of taste variables such as sweetness, sourness, saltiness, bitterness, and 'umami'. There is thus a difference between the use of a sensor array as electronic tongues or as taste sensors. A taste sensor system is used to classify the different basic taste sensations mentioned above, and the results are compared with human test panels. An electronic tongue classifies a quality of one or another kind in food, such as drinks, water, and process fluids, and the results are not necessarily compared with human sensations, but with other quality properties of the sample.

The concept of the electronic tongue and the taste sensor has developed very quickly during the last years due to its large potential. There are already commercial versions on the market [9, 10], and a number of other applications have also been reported, and are described later.

The performance of an artificial sense such as the electronic tongue can be considerably enhanced by the combination of sensors based on different technologies. The reason is, of course, that for each new measurement principal added, a new dimension of information is also added. A natural extension of this fundamental concept is the combination of different artificial senses. This is especially important when estimating the quality of food, since the guide is the impression of the human being using all five senses.

A first attempt to measure the elusive parameter 'mouthfeel' for crispy products such as potato chips or crispbread was made by the development of an artificial mouth. The intention was to collect information mimicking three human senses: olfaction, auditory, and tactile. The samples were placed in a special 'crush chamber', and, while crushed, information corresponding to three senses could be obtained: 'auditory' by a microphone, 'tactile' by a force sensor, and 'olfaction' by a gas sensor array [11, 12]. Furthermore, combinations of electronic noses and tongues have been used for quality estimation of different wines [13, 14].

A new dimension for the assessment of human-based quality evaluation is thus obtained by using the artificial analogs to all the five human senses. All information obtained from this sensor system is then fused together to form a human-like decision. Such a sensor head has been used for quality estimation of crispy products, such as crispbread and chips [15].

2.3.2 Electronic Tongues

2.3.2.1 Measurement Principles

There are several measurement principles that have the potential to be used in electronic tongues. The most important ones are based on electrochemical techniques such as potentiometry, voltametry, and conductometry, and there are a number of textbooks on the subject [16–18]. The use of electrochemical measurements for analytical purposes has found a vast range of applications. There are two basic electrochemical principles: potentiometric and voltametric. Both require at least two electrodes and an electrolyte solution. One electrode responds to the target molecule and is called the working electrode, and the second one is of constant potential and is called the reference electrode.

Potentiometry is a zero-current-based technique, in which a potential across a surface region on the working electrode is measured. Different types of membrane materials have been developed, having different recognition properties. These types of devices are widely used for the measurement of a large number of ionic species, the most important being the pH electrode, other examples are electrodes for calcium, potassium, sodium, and chloride.

In voltametric techniques, the electrode potential is used to drive an electron transfer reaction, and the resulting current is measured. The size of the electrode potential determines if the target molecules will lose or gain electrons. Voltametric methods can thus measure any chemical specie that is electroactive. Voltametric methods provide high sensitivity, a wide linear range, and simple instrumentation. Furthermore, these methods also enable measurements of conductivity and the amount of polar compounds in the solution.

Almost all electronic tongue or taste sensors developed are based on potentiometry or voltametry. There are, however, also some other techniques that are interesting to use and which have special features making them useful for electronic tongues, such as optical techniques or techniques based on mass sensitive devices.

Optical techniques are based on light absorption at specific wavelengths, in the region from ultraviolet via the visual region to near infrared and infrared. Many compounds have distinct absorption spectra, and by scanning a certain wavelength region, a specific spectrum for the sample tested will be obtained. Optical methods offer advantages of high reproducibility and good long-term stability.

Mass sensitive devices, based on piezo electric crystals are also useful. A quartz crystal resonator is operated at a given frequency, and by the absorption of certain compounds on the surface of the crystal, its frequency will be influenced [19]. For a surface acoustic wave (SAW) based device, a surface wave is propagated along the surface of the device [20], and due to adsorption of a compound in its way, the properties of this surface wave will be changed. These types of devices are very general and provide for the possibility to detect a large number of different compounds.

2.3.2.2 Potentiometric Devices

The equipment necessary for potentiometric studies includes an ion-selective electrode, a reference electrode, and a potential measuring device, as shown schematically in Fig. 2.3.1. A commonly used reference electrode is the silver-silver chloride electrode based on the half-cell reaction:

$$AgCl + e^- \longrightarrow Ag + Cl^- \qquad E^\circ = +0.22\ \text{V} \tag{1}$$

The electrode consists of a silver wire coated with silver chloride placed into a solution of chloride ions. A porous plug will serve as a voltage bridge to the outer solution.

The ion-selective electrode has a similar configuration, but instead of a voltage bridge, an ion-selective membrane is applied. This membrane should be nonporous, water insoluble and mechanically stable. It should have an affinity for the selected ion that is high and selective. Due to the binding of the ions, a membrane potential will develop. This potential, E, follows the well-known Nernst relation:

$$E = E^\circ + (RT/n\text{F}) \ln a \tag{2}$$

where E° is a constant for the system given by the standard electrode potentials, R is the gas constant, T the temperature, n the number of electrons involved in the reaction, F the Faraday constant and finally, a is the activity of the measured specie. The potential change is thus logarithmic in ionic activity, and ideally, a ten-fold increase of the activity of a monovalent ion should result in a 59.2 mV change in the membrane potential at room temperature.

In the early 1970s, ion-selective field effect transistors (ISFETs) were developed, in which the ion-selective material is directly integrated with solid-state

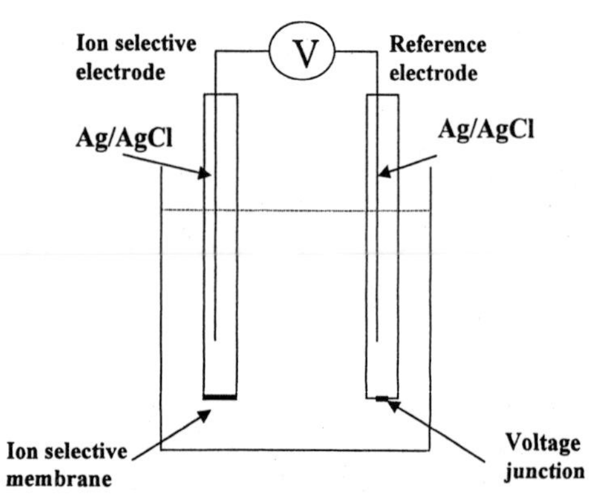

Figure 2.3.1. Schematic diagram of an electrochemical cell for potentiometric measurements.

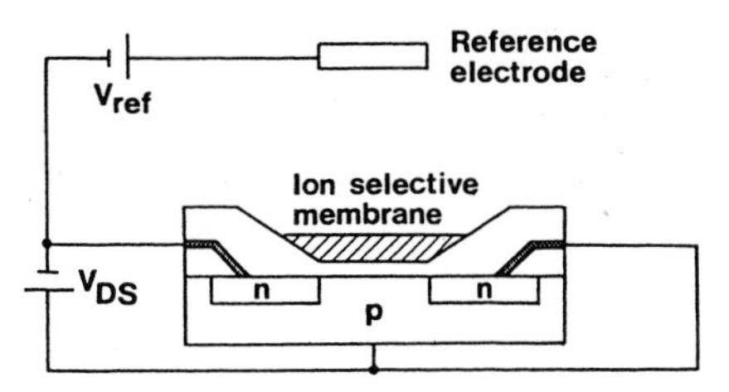

Figure 2.3.2. Schematic diagram of an ion-sensitive field effect transistor.

electronics [21]. A schematic diagram of an ISFET is shown in Fig. 2.3.2. The current between the drain and source (I_{DS}) depends on the charge density at the semiconductor surface. This is controlled by the gate potential, which in turn is determined by ions interacting with the ion-selective membrane. In the ISFET, the normal metal gate is replaced with the reference electrode and sample solution. An attractive feature of ISFETs is their small size and ability to be directly integrated with microelectronics, for example, signal processing, furthermore, if mass fabricated, they can be made very cheaply. These features make them especially valuable for use in electronic tongues.

Potentiometric devices offer several advantages for use in electronic tongues or taste sensors. There are a large number of different membranes available with different selectivity properties, such as glass membranes and lipid layers. A disadvantage is that the technique is limited to measurement of charged species only.

2.3.2.2.1 The Taste Sensor

The first concepts of a taste sensor were published in 1990 [22, 23]. It was based on ion-sensitive lipid membranes and developed to respond to the basic tastes of the tongue, that is sour, sweet, bitter, salt, and 'umami'.

The multichannel taste sensor [5, 23] was also based on ion-sensitive lipid membranes, immobilized with the polymer PVC. In this taste sensor, five different lipid analogs were used: *n*-decyl alcohol, oleic acid, dioctyl phosphate (*bis*-2-ethylhexyl)hydrogen phosphate, trioctylmethyl ammonium chloride, and oleylamine, together with mixtures of these. Altogether eight different membranes were fitted on a multichannel electrode, where each electrode consisted of a silver wire with deposited silver chloride inside a potassium chloride solution, with the membrane facing the solution to be tasted. A schematic of the multichannel electrode is shown in Fig. 2.3.3. The voltage between a given electrode and a Ag/AgCl reference electrode was measured. The setup is shown in Fig. 2.3.4. This taste sensor has been used to study responses from the five typical ground tastes, HCl (sour), NaCl (salt), quinine (bitter), sucrose (sweet), and monosodium glutamate (umami). The largest responses were obtained from the sour and bitter compounds, thereafter umami and salt, and for sucrose almost no response was obtained. For other sweet tasting substances, such as the amino acids glycine and alanine, larger responses were obtained. It was further shown that similar sub-

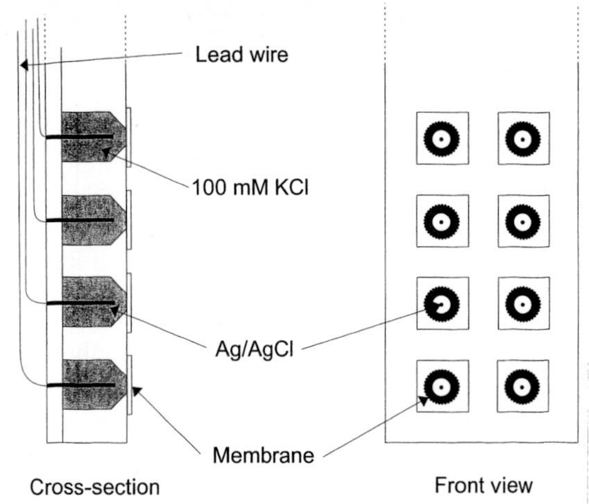

Figure 2.3.3. Schematics of the multichannel electrode with eight lipid/polymer membranes.

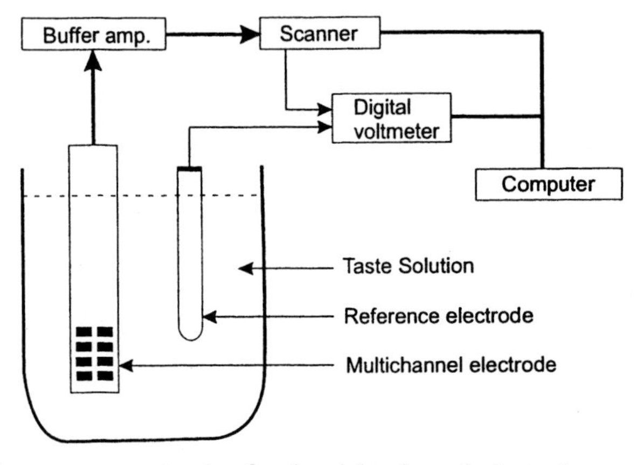

Figure 2.3.4. The measurement setup for the eight-channel electrode system.

stances, such as sour substances like HCl, acetic acid, citric acid, or salty substances such as NaCl, KCl, and KBr showed similar response patterns. The system does not respond well to nonelectrolytes, which have little effect on the membrane potential [24].

The multichannel system has been commercialized [9]. The detecting part is an eight-channel multisensor, placed on a robot arm and controlled by a computer. The samples to be tested are placed in a sample holder together with a cleaning solution as well as reference solutions. The measurements then take place in a special order: first the multisensor is cleaned by dipping into the cleaning solution, thereafter into the sample solution, and the cycle repeats. At certain intervals, the multisensor is placed in the reference solution for calibration purposes.

This taste sensing system has been used in a number of different applications. These have mainly dealt with discrimination and estimation of the taste of different drinks. In one investigation, 33 different brands of beers were studied [25]. The samples were analyzed both by using a sensory panel and by the taste sensing system. The sensory panel expressed the taste of the different beers in the parameters sharp-versus-mild and rich-versus-light. The output pattern from the taste sensor was analyzed using principal component analysis (PCA). An interesting observation was that the first principal component corresponded well to the taste parameter rich-versus-light taste, and the second principal component corresponded well to the parameter sharp-versus-mild taste.

Mineral water has also been studied using the taste sensing system [24]. A good correlation of the sensor responses to the hardness of the water could be seen in PCA plots, and also the sensor could discriminate between different brands.

Other applications involve the monitoring of a fermentation process of soybean paste [26], estimation of the taste of milk [27] or coffee [28], and the development of a monitoring system for water quality [29].

2.3.2.2.2 Ion-Selective Electrodes

The term 'electronic tongue' was first coined at the EurosensorsX conference [5, 30]. The concept had been developed as a research collaboration between an Italian group (DiNatale, Davide and D'Amico) and a Russian group (Legin, Vlasov and Rudnitskaya). This device has now been developed further, and a large number of applications have been studied, and are described in the following.

The first devices consisted of potentiometric sensor arrays of two general kinds: conventionally ones such as pH, sodium and potassium-selective electrodes, and especially designed ones. The latter ones were based mainly on chalcogenide vitreous materials. Altogether the array included 20 potentiometric sensors: glass, crystalline, PVC plasticized sensor, and metal electrodes. The sensor system was used for the recognition of different kinds of drinks such as tea, soft drinks, juices, and beers. Each sample was measured twice, and the information obtained from the sensor array was treated using PCA. The score plots showed good separation between all these samples. The detoriation of orange juice during storage was also followed, and by using an artificial neural network (ANN) on the data obtained, a model for storage-time prediction could be made.

The measurements of compounds of relevance for pollution monitoring in river water using this electronic tongue have also been reported [31]. River water was taken at three locations and artificially polluted with Cu, Cd, Fe, Cr, and Zn, all in ionic form, representing a 'common' pollution from the industries. The sensor array consisted of 22 electrodes mainly based on chalcogenide glasses and conventional electrodes. Different approaches of data analysis were performed such as multiple linear regression (MLR), projection to latent structure (PLS), nonlinear least square (NLLS), back-propagation ANN, and a self-organizing map (SOM). Two modular models were developed, the first a combination of

PCA and PLS, the second a combination of ANN and SOM, and both could predict pollutant ions well.

A similar setup of this electronic tongue has been used for qualitative analysis of mineral water and wine [32], and for multicomponent analysis of biological liquids [33]. A flow-injection system based on chalcogenide glass electrodes for the determination of the heavy metals Pb, Cr, Cu, and Cd was also developed [34]. The approach of combining flow injection analysis in combination with a multisensor system and analyzing data using multivariate data analysis appears very advantageous. The flow injection analysis (FIA) technique offers several advantages: since relative measurements are performed, the system is less influenced by sensor baseline drift, calibration samples can be injected within a measurement series, and the system is well adapted for automization. One should also remember that most electronic nose measurements are based on a gas-phase FIA technique, one reason is to compensate for the drift of the gas sensors.

2.3.2.2.3 Surface Potential Mapping Methods

A very interesting technique has been developed, in which the surface potential of a semiconductor structure is measured locally [35–37]. This is a new type of a potentiometric system that provides for a contactless sensing over a surface and is thus a convenient way to analyze a multifunctional surface. It also opens up possibilities to use gradients of different functional groups as the sensing principle. The semiconductor surface acts as the working electrode on to which the test solution is applied. Into this solution a reference electrode and an auxiliary electrode are also applied. On the backside, a light-emitting diode is applied, which can scan the surface in both x and y directions. By illuminating a certain region on the semiconductor (via the backside), a photocurrent will be generated, the size being a measure of the surface potential at that particular region.

In one application [35], five lipid membranes (oleic acid, lecithin, cholesterol, phosphatidyl ethanolamine, and dioctyl phosphate) were deposited at different areas on the semiconductor surface. First, one lipid was coated onto the whole area, the next on two thirds of the area, and the third on the last third of the area. The whole surface was rotated by 90°, and the procedure was repeated with the remaining lipids. The sensing area could thus be divided into nine different regions with varying composition and thickness of lipid layer. This sensor surface was then investigated for the basic taste substances, HCl (sour), NaCl (salt), quinine (bitter), sucrose (sweet), and monosodium glutamate (umami). The responses obtained had similar responses to the taste sensor system described earlier, that is the largest responses were obtained from the sour and bitter compounds, thereafter umami and salt, and for sucrose almost no response was obtained. The method has also been further developed [36–38].

2.3.2.3 Voltametric Devices

In voltametric devices, the current is measured at given potentials. This current is then a measure of the concentration of a target analyte. The reactions taking place at the electrode surface are:

$$\mathrm{O} + ne^{-} \longrightarrow \mathrm{R} \tag{3}$$

where O is the oxidized form and R is the reduced form of the analyte. At standard conditions, this redox reaction has the standard potential E°. The potential of the electrode, E^{p}, can be used to establish a correlation between the concentration of the oxidized (C_{o}) and the reduced form (C_{r}) of the analyte, according to the Nernst relation:

$$E^{\mathrm{p}} = E^{\mathrm{o}} + RT/n\mathrm{F}(\ln(C_{\mathrm{o}}/C_{\mathrm{r}})) \tag{4}$$

A well-known voltametric device is the Clark oxygen electrode, which operates at –700 mV, the potential at which oxygen is reduced to hydrogen peroxide on a platinum electrode. By reverting the potential, the electrode will be sensitive to hydrogen peroxide.

The use of voltametry as a sensing principle in an electronic tongue appears to have several advantages: the technique is commonly used in analytical chemistry due to features such as very high sensitivity, versatility, simplicity, and robustness. The technique also offers the possibility to use and combine different analytical principles such as cyclic, stripping, or pulsed voltametry. Depending on the technique, various aspects of information can be obtained from the measured solution. Normally, redox-active species are being measured at a fixed potential, but by using, for example, pulse voltametry or studies of transient responses when Helmholz layers are formed, information concerning diffusion coefficients of charged species can be obtained. Further information is also obtainable by the use of different types of metals for the working electrode.

When using voltametry in complex media containing many redox-active compounds and different ions, the selectivity of the system is normally insufficient for specific analysis of single components, since the single steps in the voltammogram are too close to be individually discriminated. Rather complicated spectra are therefore obtained and the interpretation of data is very difficult due to its complexity. These voltammograms contain a large amount of information, and to extract this there has been an increasing interest and use of multivariate analysis methods in the field [39–42].

Among the various techniques mentioned, pulse voltametry is of special interest due to the advantages of greater sensitivity and resolution. Two types of pulse voltametry are commonly used, large amplitude pulse voltametry (LAPV) and small amplitude pulse voltametry (SAPV). At the onset of a voltage pulse, charged species and oriented dipoles will arrange next to the surface of the working electrode, forming a Helmholz double layer. A charging nonfaradic current will then initially flow as the layer builds up. The current flow, i, is equivalent to the charging of a capacitor in series with a resistor, and follows an equation of the form:

$$i = E^{*} R_s \exp(-t/R_s^{*} B) \tag{5}$$

where R_s is the resistance of the circuit (=solution), E is the applied potential, t is the time, and B is an electrode related equivalent capacitance.

The redox current from electroactive species shows a similar behavior, initially large when compounds close to the electrode surface are oxidized or reduced, but decays with time when the diffusion layer spreads out. The current follows the Cottrell equation [16–18]:

$$i = nFADC((1/(\pi Dt)^{1/2}) + 1/rx) \tag{6}$$

where A is the area of the working electrode, D is the diffusion constant, C is the concentration of analyte and $1/rx$ is an electrode constant. At constant concentration, the equation can be simplified:

$$i = K_1(1/t)^{1/2} + K_2 \tag{7}$$

K_1 and K_2 are constants.

In LAPV, the electrode is held at a base potential at which negligible electrode reactions occur. After a fixed waiting period, the potential is stepped to a final potential. A current will then flow to the electrode, initially sharp when the Helmholz double-layer is formed. The current will then decay as the double-layer capacitance is charged and electroactive compounds are consumed, until the diffusion-limited faradic current remains, as depicted by Eqs. (5) and (7). The size and shape of the transient response reflect the amount and diffusion coefficients of both electroactive and charged compounds in the solution. When the electrode potential is stepped back to its starting value, similar but opposite reactions occur. The excitation waveform consists of successive pulses of gradually changing amplitude between which the base potential is applied.

The instantaneous faradic current at the electrode is related to surface concentrations and charge transfer rate constants, and depends exponentially on the difference of the electrode potential between the start value and the final potential.

In SAPV, a slow continuous direct current (DC) scan is applied to the electrode on to which small amplitude voltage pulses are superimposed. This DC scan causes a change in the concentration profile of the electroactive species at the surface. Since only small pulse changes in the electrode potential are considered, this will result in small perturbations in the surface concentration from its original value prior to the application of the small amplitude excitation. Normally for SAPV, the current is sampled twice, one just before the application of the pulse, and one at the end of the pulse, and the difference between these is recorded as the output. This differential measurement gives a peaked output, rather than the wave-like responses that are usually obtained.

2.3.2.3.1 The Voltametric Electronic Tongue

The first voltametric electronic tongue described used both LAPV and SAPV applied to a double working electrode, an auxiliary, and a reference electrode [6]. The double working electrode consisted of one wire of platinum and the other of gold, both with a length of 5 mm and a diameter of 1 mm. Current and current transient responses were measured by a potentiostat connected to a PC. The PC was also used for onset of pulses and measurement of current transient responses and to store data. Via two relays, the PC was also used to shift the type of working electrode (gold or platinum) used. Current responses from both LAPV and SAPV were collected and used as input data for PCA.

In a first study, samples of different orange juices, milk, and phosphate buffer were studied. A PCA plot performed on the data showed good separation of the samples, as shown in Fig. 2.3.5. This electronic tongue was also used to follow the ageing process of orange juice when stored at room temperature.

The voltametric electronic tongue has been further developed. A recent configuration is shown in Fig. 2.3.6. It consisted of five working electrodes, a reference electrode and an auxiliary electrode of stainless steel. Metal wires of gold, iridium, palladium, platinum, and rhodium used as working electrodes were embedded in epoxy resin and placed around a reference electrode in such a way that only the ends of the working electrodes and the reference electrodes were exposed. The opposite ends of the working electrodes were connected to electric wires. The arrangement was inserted in a plastic tube ending with a stainless steel tube as an auxiliary electrode. The wires from the working electrode were connected to a relay box, enabling each working electrode to be connected sepa-

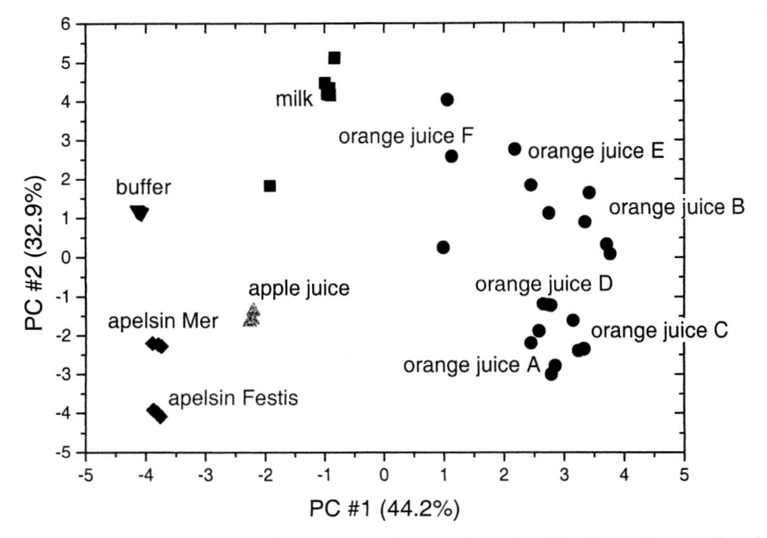

Figure 2.3.5. PCA analysis of different samples analyzed with the voltametric electronic tongue.

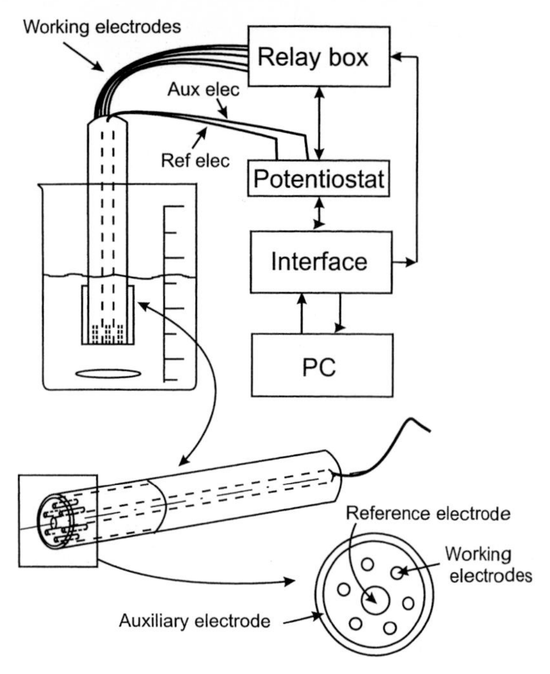

Figure 2.3.6. A recent configuration of the voltametric electronic tongue.

rately in a standard three-electrode configuration. Different types of pulsed voltametry could be applied, LAPV, SAPV and staircase. In Fig. 2.3.7, typical voltage pulses and the corresponding current responses are shown. This electronic tongue has been used to follow the deterioration of milk due to microbial growth when stored at room temperature [43]. The data obtained were treated with PCA, and the deterioration process could clearly be followed in the diagrams. To make models for predictions, projections to latent structure and ANNs were used. When trained, both models could satisfactorily predict the proceedings of bacterial growth in the milk samples.

A hybrid electronic tongue has also been developed, based on the combination of the measurement techniques potentiometry, voltametry, and conductivity [44]. The hybrid electronic tongue was used for classification of six different types of fermented milk. Using ion-selective electrodes, the parameters pH, carbon dioxide, and chloride ion concentrations were measured. The voltametric electronic tongue consisted of six working electrodes of different metals (gold, iridium, palladium, platinum, rhenium, and rhodium) and a Ag/AgCl reference electrode. The measurement principle was based on large amplitude pulse voltametry in which current transients were measured. The data obtained from the measurements were treated with multivariate data processing based on PCA and an ANN. The hybrid tongue could separate all six different types of fermented milks. Also, the composition of the microorganisms of the different fermentations was reflected in the PCA results.

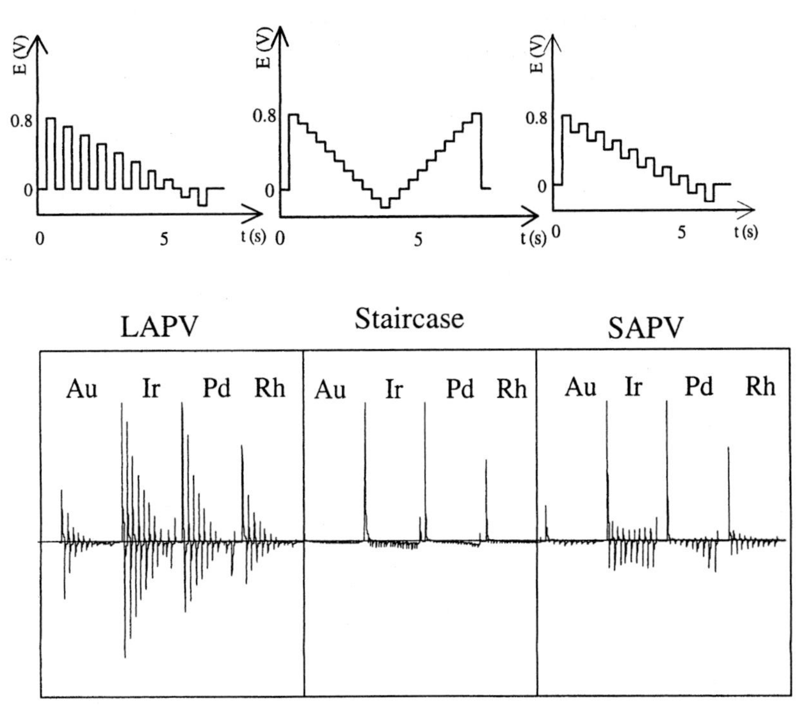

Figure 2.3.7. Three different pulsed voltametric techniques used by the voltametric electronic tongue. The upper part shows applied voltage pulses. The lower part shows the corresponding current responses for four different electrodes (gold, iridium, palladium, and platinum) due to the onset of voltage pulses.

A measurement system, based on the FIA technique applied to a voltametric electronic tongue has also been developed [45]. A reference solution was continuously pumped through a cell with a voltametric electronic tongue, and test samples were injected into the flow stream. Responses were obtained by measuring the resulting pulse height. The FIA technique offered several advantages: since relative measurements are performed, the system is less influenced by sensor baseline drift, calibration samples can be injected within a measurement series, and the system is well adapted for automization. The system was used to analyze standard solutions of H_2O_2, KCl, $CuNO_3$, $K_4[Fe(CN)_6]$, and NaCl, and results obtained were treated with multivariate data analysis. PCA showed that electrode drift was considerably decreased. The setup was also used for classification of different orange juices.

The voltametric electronic tongue has also been used for the monitoring of drinking water quality, and a review has recently been published [46].

2.3.2.3.2 Feature Extraction

To be able to correctly describe the shape of the current pulses during the voltage pulses, a large amount of variables are collected. For each pulse, up to 50 variables can be taken for the multivariate data processing. In a complete measurement series using up to 100 pulses applied to four electrodes, a total number of up to 2000 discrete values can be collected. Most of these are redundant having a low level of information.

The shape of the current responses for LAPV follows Eqs. (5) and (7) in principal, which means that constants can be calculated that express the current response. In a first attempt, constants fitting Eq. (5) were calculated, and for a given application for classification of different teas, PCA showed that a better separation was obtained using these constants compared with original data [47].

2.3.2.3.3 Industrial Applications using the Voltametric Electronic Tongue

The list of possible industrial applications for voltametric electronic tongues can be made very long. Electronic tongues are versatile in their applicability since they can give general information as well as specific information, such as pH and conductivity, about a sample [48]. In addition the construction of the voltametric electronic tongue can be made very robust – another reason that makes it suitable for many different areas of applications. One example where this quality is important is in the food industry where the use of sensors made of glass, for example, is not always acceptable.

The voltametric electronic tongue has been studied in a number of different industrial applications. One example is in the pulp and paper industry where the increasing machine speed and system closure of the papermaking process have caused an increased need to control the wet-end chemistry of the paper machine. The main challenges have been to establish knowledge of its impact on product properties as well as the most important relations between wet-end chemistry and performance of stock trades such as paper chemicals and pulp in order to improve productivity and run ability. The voltametric electronic tongue has been evaluated on pulp samples and the prediction ability of six reference parameters – pH, conductivity, chemical oxygen demand, cationic demand, zeta potential, and turbidity – was evaluated using PLS models. The results indicated that the electronic tongue studied had very promising features as a tool for wet-end control. Flexibility, fast response and wide sensitivity spectra make the electronic tongue suitable for a vast number of possible applications in the papermaking process [49].

Another example of an industrial application where the electronic tongue has been studied is as a sensor system in household appliances such as dishwashers and washing machines. The machines are today programmed to secure a good result, which often implies that the settings, such as temperature and washing time, are too high resulting in unnecessarily large consumption of energy, water, and detergent. A sensor that can give information about the water quality, type of soil loaded, and when the rinse water is free from detergents would increase the efficiency of these machines. The voltametric electronic tongue has, for example, been able to distinguish between different standardized soil types, even at high levels of detergents added to the solutions [50]. Much work remains to be done before the electronic tongue might be a conventional sensor technology in this type of machine, but these preliminary studies show its potential.

The third example of industrial applications for electronic tongues is as a monitoring device in drinking-water production plants [46]. The quality of drinking water varies due to the origin and quality of the raw water (untreated surface or ground water), but also due to efficiency variations in the drinking-water production process. Problems can be related to occurrence of, e.g., algae, bacteria, pesticides, and herbicides, and industrial contamination, in the raw water. The character of the raw water, and the biological activity at the production plant as well as in the distribution net may all cause quality problems such as bad odor/taste and/or unhealthiness. A method for monitoring variations in the raw water quality as well as the efficiency of separate process steps would therefore be of considerable value. To evaluate the voltametric electronic tongue for this purpose, water samples from each of eight parallel sand filters in a drinking-water production were collected and measured, as shown in Fig. 2.3.8. A PCA plot for the samples is shown in Fig. 2.3.9. The raw water samples are well separated from the treated water samples (slow and fast filter, and clean) in the plot. One interesting obser-

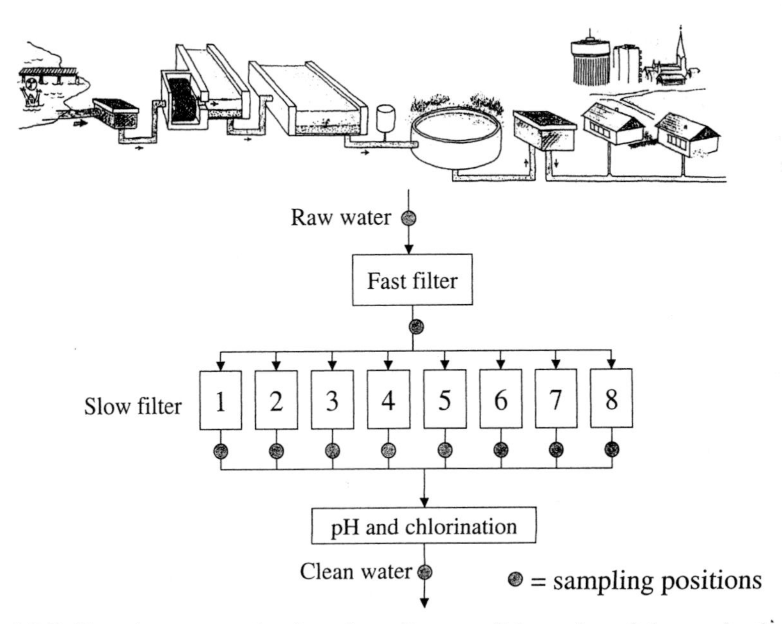

Figure 2.3.8. Top: A water production plant. Bottom: Schematics of the production plant showing the inlet of raw water, a fast filter, eight parallel sand filters and the final pH adjustment and chlorination step. The sampling positions of the electronic tongue are indicated.

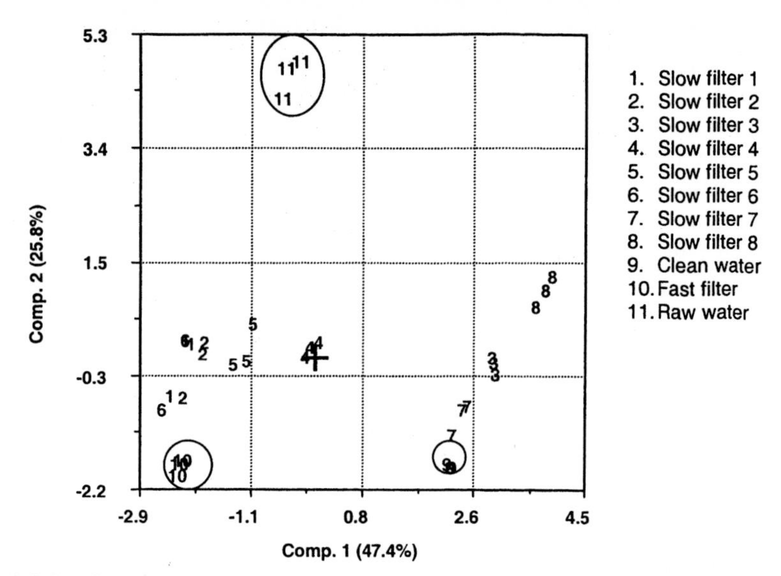

Figure 2.3.9. PCA plot of samples obtained from the water production plant.

vation is that the water quality after flowing through some of the slow filters cluster close to that after the fast filter, which indicates that the chemical composition of the two are similar. This result suggests that these slow filters are not working properly. The water quality after flowing through three of the other slow filters cluster, however, much closer to the clean water (which has also been chemically treated), which in a similar way indicates that these filters are working properly. This implies of course that the quality of the clean water is acceptable. Figure 10 demonstrates a possible use of electronic tongues (and PCA) namely to check the performance of given filters of the production plant.

The results for the drinking-water plant above suggest a possible use of the electronic tongue in continuous monitoring of the status of a given filter or other parts of the plant. After maintenance of a filter, for example, the initial position in a PCA plot of the water coming out of the filter is determined. Through a continuous measurement on the water after flowing through the filter, the time evaluation of the position in the PCA plot is followed. As long as the points cluster together within an area (determined initially by experience) the filter is performing well enough. Deviations from the 'normal' cluster indicate a malfunctioning filter (Fig. 2.3.10). To be able to associate a deviation from the 'normal' cluster to any specific parameter the reasons for malfunctioning filters must be studied. For this purpose traditional analytical chemical as well as biological methods must be used. The signals from the electronic tongue can then be correlated to these reference methods, and if there is a correlation, specific disturbances of the properties of a filter can be tracked to certain areas of the PCA plot. The possibility to detect a malfunctioning filter, regardless of the parameters causing it, is very valuable since it allows early measures to be taken against the problem.

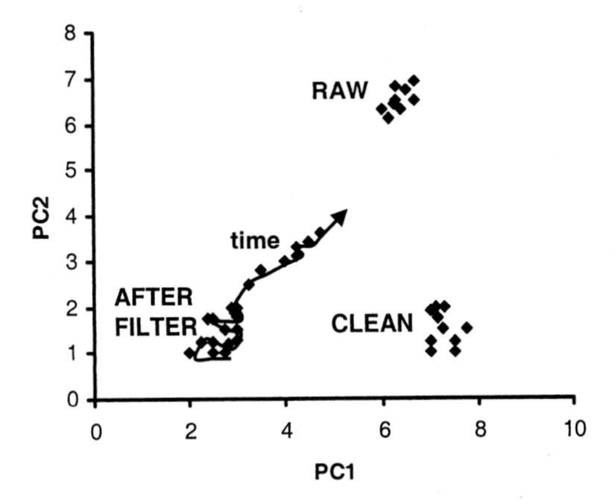

Figure 2.3.10. Schematic illustration of time-dependent PCA analysis used to detect changes in performance of a part of a plant.

Other application areas that are under study are the use of the electronic tongue for detection of microbial activity [48, 51]. One important industrial area for such applications is in the food industry where the quality of food is very much determined by its microbial status. This can be unwanted microbial occurrence like pathogenic microorganisms as well as wanted microbial growth in, for example, fermented foodstuff. Studies have shown that it is possible to follow the growth of mould and bacteria, and also to separate between different strains of moulds with the voltametric electronic tongue [48, 51].

2.3.2.3 Piezo Electric Devices

Piezo electric materials have an interesting property in that an electric field is generated by the application of pressure, and that it is distorted by the application of an electric field. The crystal will generate a stable oscillation of the electric voltage across it when an AC voltage is applied using an external oscillatory circuit. This resonance frequency is changed with the mass of the crystal according to the equation:

$$\Delta f = -cf^2(\Delta M/A) \tag{8}$$

where Δf and ΔM are the changes in resonance frequency and mass, respectively, c is a positive constant, f the resonance frequency, and A the electrode area.

Quartz crystals are widely used as sensors where the chemical sensitivity and selectivity is obtained from an adsorbent layer on the crystal. For a quartz crystal microbalance, analyte sorption on this layer will result in a frequency change [19]. Depending on the affinity properties of the adsorbing layer, different chemical compounds can be measured. Using an array based on these kinds of devices coated with hydrophilic mono- and dicarbon acids, organic and inorganic acids, and amines in drinking water could be detected [52].

A quartz resonator coated with a lipid/polymer membrane has also been investigated. The oscillation frequency showed different responses depending on taste substances and the lipid in the membrane [53].

SAW devices have also been applied for sensors in the gas and aqueous phases. For use in liquids, shear-horizontal mode SAW (SH-SAW) must be used [20, 54]. Using a 36° rotated Y-cut X-propagating $LiTaO_3$ device, a sensing system for the identification of fruit juices was developed. The device was divided in two parts: one metallized area as reference, the other area having a free surface that was electrically active. The sensor sensitivity was controlled by changing the excitation frequency. The phase difference and amplitude ratio between the reference and sensing signals were measured. A system was developed using three SH-SAW devices operated at the frequencies 30, 50 and 100 MHz, respectively, which was used to identify eleven different fruit juices [54]. In another study, a similar system was used to classify thirteen different kinds of whisky samples [55]. The device has been further studied in an application for the discrimination of four commercial brands of natural spring water [56]. Transient frequency responses were studied, and using pattern recognition based on ANNs, all four samples could be easily discriminated.

A review on recent efforts towards the development of both electronic tongues and electronic noses has been published [57], in which working principles, and the construction and performance of these systems mainly based on SAWs are discussed.

2.3.3 The Combination or Fusion of Artificial Senses

Appreciation of food is based on the combination or fusion of many senses, in fact for a total estimation all five human senses are involved: vision, tactile, auditory, taste, and olfaction. The first impression is given by the look of the food, thereafter information of weight and surface texture is gained by holding it in the hand. Thus, even before the food has come in contact with the mouth, a first conception is already made. In the mouth, additional information is given by the basic taste on the tongue, and the olfaction. Other quality parameters such as chewing resistance, melting properties, crisp sound, and temperature are added. This is often referred to as the mouth-feel, and is a very important property of the food. Individual properties correlated to special food products are especially important for their characterization, such as the crispness of crispbread or chips, the chilling properties of chocolate when melting on the tongue, or the softness of a banana.

A challenging problem in the food processing industry is maintenance of the quality of food products, and, consequently much time and effort are spent on methods for this. Panels of trained experts evaluating quality parameters are often used, which, however, entails some drawbacks. Discrepancy might occur due to human fatigue or stress, the sensory panels are time consuming, expensive, and cannot be used for online measurements. The development of replacement methods for panels for objective measurement of food products in a consistent and cost-effective manner is thus highly wanted by the food industry.

In this respect, the combination of artificial senses has great potential to at least in part replace these panels, since the outcome of such a combination will resemble a human-based sensory experience. For these purposes, both simple and more complex combinations of artificial senses have been investigated. Depending on the art of the quality parameters to be investigated, different types of artificial senses are important. For estimation of the crispness of potato chips, the human sense analogs of olfaction, auditory, and tactile would be satisfactory, but for total quality estimations, all five human sense analogs should be represented.

Applications of the combinations of artificial senses have so far only been developed for the food and beverage industry, dealing with classification and quality issues. In the future, however, it is expected that this approach also will find applications in other types of the process industry.

An important aspect is how to fuse the sense information. How a body of algorithms, methods, and procedures can be used to fuse together data of different origins and nature in order to optimize the information content has been discussed [58, 59]. The approach of abstraction level is introduced, namely the level at which the sensor data are fused together. A low level of abstraction means that the signals from the sensors are merely added together in a matrix. A high level of abstraction means that the data of each sensor system is analyzed as a stand-alone set, thus a selection of the most important features of each system can be selected and then merged together.

2.3.3.1 The Combination of an Electronic Nose and an Electronic Tongue

Various applications concerning the combination of an electronic nose and an electronic tongue have been reported. In a first study, different types of wine were classified using a taste-sensor array using lipid/polymer membranes and a smell-sensor array using conducting polymer electrodes [14]. A clear discrimination was found for the different samples. Also the effect of the aging process was studied. Later investigations performed in more detail evaluate the different information obtained from the different sensor systems, thus, in one study of wines, an electronic nose based on eight QMB sensors using different metallo porphyrins as sensing layers, and an electronic tongue based on six porphyrin-based electrodes were used [59]. The data obtained were correlated with analysis of chemical parameters. PCA loading plots showed that the artificial sensory systems were orthogonal to each other, which implies the independence of the information obtained from them.

The combination of an electronic tongue and an electronic nose for classification of different fruit juices has also been described [60]. The 'electronic nose' was based on an array of gas sensors consisting of 10 metal-oxide-semiconductor field effect transistors (MOSFETs) with gates of thin catalytically active metals such as Pt, Ir, and Pd and four semiconducting metal-oxide-type sensors. The electronic tongue was based on pulse voltametry, and consisted of six working electrodes of different metals, an auxiliary electrode, and a reference electrode. Using PCA, it was shown that the electronic nose or the electronic tongue alone was able to discriminate fairly well between different samples of fruit juices (pineapple, orange, and apple). It was also shown that the classification properties were improved when information from both sources were combined, both in the unsupervised PCA and the supervised PLS.

An original sensor fusion method based on human expert opinions about smell and taste and measurement data from artificial nose and taste sensors have been presented [12, 61]. This is achieved by a combination of ANNs and conventional signal handling that approximates a Bayesian decision strategy for classifying the sensor information. Further, a fusion algorithm based on the maximum-likelihood principle provides a combination of the smell and taste opinions, respectively, into an overall integrated opinion similar to human beings.

2.3.3.2 The Artificial Mouth and Sensor Head

Quality estimation of crispy products such as chips or crisp bread offers an intriguing problem. The human perception of crispy quality comes from the impressions collected when the product enters the mouth and is chewed. While chewing and crushing, impressions of chewing resistance, crushing vibrations, and crushing sound as well as the descriptive taste of the sample, will all contribute to give an overall quality impression. Methods developed so far only measure the crispness in terms of the hardness and brittleness of the sample. It appears that to give a better description of the crispness experienced, more subtle quality parameters referring to the 'mouth feel' should be accounted for.

A special 'artificial mouth' or 'crush chamber' has been designed, in which information corresponding to three senses could be obtained: 'auditory' by a microphone, 'tactile' by a force sensor, and 'olfaction' by a gas sensor array, thus collecting information mimicking these three human senses [11, 12]. In this artificial mouth, crispy products could be crushed under controlled conditions. The schematic of the artificial mouth is shown in Fig. 2.3.11. A piston could be moved at a constant speed by the action of a stepping motor connected to a lever. The force applied to the piston was recorded by a force sensor, and a dynamic microphone was placed at the bottom of the chamber. The chamber was thermostated to 37 °C. The sensor array consisted of 10 MOSFET gas sensors, with gates of thin, catalytically active metals such as Pt, Ir, and Pd, and four semiconducting metal-oxide type sensors.

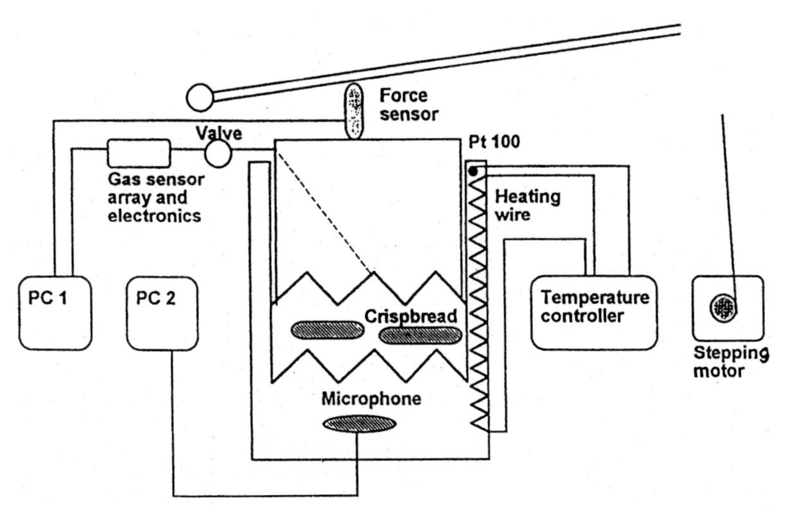

Figure 2.3.11. Schematics of the crush chamber or 'electronic mouth'.

Five types of crispbread have been investigated, one based on wheat flour, the other four based on rye flour. The information from the three information sources was first individually examined. Using information from the gas sensor, only the wheat flour based crispbread could be separated from the others. Using the sound information, a correlation to the hardness and brittleness of the samples could be obtained, and similar results were obtained from the force sensor. By combining all sense analogues, all five samples could be separated [11].

The quality of potato chips has also been investigated [12]. The aim of the study was to follow the aging process during storage. For these studies, one set of experiments was performed on potato chips stored in an opened bag, the other set in a closed bag that was opened only for sample taking. PCA analysis of data obtained from the artificial mouth showed that the information from the single information sources was not sufficient to explain the aging process, but with merged data, the aging process could be followed. A closer examination of the loading plot revealed that much of the data were strongly correlated, and from this plot, a smaller subset of data could be collected. This was used for an ANN, in which the prediction of age was modeled, and it was found that predicted values of age correlated well with true values.

To make a complete sensory evaluation, all five human senses are involved. A new approach for the assessment of human-based quality evaluation has been obtained by the design of an electronic sensor head [15]. The investigated sample enters an artificial mouth for detection of chewing resistance and recording of the chewing sound via a microphone. A video camera is used for the identification of color, shape, and similar properties of the sample. In parallel, aroma liberated during the crushing process is measured by a gas sensor array. Finally, the crushed sample is mixed with a saline solution, and an electrochemical multi-electrode arrangement analyzes the mixture. The artificial analogs to all the five human senses are therefore used for quality evaluation of the sample. All information obtained from the sensor system is then fused together to form a human-based decision. The arrangement was originally designed for quality studies of potato chips directly atline in the factory, hence it was also equipped with a robot arm, which could take out samples from the line. This sensor head has been used for quality estimation of crispy products, such as crispbread and chips.

For the chips application, it was interesting to note that vision alone could predict the quality parameters of freshness, spots, and spiciness, the olfactory analogues the amount of spiciness, and the auditory and touch analogs the freshness. The freshness of the chips can thus be determined both by change in color and by change in texture. Also, the spiciness of a chip can be determined both by the smell and by the number and color of the spices as seen by the camera. If all senses are fused together, all quality parameters could of course be correctly predicted.

2.3.4 Conclusions

Biomimetic measurement methods, as illustrated by the electronic nose and the electronic tongue, are rapidly being introduced in different applications. It is an interesting development where new achievements in both hardware and software act together to improve the performance of the sensor arrays. Some of the techniques used, such as the pulse voltametric measurements on a number of different (metal) electrodes, produce an enormous amount of data, in most cases with a large redundancy. An efficient data evaluation method is therefore necessary in order to utilize the measurements in an optimal way. The further development of algorithms is therefore an important task especially for sensor arrays based on simple, but well investigated, individual sensors. The biomimetic concept should, however, not be exaggerated. The human senses are strongly connected in the brain and give rise to associations based on an integrated previous experience. With regard to taste, the human taste sensation can, in general, not be described by one of the five simple 'basic' tastes. In olfaction, the situation is similar. One should therefore be aware of the fact that the manmade sensor arrays give responses that are only related to the taste and smell, even when they correlate with the sensation obtained by humans. Sometimes the sensor arrays do not even respond to the same molecules which give rise to the human sensation.

With this knowledge in mind, the sensor arrays are still extremely useful for quality control of products and processes as indicated in this contribution. In many applications there is no need to compare the sensor signals with sensory results, the signals themselves and their variations contain enough information. In many (industrial) applications the arrays will therefore not be calibrated against humans, but against traditional analytical techniques.

Another interesting possibility is to follow the evaluation of the data in a 'human dependent' PCA plot. In this case, process or quality monitoring can be made using references in the PCA plot itself, as discussed in correlation with the clean water production plant.

A combination of electronic noses and tongues with mechanical sensors and cameras of course increases the possibility to evaluate the properties of a given sample. The experiments made so far indicate that such a 'biomimetic sensor head' or robot has a large potential with regard to the evaluation of food, both of raw material and finished products. Such an approach will also have uses in process and product control in general.

2.3.5 References

[1] T.C. Pearce, S.S. Schiffman, H.T. Nagle, J.W. Gardner, *Handbook of Machine Olfaction*, Wiley-VCH **2003**

[2] J.W. Gardner and P.N. Bartlett, 'A brief history of electronic noses', *Sensors and Actuators* **1994** *B18–19* 211–220.

[3] F. Winquist, H. Sundgren and I. Lundström, 'Electronic Noses for Food Control', in Biosensors for Food Analysis, **1998**, A.O. Scott, Ed., The Royal Society of Chemistry, Athenaeum Press Ltd, UK.

[4] K. Toko, 'Taste sensor with global selectivity', Materials Science and Engineering **1996** *C4* 69–82.

[5] A. Legin, A. Rudinitskaya, Y. Vlasov, C. Di Natale, F. Davide, and A. D'Amico, 'Tasting of beverages using an electronic tongue based on potentiometric sensor array', *Technical digest of Eurosensors X,* Leuven, Belgium **1996** 427–430.

[6] F. Winquist, P. Wide and I. Lundström, 'An electronic tongue based on voltametry', *Analytica Chimica Acta* **1997** *357* 21–31.

[7] K. Toko, 'Taste sensor', *Sensors and Actuators* **2000** *B64* 205–215.

[8] K. Toko, 'A taste sensor', *Measurement Science and Technology* **1998** *9* 1919–1936.

[9] Taste Sensing System SA401, Anritsu Corp., Japan.

[10] The Astree Liquid & Taste Analyzer, Alfa MOS, Toulouse, France.

[11] F. Winquist, P. Wide T. Eklöv, C. Hjort and I. Lundström, 'Crispbread quality evaluation based on fusion of information from the sensor analogies to the human olfactory, auditory and tactile senses', *Journal of Food Process Engineering* **1999** *22* 337–358.

[12] P. Wide, F. Winquist and A. Lauber, 'The perceiving sensory estimated in an artificial human estimation based sensor system', *Proc. IEEE Instrumentation and Measurement Technology Conference*, Ottawa, Canada, **May 1997**.

[13] L. Rong, W. Ping and H. Wenlei, 'A novel method for wine analysis based on sensor fusion technique', *Sensors and Actuators* **2000** *B66* 246–250.

[14] S. Baldacci, T. Matsuno, K. Toko, R. Stella and D. De Rossi, 'Discrimination of wine using taste and smell sensors', *Sensors and Materials* **1998** *10(3)* 185–200.

[15] P. Wide, F. Winquist and I. Kalaykov, 'The artificial sensor head: A new approach in assessment of human based quality', *Proceedings of the Second International Conference on Information Fusion, FUSION '99. Int. Soc. Inf. Fusion*, Mountain View, CA, USA 2 **1999** 1144–1149.

[16] A.J. Bard and L.R. Faulkner 'Electrochemical Methods – Fundamentals and Applications', John Wiley & Sons, Inc. **1980**.
[17] J. Wang, 'Analytical Electrochemistry', Wiley-VCH **1994**.
[18] P.T. Kissinger and W.R. Heineman, 'Laboratory Techniques in Electroanalytical Chemistry', 2nd Edition, Marcel Dekker, Inc. **1996**.
[19] R. Lucklum and P. Hauptmann, 'The quartz crystal microbalance. Mass sensitivity, viscoelasticity and acoustic amplification', *Sensors and Actuators* **2000** *B70* 30–36.
[20] T. Yamazaki, J. Kondoh, Y. Matsui and S. Shiokawa, 'Estimation of components in mixture solutions of electrolytes using a liquid flow system with SH-SAW sensor', *Sensors and Actuators* **2000** *83* 34–39.
[21] P. Bergveld, 'The ISFET', *IEEE Trans. Biomed. Eng.* **1970** *BME-19*.
[22] K. Toko, K. Hayashi, M. Yamanaka and K. Yamafuji, 'Multichannel taste sensor with lipid membranes' *Tech. Digest 9th Sens. Symp.* **1990** 193–196.
[23] K. Hayashi, M. Yamanaka, K. Toko and K. Yamafuji, 'Multichannel taste sensor using lipid membranes', *Sensors and Actuators* **1990** *B2* 205–213.
[24] K. Toko 'Biomimetic Sensor technology', Cambridge University Press **2000**.
[25] K. Toko, 'Electronic Tongue', *Biosensors and Bioelectronics* **1998** *13* 701–709.
[26] T. Imamura, K. Toko, S. Yanagisawa and T. Kume, 'Monitoring of fermentation process of *miso* (soybean paste) using multichannel taste sensor', *Sensors and Actuators* **1996** *B37* 179–185.
[27] H. Yamada, Y. Mizota, K. Toko and T. Doi, 'Highly sensitive discrimination of taste of milk with homogenization treatment using a taste sensor', *Materials Science and Engineering* **1997** *C5* 41–45.
[28] T. Fukunaga, K. Toko, S. Mori, Y. Nakabayashi and M. Kanda, 'Quantification of taste of coffee using sensor with global selectivity', *Sensors and Materials* **1996** *8(1)* 47–56.
[29] A. Taniguchi, Y. Naito, N. Maeda, Y. Sato and H. Ikezaki, 'Development of a monitoring system for water quality using a taste sensor', *Sensors and Materials* **1999** *11(7)* 437–446.
[30] C. Di Natale, F. Davide, A. D'Amico, A. Legin, A. Rudinitskaya, B.L. Selezenev and Y. Vlasov, 'Applications of an electronic tongue to the environmental control', *Technical digest of Eurosensors X,* Leuven, Belgium, **1996** 1345–1348.
[31] C. Di Natale, A. Macagnano, F. Davide, A. D'Amico, A. Legin, Y. Vlasov, A. Rudinitskaya, and B.L. Selezenev, 'Multicomponent analysis on polluted water by means of an electronic tongue', *Sensors and Actuators* **1997** *B44* 423–428.
[32] A. Legin, A. Rudinitskaya, Y. Vlasov C. Di Natale, E. Mazzone and A. D'Amico, 'Application of Electronic tongue for quantitative analysis of mineral water and wine', *Electroanalysis* **1999** *11(10–11)* 814–820.
[33] A. Legin, A. Smirnova, A. Rudinitskaya, L. Lvova, E. Suglobova and Y. Vlasov, 'Chemical sensor array for multicomponent analysis of biological liquids', *Analytica Chimica Acta* **1999** *385* 131–135.
[34] J. Mortensen, A. Legin, A. Ipatov, A. Rudinitskaya, Y. Vlasov and K. Hjuler, 'A flow injection system based on chalcogenide glass sensors for the determination of heavy metals', *Analytica Chimica Acta* **2000** *403* 273–277.
[35] Y. Kanai, M. Shimizu, H. Uchida, H. Nakahara, C.G. Zhou, H. Maekawa and T. Katsube, 'Integrated taste sensor using surface photovoltage technique', *Sensors and Actuators* **1994** *B20* 175–179.
[36] Y. Sasaki, Y. Kanai, H. Uchida and T. Katsube, 'Highly sensitive taste sensor with a new differential LAPS method', *Sensors and Actuators* **1995** *B24–25* 819–822.

[37] M. George, W. Parak and H. Gaub, 'Highly integrated surface potential sensors', *Sensors and Actuators* **2000** *B69* 266–275.
[38] Y. Murakami, T. Kikuchi, A. Yamamura, T. Sakaguchi, K. Yokoyama, Y. Ito, M. Takiue, H. Uchida, T. Katsube and E. Tamiya, 'An organic pollution sensor based on surface photovoltage', *Sensors and Actuators* **1998** *B53* 163–172.
[39] S. Brown and R. Bear, 'Chemometric techniques in electrochemistry: A critical review', *Critical Reviews in Analytical Chemistry* **1993** *24(2)* 99–131.
[40] J.M. Diaz-Cruz, R. Tauler, B. Grabaric, M. Esteban and E. Casassas, 'Application of multivariate curve resolution to voltametric data. Part 1. Study of Zn(II) complexation with some polyelectrolytes', *Journal of Electroanalytical Chemistry* **1995** *393* 7–16.
[41] J. Menditeta, M.S. Diaz-Cruz, R. Tauler and M. Esteban, 'Application of multivariate curve resolution to voltametric data. Part 2. Study of metal-binding properties of the peptides', *Analytical Biochemistry* **1996** *240* 134–141.
[42] J. Simons, M. Bos and W.E. van der Linden, 'Data processing for amperometric signals', *Analyst* **1995** *120* 1009–1012.
[43] F. Winquist, C. Krantz-Rülcker, P. Wide and I. Lundström 'Monitoring of milk freshness by an electronic tongue based on voltametry' *Measurement Science and Technolgy* **1998** *9* 1937–1946.
[44] F. Winquist, S. Holmin, C. Krantz-Rülcker, P. Wide and I. Lundström, 'A hybrid electronic tongue', *Analytica Chimica Acta* **2000** *406* 147–157.
[45] F. Winquist, S. Holmin, C. Krantz-Rülcker and I. Lundström, 'Flow injection analysis applied to a voltametric electronic tongue', unpublished.
[46] C. Krantz-Rülcker, M. Stenberg, F. Winquist and I. Lundström, 'Electronic tongues for environmental monitoring based on sensor arrays and pattern recognition: a review', *Analytica Chimica Acta* **2001** *426* 217–226.
[47] T. Artursson, Licentiate Thesis no ,' ', Linköping University **2000**.
[48] U. Koller, Licentiate Thesis no. 859, 'The electronic tongue in the dairy industry', Linköping University **2000**.
[49] A. Carlsson, C. Krantz-Rülcker, and F. Winquist, 'An electronic tongue as a tool for wet-end control', unpublished.
[50] P. Ivarsson, Licentiate Thesis no. 858, 'Artificial senses – New technology in household appliances', Linköping University **2000**.
[51] C. Söderström, H. Borén, F. Winquist, and C. Krantz-Rülcker, 'Analysis of mould growth in liquid media with an electronic tongue', unpublished.
[52] R. Borngräber, J. Hartmann, R. Lucklum, S. Rösler and P. Hauptmann, 'Detection of ionic compounds in water with a new polycarbon acid coated quartz crystal resonator', *Sensors and Actuators* **2000** *B65* 273–276.
[53] S. Ezaki and S. Iiyama,'Detection of interactions between lipid/polymer membranes and taste substances by quartz resonator,' *Sensors and Materials* **2001** *13(2)* 119–127.
[54] J. Kondoh and S. Shiokawa 'New application of shear horizontal surface acoustic wave sensors to identifying fruit juices' *Japan Journal of Applied Physics* **1994**, *33, part I,* 3095–3099.
[55] J. Kondoh and S. Shiokawa 'Liquid identification using SH-SAW sensors', *Technical digest of Transducers 95 – Eurosensors IX,* Stockholm **1995** 716–719.
[56] A. Campitelli, W. Wlodarski and M. Hoummady 'Identification of natural spring water using shear horizontal SAW based sensors', *Sensors and Actuators* **1998** *B49* 195–201.

[57] V. Varadan and J. Gardner 'Smart tongue and nose', *Proc. SPIE-Inte. Soc. Eng.* **1999,** *3673*, 67–76. I
[58] C. Di Natale, R. Paolesse, A. Macagnano, A. Mantini, A. D'Amico, A. Legin, L. Lvova, A. Rudinitskaya and Y. Vlasov, 'Electronic nose and electronic tongue integration for improved classification of clinical and food samples', *Sensors and Actuators* **2000** *B64* 15–21.
[59] C. Di Natale, R. Paolesse, A. Macagnano, A. Mantini, A. D'Amico, M. Ubigli, A. Legin, L. Lvova, A. Rudinitskaya and Y. Vlasov, 'Application of a combined artificial olfaction and taste system to the quantification of relevant compounds in red wine', *Sensors and Actuators* **2000** *B69* 243–347.
[60] F. Winquist, P. Wide and I. Lundström, 'The combination of an electronic tongue and an electronic nose', *Sensors and Actuators* **2000** *B69* 243–347.
[61] P. Wide, F. Winquist, P. Bergsten and E. Petru, 'The human based multisensor fusion method for artificial nose and tongue data', *Proc. IEEE Instrumentation and Measurement Technology Conference,* St. Paul, Minnesota, USA **May 1998**.

List of Symbols and Abbreviations

Abbreviation	Explanation
ANN	artificial neural network
FIA	flow injection analysis
ISFET	ion sensitive field effect transistor
LAPV	large amplitude pulse voltametry
MLR	multiple linear regression
NLLS	nonlinear least square
PCA	principal component analysis
PLS	projection to latent structure
PVC	polyvinyl chloride
SAPV	small amplitude pulse voltametry
SAW	surface acoustic wave
SH-SAW	shear-horizontal mode SAW
SOM	self-organizing map

Index